Board Review in
PREVENTIVE MEDICINE AND PUBLIC HEALTH

To Miriya and Maximus

Board Review in
PREVENTIVE MEDICINE AND PUBLIC HEALTH
Questions and Answers

GREGORY SCHWAID, DO, MPH

Academic Press is an imprint of Elsevier
125 London Wall, London EC2Y 5AS, United Kingdom
525 B Street, Suite 1800, San Diego, CA 92101-4495, United States
50 Hampshire Street, 5th Floor, Cambridge, MA 02139, United States
The Boulevard, Langford Lane, Kidlington, Oxford OX5 1GB, United Kingdom

British Library Cataloguing-in-Publication Data
A catalogue record for this book is available from the British Library

Library of Congress Cataloging-in-Publication Data
A catalog record for this book is available from the Library of Congress

ISBN: 978-0-12-813778-9

Publisher: Mica Haley
Acquisition Editor: Erin Hill-Parks
Editorial Project Manager: Joslyn Chaiprasert-Paguio
Production Project Manager: Lucía Pérez
Designer: Victoria Pearson

Typeset by MPS Limited, Chennai, India

CONTENTS

About the Author *vii*
Preface *ix*
Acknowledgements *xi*

1. General Public Health **1**

1.1 General Public Health Questions 1
1.2 General Public Health Answers 7
Bibliography 19

2. Health Policy and Management **21**

2.1 Health Policy and Management Questions 21
2.2 Health Policy and Management Answers 41
Bibliography 74

3. Epidemiology and Biostatistics **79**

3.1 Epidemiology and Biostatistics Questions 79
3.2 Epidemiology and Biostatistics Answers 118
Bibliography 179

4. Environmental Medicine **187**

4.1 Environmental Medicine Questions 187
4.2 Environmental Medicine Answers 197
Bibliography 214

5. Occupational and Aerospace Medicine **217**

5.1 Occupational and Aerospace Medicine Questions 217
5.2 Occupational and Aerospace Medicine Answers 237
Bibliography 265

6. Clinical Preventive Medicine **269**

6.1 Clinical Preventive Medicine Questions 269
6.2 Clinical Preventive Medicine Answers 297
Bibliography 341

7. Infectious Disease **349**

7.1 Infectious Disease Questions 349
7.2 Infectious Disease Answers 370
Bibliography 407

8. Emergency Preparedness **413**

8.1 Emergency Preparedness Questions 413
8.2 Emergency Preparedness Answers 417
Bibliography 422

Index *425*

ABOUT THE AUTHOR

Gregory Schwaid earned his Bachelor of Science in Biomedical Science from the University of South Florida. From there he continued at USF to earn a master degree in Public Health with a concentration in Health Policy and Management. Afterwards, he attended Lake Erie College of Osteopathic Medicine in Bradenton, Florida, from where he earned his doctorate in Osteopathic Medicine. He completed his medical residency in preventive medicine with the Florida Department of Health in Palm Beach County. Upon completion of his residency, Dr. Schwaid became board certified in preventive medicine by both the American Board of Preventive Medicine and the American Osteopathic Board of Preventive Medicine.

Since completing residency, Dr. Schwaid has taken on many roles in public health and preventive medicine. These roles include practicing direct patient care and serving as faculty at both Lake Erie College of Osteopathic Medicine and the Preventive Medicine Residency at the Florida Department of Health in Palm Beach County. He also serves as Vice President and Chair of Public Health/Preventive Medicine for the American Osteopathic College of Occupational and Preventive Medicine (AOCOPM).

PREFACE

This book is intended to serve as a study resource for clinicians, medical residents, medical students, and graduate students in the fields of public health and preventive medicine. It is composed of >640 problem-based questions and answers intended to educate and reinforce public health concepts. The questions are broken into distinct sections to help the reader identify areas of weakness.

The primary audience for the book is those seeking board certification by the American Osteopathic Board of Preventive Medicine and/or the American Board of Preventive Medicine. This includes first time test-takers and those recertifying. Specialties that take the preventive medicine board exams include general preventive medicine, occupational medicine, aerospace medicine, undersea/hyperbaric medicine, correctional medicine, addiction medicine, lifestyle medicine, and bioinformatics.

The book was written by a board certified preventive medicine physician over a 2-year period with notes he took from his preparation for the preventive medicine boards, combined with notes from his masters of public health studies and real-life clinical experiences.

ACKNOWLEDGEMENTS

Dr. Schwaid would like to thank the Shoshana Levy, MD, MPH for her contributions to the completion of this book.

General Public Health

1.1 GENERAL PUBLIC HEALTH QUESTIONS

1. What are the three core functions of public health?
   A. Assessment, policy development, assurance
   B. Prevention, legislation, enforcement
   C. Epidemiology, environmental health, individual health
   D. Health education, health promotion, health care
   E. None of the above
2. What is the relation between the core public health functions and the essential public health services?
   A. Core functions are a product of federal government, while the essential services are a product of state and/or local government
   B. The core functions and the essential services are different words for the same thing
   C. The core functions fall within the essential services
   D. The essential services fall within the core functions
   E. There is no relationship between the two
3. Licensing health-care facilities is an action that falls within which public health core function?
   A. Assessment
   B. Assurance
   C. Enforcement
   D. Regulation
   E. Safety
4. The Surgeon General oversees which of the following?
   A. American Red Cross
   B. Department of Health and Human Services
   C. Physician officers in the Army
   D. Nobody, the Surgeon General is a figurehead for the health of the nation and does not have authority
   E. US Public Health Service Commissioned Corps

Board Review in Preventive Medicine and Public Health.
DOI: http://dx.doi.org/10.1016/B978-0-12-813778-9.00001-3

5. Which of the following is not true of water fluoridation?
A. After discontinuation of water fluoridation, there is an increase in missing teeth
B. After initiation of fluoridation, there is a decrease in dental caries
C. Community water fluoridation reduces dental caries across all socioeconomic status groups
D. The larger the population of those on community water fluoridation systems, the more expensive it is per individual
E. Water fluoridation reduces the number of people without a single dental cavity

6. An unemployed single mother of four children (ages 2, 4, 7, and 9) presents to the county nutrition clinic after moving from another state. How many of the children will the Woman, Infants, and Children (WIC) program directly benefit?
A. 0
B. 1
C. 2
D. 3
E. 4

7. The US PEPFAR Program targets which disease?
A. AIDS/HIV
B. Asthma
C. Diarrheal disease
D. Hepatitis
E. Malaria

8. The Health Resources and Services Administration's Ryan White Program is dedicated to helping those with which disease?
A. Acute lymphoblastic leukemia
B. Diabetes
C. HIV
D. Lung cancer
E. Pediatric obesity

9. How often are the Healthy People objectives updated?
A. Annually
B. 3 years
C. 5 years
D. 10 years
E. 20 years

10. Compared to the general population in the United States, the prison population in the United States has a lower prevalence of which ailment?
A. Diabetes mellitus
B. HIV
C. Substance abuse
D. Tuberculosis
E. None of the above

11. Which of the following is the leading cause of death in American jails?
A. Accident
B. Alcohol/drug overdose
C. Heart disease
D. Homicide
E. Suicide

12. Which model of health behavior includes perceived susceptibility?
A. Health Belief Model
B. Social Cognitive Theory
C. Theory of Reasoned Action
D. Transtheoretical Model
E. None of the above

13. Which model of health behavior proposes that the actual change in a behavior is correlated to the intention to change the behavior?
A. Health Belief Model
B. Social Cognitive Theory
C. Theory of Reasoned Action
D. Transtheoretical Model
E. None of the above

14. Which model of health behavior includes reciprocal determinism?
A. Health Belief Model
B. Social Cognitive Theory
C. Theory of Reasoned Action
D. Transtheoretical Model
E. None of the above

15. After years of deliberation, a smoker has decided to speak to his physician about quitting. The physician suggests setting a quit date. Which step in the Transtheoretical Model does setting a quit date represent?
A. Precontemplation
B. Contemplation

C. Preparation
D. Action
E. Maintenance

16. Which of the following is not considered to be one of the distinct categories of the Diffusion of Innovation Model?
A. Early adaptors
B. Early majority
C. Innovators
D. Late majority
E. Majority

17. What is the first step in completing the CDC's Community Health Assessment and Group Evaluation (CHANGE) tool?
A. Assemble the community team
B. Build the community action plan
C. Create a change summary statement
D. Gather data
E. Review all five change sectors

18. Which of the following is not one of the five Community Health Assessment and Group Evaluation (CHANGE) sectors?
A. Community-a-large sector
B. Health-care sector
C. School sector
D. Volunteer sector
E. Work-site sector

19. What does CDC's PATCH program stand for?
A. Partnerships Aimed to Create Health
B. People Against the Corruption of Health
C. Planned Approach to Community Health
D. Practitioner Alliance to Community Health
E. Practice Arrangement to Create Health

20. Which one of the following choices is not one of the Rothman community organization models?
A. Social action
B. Social planning
C. Locality development
D. Resource management
E. All of the above are Rothman models

21. The CDC's Community Preventive Services Task Force publishes how many potential grades for each recommended topic?
A. 2
B. 3
C. 4
D. 5
E. 6

22. Which of the following is the target population for which the Assessment Protocol for Excellence in Public Health (APEX PH) is intended for use?
A. Individuals
B. Local health departments
C. State health departments
D. Federal government
E. International public health emergencies

23. Mobilizing Action for Planning and Partnerships (MAPP) is best suited for addressing which of the following components of a local health department?
A. Mission
B. Organizational administrative processes
C. Organizational structure
D. Strategic plan
E. Values

24. An administrator in a publically funded teen health clinic wants to start health education program. While thinking back to her days studying public health, she remembered a popular eight-step approach to create a health program. Which of the following programs is she thinking of?
A. EMTALA
B. Ishikawa (fishbone)
C. MAPP
D. PRECEDE–PROCEED
E. SMART Objectives

25. The PATCH Model is a tool that was created within the context of which of the following?
A. APEX PH
B. MAPP
C. PACE EH
D. PDSA

E. PRECEDE–PROCEED

26. Which of the following options is not one of the "four Ps" of health-care social marketing?

A. Place

B. Price

C. Principle

D. Product

E. Promotion

27. What is the name of the program that is administered by the Centers for Disease Control and Prevention's Division of Community Health to address local racial and ethnic disparities in health status?

A. Consortium for Equal Health for All

B. Equal Opportunity for Health

C. Race and Ethnicity Health Task Force

D. Racial and Ethnic Approaches to Community Health

E. United Care for All People

1.2 GENERAL PUBLIC HEALTH ANSWERS

1. **A. Assessment, policy development, assurance**
   The three core functions of public health are assessment, assurance, and policy development. These three functions are further broken down into the 10 essential public health services, as shown in the answer for question 2 (directly below).
2. **D. The essential services fall within the core functions**
   The three core public health functions are assessment, assurance, and policy development. These three stages revolve in a continuous motion. Because the scope of these three functions is so broad, the 10 essential public health services were developed to further differentiate the stages of the public health process. The essential services framework groups public health activities into categories that can be recognized by budget officers, legislators, and the public at all levels of government. The essential services are evaluated in attaining public health accreditation.

   The essential services fit within the context of the core functions, as shown below:

3 Core Function	10 Essential Services	Example
Assessment	Monitor health status to identify health problems	Injury & disease registries Epidemiology Community needs assessment & status indicators (report cards) Vital statistics
Assessment	Diagnose and investigate health problems and hazards in the community	Injury, communicable, and chronic disease detection STD counseling and testing Outbreak investigation Environmental risk assessment Laboratory services
Policy development	Inform, educate, and empower people about health issues	Population-wide health promotion (e.g., education, programs for physical fitness) Health education Work-site health promotion

(continued)

(Continued)

3 Core Function	10 Essential Services	Example
Policy development	Mobilize community partnerships to identify and solve health problems	Forming community partnerships Community planning
Policy development	Develop policies and plans that support individual and community health efforts	Development of policies and guidelines Set the agenda Legislative activities
Assurance	Enforce laws and regulations that protect health and ensure safety	Consumer protection and sanitation Air and water quality Hazardous materials management Fluoridation services Medical examiner and forensics Enforcement agencies related to the agency's police authority
Assurance	Link people to needed personal health services and assure provision of health care	Provide health-care services in underserved communities Coordination of health services Clinical preventive services Communicable disease treatment Dental health services
Assurance	Assure a competent public health and personal health-care workforce	Maintenance of appropriate certification and licensure Mandated continuing education Maintaining labor force Employee training and programs
Assurance	Evaluate effectiveness, accessibility, and quality of personal and population-based health services	Continuous quality improvement activities (monitor and improve systems and outcomes) Facilities licensing Laboratory regulation Biomedical, clinical, and preventive investigation

(*continued*)

(Continued)

3 Core Function	10 Essential Services	Example
Combination of all three core functions	Research new insights and innovative services to health problems	Health services research Innovative technologies

3. **B. Assurance**

As described in the answer above (question 2), the 10 essential public health services were developed within the context of the 3 core public health functions: assessment, assurance, and policy development.

Licensure of health-care facilities falls within the essential service of evaluating effectiveness, accessibility, and quality of personal and population-based health services. This service is within the core function of assurance.

4. **E. US Public Health Service Commissioned Corps**

The Office of the Surgeon General sits within the Office of the Assistant Secretary for Health, a part of the Department of Health and Human Services. The Surgeon General, along with the Assistant Secretary of Health, oversees the US Public Health Service Commissioned Corps (USPHS). The Surgeon General is nominated by the President of the United States and sits for a 4-year term. In addition to overseeing USPHS, the Surgeon General is designated as the Chair of the National Prevention Council, an organization that provides leadership in prevention, wellness, and health-promotion activities. The Surgeon General also serves as a figurehead and provides Americans with the best health information available to increase health and well-being.

The mission of the USPHS is to protect, promote, and advance the safety of the nation. This mission is accomplished through rapid response to public health needs, leadership in public health, and advancement of the practice of public health. USPHS workers deploy to support public health responses to both natural and man-made events. Specific USPHS deployment activities include serving vulnerable populations, addressing disease control and prevention, supporting biomedical research, and regulating water supply.

5. **D. The larger the population of those on community water fluoridation systems, the more expensive it is per individual**

As the number of people in a community receiving water fluoridation increases, the average cost per individual decreases. This is due to economies of scale. The financial benefits of water fluoridation are enormous, as those receiving fluoridated water have less caries and missing teeth. Every dollar spent on water fluoridation has been estimated to save $38 in dental costs.

All of the other statements in the question are true. Water fluoridation serves as a primary prevention to decrease the number of dental caries across all socioeconomic groups. When fluoridation services are discontinued there is a decrease in overall dental health status in the population.

6. C. 2

Because the Women, Infants, and Children (WIC) program only benefits children aged 5 and under, only two of the children are able to directly receive WIC benefits.

WIC is operated and funded by the US Department of Agriculture, which funds local non-profit and public health agencies. Components of WIC include nutrition education, health-care referrals and financial assistance for needy families with children aged 5 years old and younger. To be eligible, families must be nutritionally at-risk and fall below a specific income.

Pregnant and breastfeeding mothers are also eligible to receive WIC benefits.

7. A. AIDS/HIV

PEPFAR stands for the US **P**resident's **E**mergency **P**lan **f**or **A**IDS **R**elief. It is the largest program within the US President's Global Health Initiative, a program aimed at saving the greatest number of lives by supporting the health infrastructure of other nations. Other programs within the Global Health Initiative include the President's Malaria Initiative and Feed the Future.

8. C. HIV

The Health Resources and Services Administration's (HRSA) Ryan White HIV/AIDS Program provides HIV-related care to those with insufficient health-care resources. The program is funded by the Department of Health and Human Services, the HRSA, and the HIV/AIDS Bureau.

9. D. 10 years

The Office of Disease Prevention and Health Promotion (ODPHP), within the Department of Health and Human Services publishes the new Health People objectives every 10 years. These science-based objectives (more than 1000) create an agenda for improving the nation's health. Each objective is outlined with baseline measures and specific goals for improvement. Government organizations, communities, and other entities often follow the Healthy People objectives to plan strategic goals.

10. E. None of the above

The United States has the largest prisoner population in the world. The average age of prisoners is advancing due to the aging baby boomer population and external political factors. With the ability to monitor and treat captive population, correctional medicine is a pure form of preventive medicine.

Across the board, prisoners experience earlier onset and increased prevalence of chronic diseases, such as hypertension and diabetes.

The prison population also has a higher prevalence of sexually transmitted diseases. Rates of HIV in prison are roughly four times as high as HIV rates found in the general population. Furthermore, an estimated 35% of the prison population carries chronic hepatitis C.

Half of all prisoners have a mental disorder. Prisons house more mentally ill persons than hospitals and mental health facilities. Substance abuse, a type of mental disorder, is also much more prevalent in the prison population than the general population.

Due interplay between the environment, host, and vector, tuberculosis (TB) infection is common in prison. TB has been found to be at least three times more prevalent within prison than outside of prison.

11. E. Suicide

Because jails typically incarcerate perpetrators for under 1 year while prisons nearly exclusively house inmates with sentences over one year, the population characteristics within jails and prisons are not the same. Because of this and over factors, health status between the two populations may differ. Prisoners have guaranteed access to care during the duration of the sentence, while jail inmates have less predictable health-care utilization while not incarcerated.

Suicide is routinely the leading individual cause of death in American jails. Nearly 1/3 of jail inmate deaths are attributed to suicide. In descending order, the cause of death in jail are illness (including heart disease, cancer, etc.), suicide, alcohol/drug intoxication, accident, and homicide. There are roughly 140 jail inmate deaths per 100,000 inmates. Most jails (~80%) do not report a single death annually.

Illness is directly responsible for 50% of deaths in jail, while illness is responsible for roughly 90% of death in prison. The two leading causes of illness-related death in prison are heart disease and cancer. After illness, the next most common causes of death amongst prisoners in descending order are suicide, homicide, alcohol/drug intoxication, and accident.

12. A. Health Belief Model

The Health Belief Model hypothesizes that an individual will make a particular health decision and take action based on their own perception of susceptibility to the illness and their ability to control it. The Health Belief Model consists of the following principles; perceived susceptibility, perceived severity, perceived benefits, perceived barriers, cues to action, and self-efficacy.

13. C. Theory of Reasoned Action

The different health belief models can be thought of as maps that provide guidance in a series of steps (constructs) to understanding health behavior. Each model emphasizes a different construct over the others.

The Theory of Reasoned Action assumes that people are rational and that their behavior is under control. Therefore, it is thought that behavioral intention leads to an actual behavioral change. The Theory of Planned Behavior is a modified version of the Theory of Reasoned Action that further explains an individual's perception of control over their own behavior.

14. B. Social Cognitive Theory

Reciprocal determinism states that there is a fluid relationship between the individual and environment, where each one reacts with the other to shape behavior. A change in any of these factors will affect the other two.

15. C. Preparation

The Transtheoretical Model is a stepwise map of intentional behavior change through the stages of precontemplation, contemplation, preparation, action, and maintenance.

Stage	Intervention
Precontemplation	Assess knowledge Discuss benefits and risks Education & research of benefits Share success stories Provide personal feedback and advise to change
Contemplation	Discuss the benefits and risks Outline reason to change behavior

(*continued*)

(Continued)

Stage	Intervention
	Review & evaluate barriers to change
	Discover resources & support network
	Discuss strategies for change
Preparation	Develop a plan
	Set a date for change
	Emphasize support
	Encourage motivation
	Set date to quit (smoking, alcohol, drugs)
Action	Affirm decision to change
	Coach through relapse triggers
	Troubleshoot problem areas
	Discuss coping strategies
	Focus on progress
	Provide follow-up
Maintenance	Reaffirm commitment
	Support change efforts
	Practice coping skills
	Relapse prevention skills

Once the decision to quit smoking has been made, the preparation stage begins. This includes setting the quit date. The action stage begins on the quit date.

16. E. Majority

The Diffusion of Innovation Model is a model of the social system that breaks down the pace at which an innovation is adapted. An innovation is new idea, practice, service, or object being introduced to the population. Consider for example new dietary recommendations. Predictably, dietary recommendations will not be adopted by everyone in the population at the same rate and time.

The Diffusion of Innovation model is broken down into six chronological categories. From first to last adaptors, the categories are innovators, early adaptors, early majority, late majority, late adaptors, and laggards. Many professionals do not recognize late adaptors as a category. The incorrect answer to this question is majority, as there is not a distinct category dedicated solely to the majority.

This model recognizes that communication is important to promote social change and bring along diffusion of innovation. The pace of adaption is influenced by perceived benefit of the change compared to the perceived risk, the ease of adaption, and whether there is evidence that the adaption works.

17. A. Assemble the community team

Community Health Assessment and Group Evaluation (CHANGE) was developed by the Healthy Communities Program of the Division of Adult and Community Health (within the CDC) to serve as a tool to help communities recognize and assess community policy, systems, and environmental changes over time. This tool has fallen out of favor with the CDC and is no longer funded by it, yet it is still used by state and local health departments across the United States.

The summary of steps in the CHANGE model are as follows:

Step 1—**Assemble the community team**

Identify and assemble a diverse team with a maximum of 10–12 individuals.

Step 2—**Develop a team strategy**

Decide whether to use the CHANGE tool as a group or divided into subgroups.

Step 3—**Review all five CHANGE sectors**

I. Community-at-large sector
II. Community institution/ organization sector
III. Health-care sector
IV. School sector
V. Work-site sector.

Step 4—**Gather data**

Gather data using multiple measures and methods to minimize bias.

Step 5—**Review gathered data**

Sit with the team to discuss what the data means in reference to the CHANGE tool.

Step 6—**Enter data**

Designate a data manager to input the information for each site.

Step 7—**Review consolidated data**

After rating each sector, complete the following steps, so the team can determine areas of improvement and subsequently develop a community action plan.

Step 7a: Create a CHANGE summary statement
Step 7b: Complete the sector data grid
Step 7c: Fill out the CHANGE strategy worksheets
Step 7d: Complete the Community Health Improvement Planning template.

Step 8—**Build the Community Action Plan**

Be specific with objectives and proposed time periods for the projects.

18. D. Volunteer sector

Community **H**ealth **A**ssessment and **G**roup **E**valuation (CHANGE) was developed by the Healthy Communities Program of the Division of Adult and Community Health, within the CDC to serve as a tool to help communities recognize and assess community policy, systems, and environmental changes over time. CHANGE has fallen out of favor with the CDC, yet it is still used by state and local health departments across the United States.

The volunteer sector is not one of the five sectors identified in Step 3 of the CHANGE community development tool. The five sectors are as follows:

1. Community-at-large sector
2. Community institution/ organization sector
3. Health-care sector
4. School sector
5. Work-site sector.

19. C. Planned Approach to Community Health

PATCH is an organizational tool used on the local level to plan, conduct, and assess health programs. It is an acronym for **P**lanned **A**pproach **t**o **C**ommunity **H**ealth and was developed in the 1980s as a joint effort between the CDC and state/local health departments to create a local-based process based on current knowledge in health theory, promotion, education and community development. It was created within the context of the PRECEDE model. Because PATCH is community-based, each individual community is able to tailor fit the PATCH process to fit their unique locale.

The phases of PATCH are as follows:

1. Mobilizing the community
2. Collecting and organizing data
3. Choosing health priorities
4. Developing a comprehensive intervention plan
5. Evaluating PATCH

20. D. Resource Management

In 1967, Jack Rothman presented an article identifying three distinct community organization models; social action, social planning, and locality development. All three of these models have become frameworks used in social planning.

21. **B. 3**

The CDC's Community Guide reviews public health interventions to analyze which interventions have a positive net impact. The Community Preventive Services Task Force uses the information obtained from the Community Guide to issue evidence-based recommendations to public health organizations. It may be thought of as analogous to the US Preventive Services Task Force, but the recommendations are for communities rather than primary care clinicians. There are only three types of stances: recommended, recommended against, and insufficient evidence.

22. **B. Local health departments**

Assessment **P**rotocol for **Ex**cellence in **P**ublic **H**ealth (APEX PH) is a three-step process that helps local health departments (LHDs) assess their internal capacity, better understand local health issues, and create action plans.

The APEX PH was created as a group effort between the American Public Health Association, the Association of State and Territorial Health Officials, the National Association of County Health Officials, the National Association of County and City Health Officials and, the CDC to provide LHDs with a tool to increase organizational capacity and strengthen their role within the community.

23. **D. Strategic plan**

Mobilizing for **A**ction through **P**lanning and **P**artnership (MAPP) was developed by the National Association of County and City Health Officials with support from the CDC to perform community assessment and planning, with the vision to assist in communities achieving improved health and quality by mobilizing partnerships and taking strategic action.

MAPP consists of community partnership development, visioning, continuous assessments, identifying strategic issues, formulations of strategic objectives, and implementing (while assessing) the process.

Unrelated to this question, although relevant to MAPP is the Protocol for Assessing Community Excellence in Environmental Health assessment. It similar to MAPP, but focuses exclusively as being an assessment tool to create an operational plan (as opposed to strategic plan) tailored for environmental health.

24. D. PRECEDE–PROCEED

The PRECEDE–PROCEED Model is the predominant model used to plan a health-education program. PRECEDE stands for **pre**-disposing, **r**einforcing, and **e**nabling **c**onstructs in **e**ducational/**e**nvironmental **d**iagnosis and **e**valuation. PROCEED stands for **p**olicy, **r**egulatory, and **o**rganizational **c**onstructs in **e**ducational and **e**nvironmental **d**evelopment. This model can be used to provide guidance for any health education program.

There are eight stages of the PRECEDE–PROCEED Model:

1. Social assessment
2. Epidemiological assessment
3. Educational and ecological assessment
4. Administrative and policy assessment and intervention alignment
5. Implementation
6. Process evaluation
7. Impact evaluation
8. Outcome evaluation

EMTALA stands for **E**mergency **M**edical **T**reatment and **A**ctive **L**abor **A**ct. It is federal legislation dictating that hospitals receiving Medicare dollars must screen and stabilize all patients the come through the emergency room.

An Ishikawa diagram (aka fishbone diagram, cause and effect diagram) is a quality-management tool used to identify the root cause of a problem and identify opportunities for improvement.

MAPP stands for **M**obilizing **A**ction for **P**lanning and **P**artnerships. It is a tool used by communities to perform community assessment and planning.

SMART objectives is an acronym for creating goals that are **s**pecific, **m**easurable, **a**ttainable, **r**ealistic and **t**imely.

25. E. PRECEDE–PROCEED

The **P**lanned **A**pproach **t**o **C**ommunity **H**ealth (PATCH) Model was created within the context of the PRECEDE model. PATCH is a community health planning model to increase the capacity of health agencies to plan, implement, and evaluate community health promotion programs.

APEX PH is an acronym for **A**ssessment **P**rotocol for **Ex**cellence in **P**ublic **H**ealth. It provides local health departments with a tool to increase organizational capacity and strengthen their role within the community.

MAPP stands for **M**obilizing **A**ction for **P**lanning and **P**artnerships. It is a tool used by communities to perform community assessment and planning.

PACE EH stands for **P**rotocol for **A**ssessing **C**ommunity **E**xcellence in **E**nvironmental **H**ealth. It is an assessment tool used to create an operational plan tailored to improving environmental health.

PDSA is an acronym that stands for **p**lan, **d**o, **s**tudy, **a**ct. It represents a continuous quality improvement cycle.

26. C. Principle

The "four Ps" of social marketing in health-care include place, price, product, and promotion. The health-care service should be in a place that is accessible and appropriate. The price component includes social, environmental, monetary, and environmental costs. The product should be attractive and beneficial to the recipient. Finally, the product should be promoted to the target audience and information disseminated to the appropriate target.

Some experts also believe that positioning should be included as the fifth P in health-care social marketing. Positioning would entail framing an issue so that the target population relates to it. Meanwhile, in the for-profit world, the fifth P stands for profit.

27. D. Racial and Ethnic Approaches to Community Health

The CDC's Division of Community Health operates the Racial and Ethnic Approaches to Community Health (REACH), a program designed to reduce racial and ethnic disparities in health. The REACH program provides monetary awards to community-based programs that administer programs intended to increase the health status of Blacks, American Indians, Hispanics, Asians, Alaska Natives, and Pacific Islanders. These programs typically focus on providing education and intervention on proper nutrition, physical activity, tobacco use, and chronic diseases, such as diabetes.

Outside of REACH, the other question options do not exist.

Other Division of Community Health (DCH) programs include Partnerships to Improve Community Health (PICH) and National Implementation and Dissemination for Chronic Disease Prevention. These programs (including REACH), as well as others, typically expire and renew according to times of need and political climate. The DCH aims to be the national leader in advancing the practice of community health and making healthy living easier. Specific DCH principles include maximizing public health impact, advancing health equity, using evidence-based practices, and engaging the community.

BIBLIOGRAPHY

[1–3] Shi L, Johnson JA. Novick & morrow's public health administration principles for population-based management. 3rd ed Sudbury, MA: Jones and Bartlett Publishers; 2013. p. 56–63.

[4a] USPHS: About us. ⟨http://www.usphs.gov/aboutus/⟩ [accessed 05.10.2016].

[4b] About the Office of the Surgeon General. ⟨http://www.surgeongeneral.gov/about/index.html⟩ [accessed 05.10.2016].

[5] Guide TC. Summary—oral health, dental caries: community water fluoridation. ⟨http://www.thecommunityguide.org/oral/fluoridation.html⟩ [accessed 05.10.2016].

[6] Marotz LR. Health, safety, and nutrition for the young child. 9th ed Stamford, CT: Wadsworth Publishing Co.; 2014. p. 517.

[7] ⟨http://www.pepfar.gov/about/index.htm⟩ [accessed 05.10.2016].

[8] HRSA HIV/AIDS programs. ⟨http://hab.hrsa.gov/⟩ [accessed 05.10.2016].

[9] About healthy people. ⟨https://www.healthypeople.gov/2020/About-Healthy-People⟩ [accessed 05.10.2016].

[10] Lorry CKM, Schoenly L, Knox CM. Essentials of correctional nursing. New York: Springer Publishing Co; 2012. p. 9–16.

[11] Noonan M, Rohloff H, Ginder S. Mortality in local jails and state prisons, 2000–2013—statistical tables. Bureau of Justice Statistics; 2015.

[12–15] Glanz K, Lewis FMM, Rimer BK, McGinnis MJ. Health behavior and health education: theory, research, and practice. 2nd ed San Francisco, CA: Jossey-Bass Inc., US; 2002.

[16] Skolnik RL. Global health 101. 2nd ed Sudbury, MA: Jones and Bartlett Publishers; 2011. p. 126–7.

[17–18] Centers for Disease Control and Prevention. Community Health Assessment and Group Evaluation (CHANGE) action guide: building a foundation of knowledge to prioritize community needs. Atlanta, GA: U.S. Department of Health and Human Services; 2010.

[19a] Kreuter MW. Community health promotion: the agenda for the '90s, PATCH. J Health Educ 1992;23(3):135–9.

[19b] U.S. Department of Health and Human Services. Planned approach to community health: guide for the local coordinator. Atlanta, GA: U.S. Department of Health and Human Services, Department of Health and Human Services, Centers for Disease Control and Prevention National Center for Chronic Disease Prevention and Health Promotion http://www.lgreen.net/patch.pdf.

[20] Weil M. Community practice: models in action. New York: Haworth Press; 1997. p. 28.

[21] Systematic review methods. ⟨http://www.thecommunityguide.org/about/methods.html⟩ [accessed 05.10.2016].

[22–23] Rowitz L. Public health leadership: putting principles into practice. 3rd ed Sudbury, MA: Jones & Bartlett Publishers; 2014. p. 190–4.

[24–25] Sharma M, Romas JA. Theoretical foundations of health education and health promotion. 2nd ed Sudbury, MA: Jones and Bartlett Publishers; 2012. p. 43–8.

[26] Corcoran N, editor. Communicating health: strategies for health promotion. 2nd ed Thousand Oaks, CA: SAGE Publications; 2013.

[27] CDC. About Division of Community Health (DCH). Division of Community Health (DCH): making healthy living easier. ⟨https://www.cdc.gov/nccdphp/dch/about/index.htm⟩ [accessed 05.10.2016].

CHAPTER TWO

Health Policy and Management

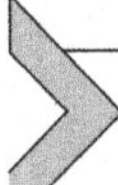

2.1 HEALTH POLICY AND MANAGEMENT QUESTIONS

1. Commercial insurance most often executes payment from ______ to ______ entities.
   A. Private to private
   B. Public to public
   C. Private to public
   D. Public to private
   E. None of the above
2. Which option best describes The Joint Commission (TJC)?
   A. Government organization
   B. Not-for-profit, contracted by government
   C. Not-for-profit, independent of government
   D. For-profit, contracted by government
   E. For-profit, independent of government
3. Which of the following measures patient perception of their healthcare?
   A. Agency for Healthcare Research and Quality (AHRQ)
   B. Hospital Consumers Assessment of Healthcare Providers and Systems (HCAHPS)
   C. Prospective Payment System (PPS)
   D. Public Health Accreditation Board (PHAB)
   E. The Joint Commission (TJC)
4. What is the source of the majority of funding for graduate medical education (GME) for physicians in the United States?
   A. Health Resources and Services Administration
   B. Medicare
   C. Medicaid
   D. Medical malpractice lawsuits
   E. Veterans Health Administration

Board Review in Preventive Medicine and Public Health.
DOI: http://dx.doi.org/10.1016/B978-0-12-813778-9.00002-5

5. Which organization pays the highest percentage of long-term care expenses in the United States?
   - A. Medicare
   - B. Medicaid
   - C. Self-Pay
   - D. Private Insurance
   - E. American Association of Retired Persons
6. Where do the largest percentage of Americans receive their health insurance from?
   - A. Centers for Medicare and Medicaid Services
   - B. Veterans Affairs
   - C. Individual private insurance
   - D. Employer private insurance
   - E. Self-insured (pay out-of-pocket)
7. Which payer contributes the most to national healthcare expenditures?
   - A. Government financed programs
   - B. Individual private insurance
   - C. Employer private insurance
   - D. Self-insured (pay out-of-pocket)
   - E. Other
8. Patients enrolled in the Early and Periodic Screening, Diagnostic, and Treatment (EPSDT) program may benefit from these services until what age?
   - A. 18
   - B. 21
   - C. 26
   - D. 65
   - E. Indefinitely
9. Primary care physician health professional shortage areas (HPSAs) are defined by having less than one primary care physician per how many residents in a geographic population?
   - A. 1000
   - B. 3500
   - C. 5000
   - D. 15,000
   - E. 30,000
10. Which of the following is NOT true of Federally Qualified Health Centers (FQHCs)?

A. FQHCs receive funding from the Health Resources and Service Administration
B. FQHCs primarily provide outpatient health services
C. FQHCs may also be approved as a Rural Health Center
D. FQHCs must provide a sliding fee scale to persons with an income below 200% of the federal poverty level
E. FQHCs may be located in urban or rural settings

11. What is the name of a hospital that receives special designation for being located in a rural area, is far away from other hospitals, has a 24 hour emergency room, and may only maintain an average inpatient length of stay for 96 hours to maintain designation?
A. Critical access hospital
B. Community support hospital
C. Isolated infrastructure hospital
D. Rural care hospital
E. Underserved regional hospital

12. Approximately what percentage of Medicare dollars is spent on a beneficiary's last year of life?
A. 5%
B. 15%
C. 25%
D. 35%
E. 45%

13. Which of the following organizations certifies Patient-Centered Medical Homes (PCMHs)?
A. Agency for Healthcare Research and Quality (AHRQ)
B. Healthcare Effectiveness Data and Information Set (HEDIS)
C. National Committee for Quality Assurance (NCQA)
D. Health Resources and Service Administration (HRSA)
E. Department of Health and Human Services (DHHS)

14. Which of the following actions is incentivized by hospitals receiving payments through diagnostic-related groups (DRGs)?
A. Conduct internal utilization review
B. Increase length of stay
C. Increase the number of nurses per patient
D. Order more tests and procedures
E. Use nongeneric medications

15. Tired of contracting directly with health maintenance organizations (HMOs) individually, a group of physicians decides to form an organization to negotiate with HMOs as a group.

What is the name of the structure of this organization?

A. Accountable Care Organization (ACO)
B. Diagnostic-related group (DRG)
C. Independent Practice Association (IPA)
D. Physician Hospital Organization (PHO)
E. Preferred Provider Organization (PPO)

16. Which agency provides the US Preventive Services Task Force (USPSTF) with the resources necessary to create recommendations?

A. Agency for Healthcare Research and Quality
B. Healthcare Effectiveness Data and Information Set
C. National Committee for Quality Assurance
D. Health Resources and Service Administration
E. Department of Health and Human Services

17. A 27 year-old experiences an episode of anaphylaxis immediately after receiving a flu shot and is transported to the emergency room via ambulance. When looking to recover financial losses, his attorney suggests that he seek to recover from which entity?

A. National Vaccine Injury Compensation Program
B. Nurse that administered the shot
C. Private Insurance Company
D. Vaccine manufacturer
E. Nobody. Immunizations are exempt from financial liability.

18. Charging a higher premium for insuring a woman is an example of what type of rating?

A. Health status rating
B. Experience rating
C. Demographic rating
D. Industry rating
E. Durational rating

19. In health economics, which of the following is a way to combat adverse selection?

A. Pay-for-performance compensation for physicians
B. Mandated purchase of health insurance
C. Electronic medical records
D. Randomization
E. Utilization review

20. Which is the method used to evaluate for appropriate use of health resources in the process of providing healthcare?

A. Utilization management
B. Evidence based medicine

C. Electronic medical records
D. Insurance authorization
E. Reimbursement of services

21. What is the name of the legislation that prevents a physician from referring his/her patients to a facility that he/she has a financial interest in?
A. Flexner Act
B. Hill–Burton Act
C. The Physician Self-referral Act (also known as the Stark Law)
D. The name of the law varies depending on the state
E. There are no laws preventing this action

22. Which part of Medicare is responsible for payment of physicians?
A. Part A
B. Part B
C. Part C
D. Part D
E. Part E

23. Which Medicare program allows its beneficiaries to enroll in a private managed care plans?
A. Medicare A
B. Medicare B
C. Medicare C
D. Medicare D
E. Medicare E

24. What is the Federal Medical Assistance Percentage (FMAP)?
A. The percentage of gross domestic product the United States pays towards medical expenses
B. The amount of Medicare dollars spent annually
C. Money spent for healthcare in the Indian Health Services
D. Tax relief for healthcare organizations with nonprofit status
E. The dollars matched by the federal government for state Medicaid expenses

25. The chief financial officer (CFO) of a large HMO is disregarding evidence-based medicine and is performing cost-minimization analysis (CMA) to determine the company's preferred method of treating a specific disease. In doing so, the CFO wishes to discount the price of necessary supplies. If $1000 worth of supplies depreciates at 5% annually, approximately what will they be worth in three years?
A. $950
B. $922

C. $900
D. $862
E. $810

26. Which of the following has the least favorable cost-effectiveness analysis (CEA)?
A. Health promotion
B. Primary prevention
C. Secondary prevention
D. Tertiary prevention
E. Need more information to answer

27. A hospital administrator has asked his economist to tell him whether or not he should stay with the old, time-tested surgical treatment or switch to a newer one. After careful cost-effectiveness analysis (CEA), the administrator's economist tells him that the old surgical approach is dominated, while the new approach dominates.
Because the administrator does not want to feel embarrassed in front of the economist, he comes to your office to ask what this means and what he should do.
Which is the correct response?
A. The old approach is more effective and costs less
B. The old approach is less effective and costs less
C. The new approach is more effective and costs less
D. The new approach is more effective and costs more
E. There is not enough information to tell

28. For cost-minimalization analysis (CMA) calculations to be accurate, which of the following must hold true?
A. Outcome from the alternative options should be measured in dollars
B. Outcome of alternative options should be similar to each other
C. The same approach should be used for all options
D. Both the inputs and the outputs of the process should vary from the alternative options
E. None of the above

29. A charitable organization donated a large sum of money to a local health department to increase the health status of the local population. The health department has a list of projects it would like to spend the money on. These projects include creating tobacco-free zones, teaching free obesity prevention classes and operating an after school clinic to discourage teen pregnancy. Because the programs are so different from each other, the health department decided to

estimate potential outcomes in dollars. Which economic evaluation tool would be most appropriate to prioritize which projects the grant money should be used for?

A. Cost-benefit analysis (CBA)
B. Cost-effectiveness analysis (CEA)
C. Cost-minimization analysis (CMA)
D. Cost-utility analysis (CUA)
E. Sensitivity analysis

30. Cost, access, and _____ are three elements that dictate nearly all healthcare decisions. These three variables have been referred to as the "iron triangle" and the "three-legged stool."

Which option below is the third element?

A. Availability
B. Care
C. Price
D. Quality
E. Time

31. Healthcare quality measures can be categorized as either ________, process, or outcome.

Which of the variables below belongs in this statement?

A. Business
B. Money
C. Organization
D. Structure
E. Time

32. All health outcomes can be further categorized as clinical, ______, or humanitarian.

A. Economic
B. Healthy
C. Nosocomial
D. Preventive
E. Timely

33. Which of the following is not an example of cost-sharing?

A. Deductible
B. Flexible spending account
C. Copayment
D. Coinsurance
E. All of the above are examples of cost-sharing

34. Which USPSTF recommendation grades receive coverage from insurers without the patient paying money out-of-pocket, copayments, coinsurance, or deductibles?
A. A
B. A, B
C. A, B, C
D. A, I
E. I

35. What is the name of the group created by the Affordable Care Act (ACA) to reduce Medicare spending?
A. Diagnostic-related group
B. Healthcare Expenditure Review Committee
C. Independent Payment Advisory Board
D. Medicare Cost Advisory Committee
E. Medicare Finance Advisory Panel

36. What type of physician payment reimburses physicians for caring for a patient during a specified period of time?
A. Capitation
B. Fee-for-service
C. Diagnostic-related group
D. Resource based relative value scale
E. Salary

37. A young physician is completing residency and looking for employment. She is looking for the best fit for her and is exploring all the different types of physician compensation systems. Which type of compensation system would give her the most incentive to perform more procedures and order more tests?
A. Capitation
B. Diagnostic-related group
C. Fee-for-service
D. Pay-for-performance
E. Paid salary

38. Which form of cost-sharing requires the patient to pay a set percentage of all dollars spent?
A. Coinsurance
B. Copayment
C. Deductible
D. Premium
E. Utilization

39. A terminally ill patient is given a $20,000 treatment that will make her more comfortable. This is an example of which of the following types of health outcomes?
  A. Clinical
  B. Economic
  C. Humanitarian
  D. Donation
  E. Charity

40. Which is a unit of measure in cost-utility analysis (CUA)?
  A. Dollars saved
  B. Quality adjusted life years
  C. Heart attacks averted
  D. Body mass index (BMI) of heart attack patients
  E. None of the above

41. The National Healthcare Safety Network (NHSN) is part of which governmental organization?
  A. Centers for Disease Control and Prevention (CDC)
  B. Food and Drug Administration (FDA)
  C. Department of Justice (DOJ)
  D. Health Resources and Services Association (HRSA)
  E. National Institutes of Health (NIH)

42. The National Practitioner Data Bank (NPDB) is housed within which organization?
  A. Agency for Healthcare Research and Quality
  B. Health Resources and Services Administration
  C. Kaiser Family Foundation
  D. Department of Homeland Security
  E. Department of Justice

43. Which US executive department does the Occupational Safety and Health Administration (OSHA) reside within?
  A. Department of Agriculture
  B. Department of Health and Human Services
  C. Department of Justice
  D. Department of Labor
  E. Department of Veterans Affairs

44. Which US executive department does the administration of Women, Infants, and Children (WIC) benefits fall within?
  A. Department of Agriculture
  B. Department of Health and Human Services

C. Department of Justice
D. Department of Labor
E. Department of Veterans Affairs

45. Which organization administers the Supplemental Nutrition Assistance Program (SNAP)?
A. US Department of Agriculture
B. US Department of Health
C. Individual State Departments of Health
D. Individual State Departments of Agriculture
E. Private not-for-profit organizations

46. Which entity holds the most power and influence in a centralized health department?
A. Town
B. City
C. Region
D. State
E. Country

47. Which of the following is a national organization in the United States composed of local health departments (LHDs) with a mission "to be a leader, partner, catalyst, and voice for local health departments in order to ensure the conditions that promote health and equity, combat disease, and improve the quality and length of all lives"?
A. American Consortium of Local Health Departments (ACLHD)
B. Federal Organization of Local Health Departments (FOLHD)
C. Local Health Department Federal Coalition (LHDFC)
D. National Association of County and City Health Officials (NACCHO)
E. Representative Local Health Organization (RLHO)

48. Which of the following statements regarding local health departments (LHDs) is true?
A. LHDs may only serve counties
B. Centralized LHDs are mainly run by the local jurisdiction
C. Decentralized LHDs are mainly run by the state
D. The number of decentralized and centralized health departments is static and can't change
E. All LHDs are included under the state public health agency/state health department

49. The National Institute for Occupational Safety and Health (NIOSH) is part of which organization?

A. American Conference of Governmental Industrial Hygienists
B. Centers for Disease Control and Prevention
C. Department of Labor
D. Occupational Safety and Health Administration
E. World Health Organization

50. Which of the following functions is NOT performed by the US Department of Agriculture (USDA)?
A. Administering the Woman, Infants, and Children (WIC) Supplemental food assistance program
B. Ensuring safety of poultry
C. Ensuring safety of tobacco products
D. Regulate genetically engineered foods
E. All of these functions are performed by the USDA

51. IHS is a federal health program that falls under which organization?
A. Centers for Disease Control and Prevention (CDC)
B. US Department of Health and Human Services (DHHS)
C. Agency for Healthcare Research and Quality (AHRQ)
D. National Rural Health Association (NRHA)
E. American Public Health Association (APHA)

52. Which US executive department does the Agency for Toxic Substances and Disease Registry (ATSDR) fall within?
A. Department of Agriculture
B. Department of Health and Human Services
C. Department of Justice
D. Department of Labor
E. Department of Veterans Affairs

53. Which US executive department does the Veterans Benefits Administration fall within?
A. Department of Agriculture
B. Department of Health and Human Services
C. Department of Justice
D. Department of Labor
E. Department of Veterans Affairs

54. Which US executive department supplies clinicians with Drug Enforcement Agency (DEA) numbers?
A. Department of Agriculture
B. Department of Health and Human Services
C. Department of Justice
D. Department of Labor
E. Department of Veterans Affairs

55. Which Title of the Social Security Act (SSA) primarily addresses Medicare?
A. 17
B. 18
C. 19
D. 20
E. 21

56. Which Title of the Social Security Act (SSA) primarily addresses Medicaid?
A. 17
B. 18
C. 19
D. 20
E. 21

57. Which of the following is not a police power?
A. Approval of new drugs
B. Licensure of physicians
C. Fluoridation of drinking water
D. Inspection of restaurants
E. Quarantine sick patients

58. Which entity licenses dental hygienists?
A. City
B. State
C. Federal government
D. Private organization
E. No license is required to be a dental hygienist

59. Which of the following programs is sponsored by the Office of Disease Prevention and Health Promotion (ODPHP)?
A. American Medical Association
B. National Association of City and County Health Officials
C. National Health Information Center
D. National Institutes of Health
E. WIC Program

60. What is SWOT analysis primarily used for?
A. Strategic planning
B. Financial analysis
C. Workers' compensation
D. Legal analysis
E. Physician performance evaluation

61. In addition to a SWOT analysis, public health agencies may also conduct a PEST analysis. What does the 'T' in PEST analysis stand for?
A. Team
B. Technological
C. Temporary
D. Timely
E. Together

62. What does the 'A' stand for in the PDSA cycle?
A. Act
B. Add
C. Alert
D. Always
E. Assert

63. The acronym DMAIC is part of which quality improvement tool?
A. Continuous quality improvement
B. Ishikawa diagram
C. Six sigma
D. Total quality improvement
E. Value stream mapping

64. Which of the following S.M.A.R.T. objectives is incorrect?
A. Specific
B. Measurable
C. Attainable/Achievable
D. Reliable
E. Time-oriented

65. Under the Secure and Responsible Drug Disposal Act and the Controlled Substances Act (CSA), which of the following medications can the Drug Enforcement Agency accept in a take back event?
A. Heroin
B. Hydrocodone cough syrup
C. Medications containing iodine
D. Used insulin needles
E. All of the above

66. A surgery center is experiencing a high rate of postsurgical infections. Management determines that the cause is due to members of the surgical team forgetting to sanitize appropriately because they are too fixated on making sure they operate on the correct side of the body.

One administrator suggests using a flowchart, so the surgical team can follow the correct steps in preparation for surgery.

Assuming everything else within the organization remains the same, which of the following statements accurately summarizes the administrator's thought process?

A. Changing the structure will change the process
B. Changing the structure will change the outcome
C. Changing the process will change the structure
D. Changing the process will change the outcome
E. The outcome will change once the structure is changed

67. Which tool is best equipped to compare performance indicators of one healthcare plan with another?

A. Healthcare Effectiveness Data and Information Set
B. National Guideline Clearinghouse
C. National Quality Measures Clearinghouse
D. Records from the Department of Vital Statistics
E. Records from The Hospital Quality Initiative

68. An Ishikawa diagram (also known as fishbone diagram) may be used by healthcare organizations for which purpose?

A. Create a project schedule
B. Evaluate meta-analysis studies to evaluate for publication bias
C. Map the decision-making process
D. Determine the root cause of a problem
E. Graphically demonstrate how the components of a whole variable are divided

69. An ICU physician questions the competency of his patient. Although the patient is capable of eating, she has chosen not to. Despite her stating she does not want a feeding tube, the physician makes sure that one is placed. The physician believes that his beliefs are more appropriate than the patient's, given the situation. The patient makes a full recovery from her ailment and seeks legal action against the physician.

Which principle is most likely the physician's best legal defense?

A. Autonomy
B. Beneficence
C. Justice
D. Medical paternalism
E. Nonmaleficence

70. Which principle states that physicians have an obligation to provide guidance and act in the best interest of the patient?

A. Beneficence
B. Justice
C. Nonmaleficence
D. Autonomy
E. Utilitarianism

71. In which of the following cases is it a violation of HIPAA to share any part of a patient's health information?
A. Give a diagnosis to workers' compensation nurse
B. Report a disease to the local health department
C. Respond to a request by law enforcement for address and blood type
D. A news reporter calls to speak with the treating physician
E. A physician may contact a patient's family member if he believes that the patient is a potential harm to herself.

72. What does HIPAA stand for?
A. Health Information Privacy and Accountability Act
B. Health Insurance Portability and Accountability Act
C. Historical Individual Privacy Awareness Act
D. Honor Individual Protected Ailment Act
E. Human Information and Personal Awareness Act

73. What is the Hospital Consumer Assessment of Healthcare Providers and Systems (HCAHPS) based upon?
A. Objective health information such as blood pressure control
B. Patient perception of care provided
C. Healthcare worker perception of care provided
D. Patients and healthcare workers feel about the care provided
E. Comparison between different types of healthcare delivery systems

74. Which of the following legislative acts protects the minimum standards for health plans within the private workforce?
A. Consolidated Omnibus Budget Reconciliation Act
B. Employee Retirement Income Security Act
C. Hill–Burton Act
D. Health Insurance Portability and Accountability Act
E. Patient Self-Determination Act

75. Which legal principle allows the government to take action to protect an individual when they are unable to do it themselves?
A. Advanced directive
B. Adverse selection
C. Community rating
D. Entitlement program
E. Parens patriae

76. The Age Discrimination in Employment Act (ADEA) provides protection for which age group?
A. All ages
B. ≤ 30 years old
C. ≥ 40 years old
D. ≥ 50 years old
E. ≥ 60 years old

77. The length of COBRA benefits varies by specific situation. What is the longest time period that COBRA benefits can last?
A. 6 months
B. 12 months
C. 24 months
D. 36 months
E. 48 months

78. Which of the following is not true of the ACA of 2010?
A. The ACA establishes an employer mandate to provide insurance
B. The ACA establishes an individual mandate to obtain insurance
C. The ACA allows children up to age 26 to be insured under their parents' health insurance
D. The ACA allows health insurance exclusion due to preexisting conditions
E. The ACA limits the percentage of premiums that can be retained by an insurance company

79. Which act enabled the creation of Medicare and Medicaid?
A. Balanced Budget Act
B. Consolidated Omnibus Budget Reconciliation Act
C. Health Information Portability and Accountability Act
D. Hill–Burton Act
E. Social Security Act

80. Which hospitals must provide care under the Emergency Medical Treatment and Active Labor Act (EMTALA)?
A. All hospitals must provide care under EMTALA
B. Only critical access hospitals (CAHs)
C. Only hospitals that are certified by The Joint Commission
D. Only hospitals that receive Medicare reimbursement
E. Only hospitals with level one and level two trauma centers

81. A primary care physician has received notification that one of his former patients plans to bring on a lawsuit for failure to diagnose and treat bacterial sinusitis. The patient claims that he was told his condition would resolve without antibiotics. However, he did not get better within a four days after seeing the physician and had to go to a different clinic to get antibiotics.

Which of the following elements must a plaintiff successfully prove in a malpractice claim?

A. A breach of duty contributes to harm of the patient
B. A breach of duty, where the clinician does not provide standard of care
C. A patient suffers harm or injury
D. There is a doctor—patient relationship and a duty for the health professional to act reasonably and appropriately
E. All of the above elements must be present for a successful malpractice suit.

82. What percentage of the overall national health expenditure is spent on dentistry?

A. 0.5%
B. 2%
C. 4%
D. 8%
E. 15%

83. What is the name of the laboratory regulatory standards monitored by the CMS?

A. Clinical Laboratory Improvement Amendments (CLIA)
B. Evaluating Laboratory Specimen Measures (ELSM)
C. Human Laboratory Testing Guidelines (HLTS)
D. Laboratory Response Network (LRN)
E. US Laboratory Criteria (USLC)

84. Which organization is responsible for approval of genetically modified organisms (GMOs) consumed as food in the United States?

A. Centers for Disease Control and Prevention
B. Food and Drug Administration
C. Genetically Modified Organism Approval Committee
D. GMOs do not need to be approved in the United States
E. Individual states may approve GMOs sold and produced in their jurisdiction

85. What is the most common type of healthcare-associated infection (HAI; also known as nosocomial infection)?

A. Bacteremia
B. Deep venous thrombosis

C. Pneumonia
D. Skin infection
E. Urinary tract infection

86. What is the best way to reduce nosocomial infections?
A. Contact precautions
B. Hand washing
C. Frequent antibiotic use
D. Influenza vaccination
E. UV lighting

87. Roughly, how many Americans die annually due to medical errors?
A. <20,000
B. 20,000–40,000
C. 40,000–60,000
D. 60,000–80,000
E. >100,000

88. A clinician is reimbursed if 70% of his patients meet the Eighth Joint National Committee (JNC 8) guidelines. What type of reimbursement concept is this an example of?
A. Fee-for-service
B. Capitated
C. Pay-for-performance
D. Diagnostic-related group
E. Salary

89. A 26-year-old man presented to your clinic, where he disclosed that he had intercourse while on vacation two weeks prior with a partner outside of his sexually exclusive relationship. Reluctantly, the patient agrees to wait for the results of his gonorrhea and chlamydia tests before receiving treatment. When the results returned negative three days later, the patient still requests treatment for both gonorrhea and chlamydia.

Which of the following best describes this vignette?
A. The patient has a need for treatment
B. The patient has a demand for treatment
C. The patient has not utilized health resources until he has received treatment for his concerns
D. The patient does not have a need or demand for treatment
E. More than one of the answers above is correct

90. The Strategic Advisory Group of Experts (SAGE) on Immunization falls within which organization?

A. Centers for Disease Control and Prevention
B. Department of Health and Human Services
C. Health Resources Services Administration
D. The Joint Commission
E. World Health Organization

91. Which entity ultimately determines when healthcare organization are able to build new hospitals?
A. Federal government
B. Local government
C. State government
D. Individual healthcare organizations determine their own growth
E. Stock exchanges

92. Which of the following programs is considered to be means-tested?
A. Employer-based health insurance
B. Medicare
C. Medicaid
D. Social security
E. All of the above

93. Which of the following parties determines the schedule of controlled substances?
A. Attorney General of the United States
B. Centers for Disease Control and Prevention
C. Food and Drug Administration
D. Health Resources and Services Administration
E. National Academy of Medicine

94. Heroin and other illicit drugs are classified under which controlled substance schedule?
A. I
B. II
C. III
D. IV
E. V

95. A prescription drug monitoring program is intended to serve which purpose?
A. Allow clinicians and law enforcement to view a person's history of certain prescription medications
B. Allow insurers to screen for prescription drug fraudulent activity
C. Collect data of prescribing habits for statistical purposes
D. Create a tool for discounting expensive out-of-pocket payments

E. All of the above

96. The Merit-Based Incentive Payment System (MIPS) pays providers by using elements of which of the following payment strategies?

A. Capitation
B. Cash directly from the patient
C. Diagnostic-related group
D. Overtime pay
E. Pay-for-performance

97. Which of the following organizations operates the Physician Quality Reporting System (PQRS)?

A. Agency for Healthcare Research and Quality (AHRQ)
B. Centers for Disease Control and Prevention (CDC)
C. Center for Medicare and Medicaid Services (CMS)
D. Federation of State Medical Boards (FSMB)
E. Health Effectiveness Data and Information Set (HEDIS)

2.2 HEALTH POLICY AND MANAGEMENT ANSWERS

1. **A. Private to Private**
   In a simplified explanation of the American healthcare marketplace, there are two main parties: Either an entity is public (government) or private (nongovernment). Commercial insurance in the United States is privately owned. This includes for-profit and not-for-profit commercial insurance. Commercial insurance typically pays private recipients (hospitals/practitioners/pharmacies). Rare exceptions include payments to the VA Health System and LHDs. On the contrary, Medicare is always paid by the federal government. The vast majority of the time, it pays private organizations (hospitals/practitioners/pharmacies). Exceptions include payments to VA hospitals.
2. **C. Not-for-profit, Independent of Government**
   TJC is a not-for-profit private organization that is independent of any government organization, with a mission is to improve healthcare for the public. It does so by emphasizing clinical practice guidelines and providing framework for organization structure in order to establish a consistent approach to care across the >20,000 healthcare organizations it accredits in the United States. Achieving TJC accreditation validates an organization's efforts and provides bragging rights for an organization. More importantly, TJC accreditation is frequently linked to insurance reimbursement (including Medicare and Medicaid) and state regulatory requirements.

   TJC accreditation lasts three years and is paid for by the healthcare organization that is being evaluated.
3. **B. Hospital Consumers Assessment of Healthcare Providers and Systems**
   HCAHPS is a nationally standardized survey of patient perception of healthcare. By increasing health transparency, HCAHPS allows healthcare organizations to better address problems. It also allows for a national comparison of healthcare facilities based off of patient perception. It is often used in payment and reimbursement.

   The AHRQ is a federal agency tasked with improving the quality, safety, efficiency, and effectiveness of healthcare in the United States. The AHRQ hosts the USPSTF.

   The PPS is a reimbursement strategy of paying for an episode of care. DRGs are an example of a PPS.

The PHAB is a not-for-profit organization dedicated to protecting the health of the public by increasing health department performance through nationally standardized accreditation of state, local, and tribal health departments.

TJC is a not-for-profit private organization that is independent of any government organization with a mission is to improve healthcare for the public. It accredits organizations, allowing them to enjoy bragging rights, insurance reimbursements, and satisfy state regulations.

4. **B. Medicare**

GME is the process by which physicians become specialty trained through residency training. The GME system in the United States is complex.

Medicare is the largest contributor to GME funding and Medicaid is a distant second, followed by the VHA and HRSA. Many residency programs receive funding from private industry, but these monies are harder to quantify. Private funding may come from a variety of sources, including donations and increased payments that private insurance companies must make to hospitals with residency programs. States contribute to GME through Medicaid payments and some pay directly into programs. Other state-based programs include loan repayment incentives to assure physician coverage in healthcare shortage areas.

Children's hospitals receive a large amount of funding from HRSA, as they do not receive as many Medicare dollars.

5. **B. Medicaid**

Long-term care includes healthcare provided by informal caregivers, home health services, and nursing homes. While Medicare pays the highest percentage of home health services (with Medicaid being second), Medicaid pays the highest percentage of all long-term care services in the United States. In declining order of payment of long-term care expenses after Medicaid is Medicare, out-of-pocket expenses, private insurance, and miscellaneous other forms of payment.

When it comes to long-term care Medicare typically covers "skilled care" expenses, while Medicaid typically covers "custodial care" expenses. Custodial care episodes usually last much longer than skilled care episodes.

Out-of-pocket (self-pay) expenses are typically highest in the beginning of long-term care. After personal finances have been spent down to poverty, Medicaid picks up the payments.

6. **D. Employer Private Insurance**

Nearly 50% of Americans receive their health insurance coverage from their employer, where as a much smaller percentage receive their private insurance through other routes. Private insurance represents the largest mode of financing healthcare. Although government financing is the second most common form of insurance coverage in the United States, it represents the largest source of payment.

7. **A. Government Financed**

Nearly 50% of the American population is insured through employment-based private insurance. However, the largest source of healthcare dollars spent is by government funding. Of the two major government payers, Medicare spends more money than Medicaid. The two largest spenders of Medicare are Medicare A and Medicare B, which spend roughly equivalent amounts. On top of Medicare and Medicaid, the federal government also finances healthcare through the Veterans Health Services, IHS, and tax breaks for healthcare expenditures. Employment-based insurance contributes the second highest amount of money.

8. **B. 21**

The EPSDT program, as specified in Section 1905(r) of the SSA, provides a comprehensive array of treatment services for low-income individuals up to age 21. These services include preventive and diagnostic treatments designed to assure that children under age 21 receive early detection and care.

States have flexibility in determining how to ensure EPSDT services though their Medicaid program. They may choose to provide the services themselves or outsource them. Resources potentially covered by EPSDT include medical supplies and equipment, mental health, hearing and vision services, dentistry, physical and occupational therapy, home health services, and physician care.

9. **B. 3500**

By the HRSA identifying HPSAs, regions of extraordinary clinical need may be identified and potentially helped. Meeting the designation of a HPSA allows for incentives such as physician bonuses to be used to entice physicians to move to the area.

Primary care HPSAs are defined by a physician to population ratio of one physician to a population of 3500. A population may be considered a town, county, group of counties, etc. This formula does not take into account physician extenders, such as nurse practitioners and physician assistants. Roughly, 20% of the population of the United States meets this criteria.

Dental HPSAs are defined by a dentist to population ratio of one dentist to a population of 5000.

Mental health HPSAs are defined by a psychiatrist to population ratio of one Psychiatrist to a population of 30,000.

Medically underserved areas (MUAs) and medically underserved populations (MUPs) are identified by number of physicians per population, poverty, cultural differences, language differences, and infant mortality rate (IMR). These factors all play into the formula to determine the amount of assistance from the federal government.

10. C. FQHCs May Also be Approved as a RHC

FQHCs are healthcare facilities that create a safety net for patients with barriers to care. They receive funding from the HRSA and are largely engaged in providing primary care outpatient health services. To reduce barriers to care, FQHCs offer a sliding fee scale to people with incomes below 200% of the federal poverty level.

Services provided by FQHCs include care and supplies from encounters with physicians, nurse practitioners, physician assistants, midwifes, psychologists, nutritionists, and social workers. Many of these clinical encounters are to provide preventive health services.

Organizations that are eligible to collect Medicare payment as FQHCs include community health centers, migrant health centers, healthcare for the homeless health centers, public housing primary care centers, and organizations operated by tribal associations.

By statue, FQHCs must service MUAs or MUPs. These regions include both rural and urban areas. However, FQHCs may not be concurrently approved as a RHC. RHCs are clinics that receive different streams of funding and are designated as rural areas by the Bureau of the Census and as medically underserved by the Secretary of the DHHS.

11. A. Critical Access Hospitals

The CAHs are hospitals that receive special payment incentives to provide care to isolated rural communities. This financial support allows hospitals to operate in areas of need, with less fear of losing money. These hospitals are at least 35 miles away from other hospitals (or 15 miles when there is mountainous terrain). To maintain the CAH designation, a hospital must have no more than 25 inpatient beds, with an average inpatient length of stay of no more than 96 hours. These hospitals must also provide emergency services 24 hours a day for 7 days a week.

The limited size and length of stay is intended to encourage the CAHs to provide care for common conditions and defer more complicated care to larger hospitals.

12. C. 25%

Roughly, 27% of all Medicare dollars are spent on beneficiaries in their final year of life. Nearly half of this amount is spent in the final two months.

13. C. National Committee for Quality Assurance (NCQA)

The NCQA is a private not-for-profit organization that certifies and accredits organizations based off quality standards and performance management. HEDIS is a data set of health performance measures maintained by the NCQA, which turns data into information that can be used to directly compare one healthcare plan to another.

The HRSA and the AHRQ are divisions within the DHHS.

14. A. Conduct Internal Utilization Review

Reimbursement through a DRG is a form of reimbursing healthcare facilities per episode of hospitalization. With a DRG, all expenditures accrued within a single hospitalization are included in a lump sum assigned to the patient specific diagnosis. Healthcare facilities have incentive to reduce expenditures, which is done by decreasing patient length of stay, reducing the number of nurses per patient, ordering less tests and using generic medications. Hospitals may conduct internal utilization reviews to monitor excessive spending and find opportunities to save money.

A hospital may apply for outlier payments if a patient experiences medically necessary extra costs outside of the normal episode of care for a specific DRG.

15. C. Independent Practice Association (IPA)

A group of practitioners that band together to negotiate with third party payers is known as an independent practice association (IPA). For practitioners, benefits of becoming part of an IPA include increased negotiating power, organizational support and distribution of financial risk. Many IPAs take measures to control costs and conduct internal reviews.

Accountable care organizations (ACOs) are a coordinated group of healthcare providers and facilities that share resources and work together to provide coordinated care to patients.

DRGs are part of a prospective payment system where hospitals are paid a set amount per episode for the entire hospitalization. The amount paid depends on the diagnosis leading to hospitalization. This payment structure incentivized hospitals to reduce costs.

Physician hospital organizations (PHOs) are similar to IPAs. Only in this organization, the physicians also band together with hospitals to makes contracts with the payers.

Preferred provider organizations (PPOs) are organizations where insurers contract with providers and healthcare facilities on an individual basis or [rarely] through an IPO to provide discounted fee-for-service payments.

16. A. Agency for Healthcare Research and Quality (AHRQ)

The USPSTF is an independent group of experts in primary care and preventive medicine that has been authorized by congress to convene through the AHRQ. The AHRQ supports the USPSTF by assisting with daily operations, coordination of gathering evidence reports and assisting in the dissemination of USPSTF recommendations.

17. A. National Vaccine Injury Compensation Program

The VICP was created after lawsuits against vaccine companies threatened to cause vaccination shortages, potentially leading to a reduction in vaccination rates. It is a no-fault program that provides a financial settlement to those that have been injured by a VICP-covered vaccine (as chosen by the Advisory Committee on Immunization Practices).

The Vaccine Injury Compensation Trust Fund maintains funding for VICP, receiving $0.75 from each disease-specific immunization. For example, $0.75 is paid into the fund for a pneumococcal immunization, while $2.25 is paid into the fund for an MMR immunization. The program aims to ensure an adequate supply of vaccines and stabilize vaccine costs. The US DHHS hosts VICP and the US DOJ represents DHHS in court.

18. C. Demographic Rating

Demographic ratings are based off individual factors such as age, gender and zip code.

In health status rating, a new patient is given a new issue rate. An experience rating is based off current health status and past medical history. Industry rating is applied to groups of workers in specific occupations that have increased or decreased use of insurance services. Durational rating is when premiums raise over the time that the patient is insured by the same company.

19. B. Mandated Purchase of Health Insurance

Adverse selection occurs when people with a higher likelihood of using healthcare resources purchase health insurance, while those that are less likely to use healthcare resources forego purchasing health insurance. Without dollars contributed to the risk pool from the low utilizers of healthcare, the overall cost of healthcare is higher for health insurance policy holders.

In an unregulated market, insurance companies can combat adverse selection by excluding preexisting conditions, or charging more money to patients with these conditions. The ACA made it illegal for insurances to refuse coverage based on preexisting conditions. To help pay for the increase in costs for the substantial utilizers of healthcare resources, the ACA also generated an individual mandate, which taxes people that do not have health insurance.

The other options do not relate to adverse selection. Physicians receiving pay-for-performance may see incentive in only caring for those with optimal health. Patients with suboptimal health are more likely to have poor health outcomes, which translates into less money for the physician. Randomization is a tool used in research to minimize bias. Utilization review is a process to analyze the use of healthcare resources.

20. A. Utilization Management

Utilization management, also called utilization review, is a method of using evidence based medicine to evaluate the underuse or overuse of appropriate medical services and contain costs. The party holding the most financial risk performs utilization management, because this party has the incentive to reduce cost.

21. C. The Physician Self-referral Act (Stark Law)

The Physician Self-referral Act (Stark Law) is a part of a series of federal laws aimed to prevent physicians from referring to facilities that they have financial interest in.

An example of violation of Stark would be for a physician to refer patients to an imaging facility for x-rays that the physician holds ownership in. If the physician was allowed to refer patients to his/her own facility, this would increase incentive to order unnecessary tests.

22. B. Part B

Medicare A is for hospital stays, skilled nursing facilities, and hospice care. It is financed through the social security system and monthly premiums for those not enrolled in social security.

Medicare B is for clinician services, medical supplies, and preventive services. It is funded by general federal revenues (taxes) and monthly premiums.

Medicare C is for contracting with managed care plans. It typically offers lower out-of-pocket expenses, but the beneficiaries may have less freedom in the resources available to them.

Medicare D is for prescription care. It is primary financed through tax revenues.

Medicare A and Medicare B both have a similar number of enrollees and spend a similar amount of money.

23. C. Medicare C

Refer to the answer from Answer #22 (directly above) for an explanation of the types of Medicare.

24. E. The Dollars Matched by the Federal Government for State Medicaid Expenses

Medicaid, funded by title 18 of the SSA, is a means-tested program jointly operated by the states and federal government to provide healthcare for those without resources. The portion that the federal government pays is known as the FMAP and almost always exceeds 50% of healthcare costs.

25. D. $862

One dollar at current time is worth more than that same dollar in the future. This is partly because if the dollar were invested initially, it would accumulate value. Thus, money spent at present time can be seen as a loss of investment. The current value of money is adjusted for the future [discounted] value of money by using the following formula:

$$\textit{Discounted price} = \frac{\textit{Original value}}{(\text{Discounted percent} + 1)^{\text{Number of years}}}$$

$$\textit{Discounted price} = \frac{\$1000}{(0.05+1)^3} = \frac{\$1000}{(1.05)^3} = \frac{\$1000}{1.16} = \$862$$

26. D. Tertiary Prevention

Cost-effectiveness analysis (CEA) represents how much money is used to obtain one natural unit of change, when comparing one intervention to another. For example, one may wish to study the amount of heart attacks after using dollars spent to pay for one medication versus the same amount of money spent on another intervention. Tertiary prevention occurs once the disease has already manifested. With less opportunity for prevention, tertiary prevention is typically the least cost effective type of prevention.

27. C. The New Approach is More Effective and Costs Less

Cost-effectiveness analysis (CEA) provides a guideline that is helpful for implementing the most efficient program. Administrators and their ancillary staff may choose to utilize an incremental cost-effectiveness ratio (ICER) to directly compare programs.

$$ICER = \frac{\text{Cost of option B} - \text{Cost of option A}}{\text{Effectiveness of option B} - \text{Effectiveness of option B}}$$

$$= \frac{C_B - C_A}{E_B - E_A}$$

ICER may be graphically depicted in the following diagram. The upper left quadrant is more costly and less effective. It is always considered the less desirable option, and is, therefore, said to be dominated. The lower right corner is less costly and more effective. It is always the more desirable option and is considered to be dominant. The two other quadrants are not as simple to decide upon and require more analysis. A useful tool within ICER is a cost-effectiveness threshold, wish can be used to delineate between efficient and less efficient options.

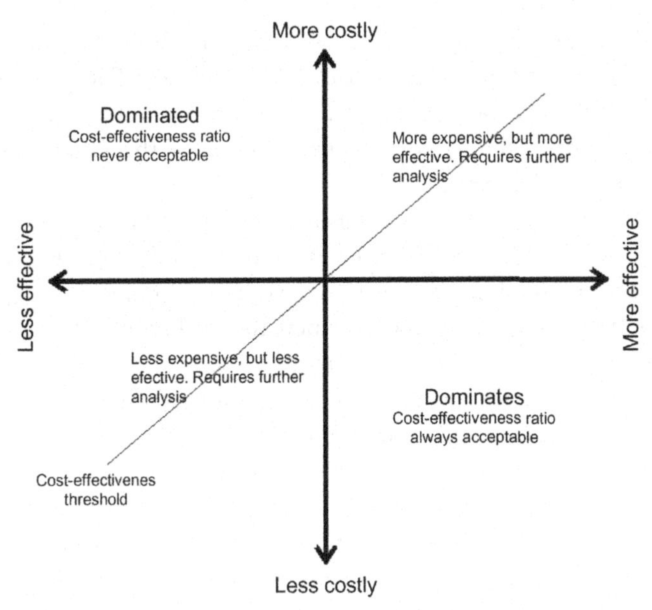

28. B. Outcome of Alternative Options Should be Similar to Each Other

CMA assumes that the outcomes from all of the options are equivalent. The goal of CMA is to find the cheapest approach to achieving an outcome. For example, comparing the price of generic drugs of the same compound based on the cheapest price is a form of CMA. If the same target blood pressure is reached with the same agent, the healthcare plan will choose to supply the cheapest form of that agent.

29. A. Cost-benefit Analysis (CBA)

CBA, CEA, CMA, and CUA are the ways of measuring and comparing outputs from inputs.

CBA measures outputs in dollars. It is useful for comparing and prioritizing nonrelated processes. CBA does this by measuring which program yields the most desirable financial benefit. For example, the chief medical officer (CMO) of an insurance plan may want to know if his company will save more money by providing free Shingles vaccinations or offering free smoking cessation tools.

CEA measures units in natural units. For CEA, consider that the same CMO wants to know how many heart attacks will be averted by using $100,000 for pharmacologic treatment of coronary heart disease compared to the surgical option, cardiac catheterizations.

CMA assumes that the outcomes from all of the options are equivalent; therefore, it is the easiest to calculate. The goal of CMA is to find the cheapest approach to achieving an outcome. For example, comparing the price of generic drugs of the same compound based on the cheapest price is a form of CMA.

CUA measures the humanistic and, often times, less objective outputs of healthcare. To make it more objective, CUA often uses disability adjusted life years (DALYs). For example, will spending $60,000 on a renal transplant increase quality of life more than spending $60,000 for years of dialysis?

Sensitivity analysis is the process of examining how results change when they're placed under different assumptions.

30. D. Quality

Cost, access, and quality are the three variables that contain all other elements of healthcare. If one of these variables increases, at least one other will decrease. For example, if the cost of healthcare increases, the access to this care will decrease.

31. D. Structure

The three main stages of healthcare delivery include structure, process, and outcome. Each of these stages can be measured for quality assessment and improvement.

The structure of a healthcare system describes the way in which healthcare is provided. Examples of healthcare structure include organizational characteristics and material resources.

The process of a healthcare system includes the activities that constitute healthcare. This includes medical decision-making, diagnosis, and treatment.

The outcome of a healthcare system measures the effectiveness of healthcare. This can be measured on both the personal and population levels. Types of outcomes include clinical, economic, and humanitarian results.

32. A. Economic

Health outcomes are measured as being clinical, economic, or humanitarian.

Clinical outcomes include results of diagnosis and treatment. For example, the resolution of acute otitis media from amoxicillin would be a clinical outcome.

Economic outcomes include the financial results of treatment. For example, the money spent on re-treating a patient that did not respond to amoxicillin for acute otitis media would be an economic outcome.

Humanitarian outcomes include healthcare results that cannot be measured in clinical or economic outcomes. For example, a patient may feel better taking acetaminophen in addition to amoxicillin for otitis media, but it does not influence clinical or economic results.

33. B. Flexible spending account (FSA)

The FSA allows a worker to set money aside tax-free at the beginning of the year for healthcare expenses. By giving a tax break to the FSA users, it reduces barriers to care.

Cost-sharing mechanisms are methods to have the insured member pay for part of their own care. They're often put in place to create a small barrier to care and prevent overutilization of services. There is wide debate over whether these mechanisms represent or create too large of a barrier and limit the use of appropriate services.

A deductible represents a set amount of money that a person must pay out-of-pocket before their insurance begins to pay for expenses.

A copayment is a set fee that a person must pay out-of-pocket before their insurance begins to pay for expenses.

Coinsurance is a set percentage of expenses from a medical bill that a person must pay out-of-pocket.

34. B. A, B

Under the ACA, USPSTF grade A and B recommendations receive first dollar coverage. This means that any recommendations given a grade A or B will be covered by the insurer, without the insured having to pay any money out of his or her pocket. Insurers are not forbidden from covering grade C, D, and I recommendations. Rather, they are encouraged to develop their own policy in paying for these lesser recommended tests.

One notable exception to this rule is mammography. Although the USPSTF does not recommend screening mammography until age 50 (grade B recommendation), first dollar coverage is provided for those over age 40 (grade C recommendation).

35. C. Independent Payment Advisory Board (IPAB)

The ACA created the IPAB to control Medicare expenditures. The IPAB is composed of a panel of 15 healthcare experts (plus the HHS Secretary along with CMS and HRSA nonvoting members) appointed by the president and confirmed by the senate. It is housed within the executive branch, further out of the reach of the lobbying external stakeholders with access to congress.

Every year, the Chief Actuary of CMS must calculate the projected Medicare growth rate. If this rate is higher than the future targeted growth rate, the actuary is then to determine how much Medicare will have to trim its spending growth. The IPAB will take these numbers to recommend ways to cut spending. The IPAB then submits a draft proposal to the Medicare Payment Advisory Commission (MedPac) and HHS. Congress must then adopt these recommendations or enact savings of similar size in Medicare. If congress fails to act, the HHS Secretary must pass these recommendations. The IPAB may not recommend changes to premiums, benefits, eligibility, taxes, or changes that would ration healthcare.

36. A. Capitation

Capitated payment is a payment made to a healthcare provider for each patient cared for over a specified period of time. In this model, the clinician undertakes some of the risk previously carried by the insurer. Clinicians have incentives to order less services, which cut into their bottom line.

In a fee-for-service reimbursement system, the clinician is reimbursed for each procedure performed. This creates an incentive to order more services than are necessary.

Reimbursement through a DRG is a form of reimbursing healthcare facilities per episode of hospitalization. With a DRG, all services used within a single hospitalization are included in a lump sum assigned to the diagnosis. Healthcare facilities have incentive to reduce expenditures and typically monitor this through internal utilization review. A hospital may apply for outlier payments if a patient experiences medically necessary extra costs outside of the normal episode of care for a specific DRG.

RBRVS is a component of fee-for-service payments, where fees are set by estimating components (time, effort, level of judgment, and technical skill) inherent to that service.

Physicians receiving a salary are typically employed by an organization and are paid for their time. Salaried physicians have less external influence from financial risk and balancing the appropriateness of services. However, salaried physicians often times experience less freedom in terms of their individual ability to control their patient load.

37. C. Fee-for-service

In a fee-for-service arrangement, clinicians are reimbursed for every service provided. This creates an incentive to work harder and perform more procedures to increase income.

The capitation reimbursement platform works in sharp contrast to the fee-for-service arrangement. Under capitation, providers are paid for each person's healthcare over a period of time, regardless of the number of procedures performed.

DRGs are a method of prospective payment, where hospitals are reimbursed per episode of illness.

The pay-for-performance model reimburses clinicians based on healthcare outcomes and patient perception of care.

Physicians that receive a paid salary are typically not reimbursed directly for patient seen or number of procedures ordered. However, salaried physicians may receive other forms of compensation, such as from the resource-based relative value scale (RBRVS).

38. A. Coinsurance

Cost-sharing is when patients must pay for a part of their healthcare out-of-pocket. The three main types of cost-sharing (coinsurance, copayment, and deductibles) may be used individually or together.

Patients with a coinsurance must pay a set percentage of healthcare expenditures out-of-pocket. For example, a patient may have an insurance plan that agrees to pay 80% of healthcare expenditures if they pay 20% out-of-pocket.

Copayments are set amounts that a patient must pay to access a service. The set amount typically varies by the type of service being provided. For example, one plan may charge $10 to fill each prescription and $100 for an emergency room visit.

Deductibles are a set amount of money that a person must pay out-of-pocket as a prerequisite for their insurance to begin reimbursing for their healthcare. A plan may decide that each member must pay $10,000 out-of-pocket before contributing 80% (coinsurance) to the rest of their care costs.

Many critics view cost-sharing mechanisms to be controversial, as they may place barriers to care. Studies on this topic show that patients are typically uninformed consumers of healthcare and cost-sharing acts as a barrier that decreases both appropriate and inappropriate use of medical services. While cost-sharing reduces patient healthcare usage, utilization review reduces provider healthcare usage and is performed by whoever is at financial risk.

39. C. Humanitarian

There are three main outcomes for healthcare interventions. Healthcare outcomes may be categorized as either clinical, economic, or humanitarian. Humanitarian outcomes are often measured by cost-utility analysis (CUA), which considers the quality of life. Cost-effectiveness analysis (CEA) can best measure both clinical and humanitarian outcomes because it uses a natural unit (i.e., months of life gained). Cost-benefit analysis (CBA) is best used for economic outcomes, as it compares dollars spent to dollars benefited.

40. B. Quality Adjusted Life Years (QALY)

CUA, CBA, and CEA are ways of measuring and comparing outputs from inputs. Each formula measures a different type of output. CUA measures the humanistic and, often times, less objective outputs of healthcare. To make it more objective, CUA often uses disability adjusted life years (DALYs). An example of using CUA would be calculating DALYs to determine if spending $60,000 on a renal transplant will increase quality of life more than spending $60,000 for 10 years of dialysis.

CBA measures outputs in dollars. For example, the chief medical officer of an insurance plan may want to know if his company will save more money by providing Pap Smears rather than waiting to use a surgical approach in those that develop cervical cancer.

CEA measures units in natural units. For CEA, consider that the same chief medical officer wants to know how many heart attacks will be averted by using $100,000 for pharmacologic treatment of coronary heart disease compared to the surgical option, cardiac catheterizations.

41. A. Centers for Disease Control and Prevention (CDC)

The NHSN falls within the CDC. It is the most prominent healthcare associated infection (HAI) tracking system in the United States. The NHSN provides support in data collection, reporting and analysis regarding patient safety, healthcare worker safety, biovigilence, and long-term care. It turns this data into useful information for facilities, states, regions, and the country to identify problem areas, measure progress of prevention efforts, and reduce HAIs. The NHSN has become the preferred avenue for healthcare facilities to comply with reporting requirements set forth by the CMS. NHSN information is available to the public.

The CDC, FDA, HRSA, and NIH are all part of the DHHS.

42. B. Health Resources and Services Administration (HRSA)

The NPDB is a national clearinghouse used as a comprehensive review of professional credentials for healthcare providers and entities. NPDB has been assigned several roles in monumental health legislation. Title IV of Public Health Law of the Healthcare Quality Improvement Act established NPDB as the record keeper of medical malpractice and clinician misconduct. The SSA further expands NPDB to include adverse actions taken by the federal government, state licensing authorities, Medicaid fraud control units, and several other organizations.

43. D. Department of Labor (DOL)

The OSHA, Mine Safety and Health Administration (MSHA) Department of Mine Safety, and Equal Employment Opportunity Commission (EEOC) all fall within the DOL.

44. A. Department of Agriculture

Child nutrition programs (including school meals), the Supplemental Nutrition Assistance Program (SNAP), the WIC Program, and food security programs are all public health efforts made through the Food and Nutrition Service (FNS) sector within the USDA.

45. A. US Department of Agriculture (USDA)

The Food and Nutrition Service (FNS) program within the USDA is dedicated to providing needy people with nutrition assistance by providing nutrition education and access to food. FNS programs include the Supplemental Nutrition Assistance Program (SNAP), supplemental nutrition for WIC, as well as free meals at public schools.

46. D. State

In a centralized health department classification system, the state government holds the authority to make fiscal decisions for the municipalities. In a shared governance, LHDs may be led by state or local officials. In decentralized systems, the LHDs retain authority over local decisions.

47. D. National Association of County and City Health Officials (NACCHO)

The NACCHO is a national organization composed of 2800 LHDs across the United States. NACHO's visions is "health, equity, and security for all people in their communities through public health policies and services." The organization act as a voice for local health departments and promote health in areas of community health, environmental health, public health preparedness, and public health infrastructure and systems.

48. E. All LHDs are Included Under the State Public Health Agency/SHD

LHDs may serve counties, cities, towns, regions, or any combination of the above. They are components of the state public health agency/SHD. As the political landscape changes in the state, the LHD's relationship to the SHD may undergo changes.

There are four main types of relationships between SHDs and LHDs.

Type of relationship	Description of relationship
Centralized	The LHD is operated by the SHD or state board of health
Decentralized	The LHD is largely independent of the SHD
Shared authority	The LHD operates under the SHD, state board of health, and local government
Mixed authority	LHD services are provided by blends of the SHD, state board of health, and local government

49. B. Centers for Disease Control and Prevention (CDC)

The National Institute for Occupational Safety and Health (NIOSH), is a division within the CDC, which is within the DHHS. NIOSH researches occupational problems to provide solutions and recommendations to protect the health of workers. NIOSH creates recommended exposure limits (RELs) to guidance workers on how much exposure to a chemical is acceptable. NIOSH does not have the power to enforce regulations.

The ACGIH is an independent member-based organization dedicated to advancing occupational and environmental health. This organization publishes both threshold limit values (TLVs) and biological exposure indices (BEIs) as guidelines to use for well-informed decision-making in the occupational arena. ACGIH cannot enforce regulations.

The OSHA falls within the DOL.

OSHA is tasked with regulating the work environment, to make it safe for employees. It publishes permissible exposure limits (PELs), based off NIOSH's recommended exposure limits (RELs) and has the power to enforce legislation.

The WHO has been tasked by the United Nations to coordinate national health efforts.

50. C. Ensuring Safety of Tobacco Products

The Center for Tobacco Products within the FDA regulates the manufacturing, distribution, and marketing of tobacco products.

The USDA administers the WIC Program, ensures safety of eggs, poultry, and beef and regulates genetically engineered foods. The USDA is focused on providing leadership on agriculture, food, natural resources, nutrition, and rural development. Some of the USDA divisions are as follows:

Animal and Plant Health Inspection Service (APHIS): Protects agricultural resources from diseases and regulates genetically engineered organisms.

Agriculture Research Service (ARS): Produces research in food production and protection, human nutrition, food safety, and food bioengineering.

Center for Nutrition Policy and Promotion (CNPP): Establishes federal nutrition policies and standards. It also disseminates the Dietary Guidelines for Americans and the MyPlate food guidance system.

Food and Nutrition Service (FNS): Provides access to food and nutritional resources to the needy. Programs include the SNAP, WIC, and free meals to public schools.

Food Safety and Inspection Service (FSIS): Monitors the nation's commercial supply of meat, poultry, and processed egg products.

51. **B. US Department of Health and Human Services (DHHS)**
The IHS is one of 11 operating divisions within the DHHS. It is responsible for providing healthcare to American Indians and Alaska Natives.

The 11 operating divisions within DHHS are:

- Administration for Children and Families (ACF)
- Administration for Community Living (ACL)
- Agency for Healthcare Research and Quality (AHRQ)
- Agency for Toxic Substances and Disease Registry (ATSDR)
- Centers for Disease Control and Prevention (CDC)
- Centers for Medicare and Medicaid Services (CMS)
- Food and Drug Administration (FDA)
- Health Resources and Services Administration (HRSA)
- Indian Health Services (IHS)
- National Institutes of Health (NIH)
- Substance Abuse and Mental Health Services Administration (SAMHSA)

The NRHA is a national nonprofit organization with a mission to provide leadership on rural health issues.

The APHA is a nonprofit organization that strives to strengthen the health of all people, while advocating to strengthen public health as a profession.

52. **B. Department of Health and Human Services (DHHS)**
The ATSDR falls within the DHHS. See the answer for Question #51 (directly above) for a full listing of the divisions within DHHS.

53. **E. Department of Veterans Affairs**
The Veterans Benefits Administration falls within the Department of Veterans Affairs.

54. **C. Department of Justice (DOJ)**
Of public health interest, the DOJ houses the Bureau of Tobacco, Alcohol, Firearms, and Explosives and the Drug Enforcement Administration (DEA).

55. **B. 18**

Title of SSA	Content
17	Grants for planning comprehensive action to combat intellectual disability
18	Health insurance for the aged and disabled, (Medicare)
19	Grants to states for medical assistance programs (Medicaid)
20	Block grants to states for social services
21	State children's health insurance program (SCHIP)

56. C. 19

See the explanation in Answer #55 (directly above) for a full listing of the relevant titles within the SSA.

57. A. Approval of New Drugs

The Reserved Powers Doctrine holds that the states inherent government powers that are not explicitly granted to the federal government. Two of these powers include the parens patriae power (protecting interests of minors and incompetent persons) and police powers, which include the authority of the state to enact laws and create regulations to protect, preserve, and promote the general health and safety of the people. States may choose to pass on their police powers to local municipalities. Often times, the state and local governments are better able to understand and address their needs than the federal government.

Police powers are not addressed in the constitution. Some of the police powers are include: Licensure of healthcare professionals, fluoridation of drinking water, inspection of eating establishments, regulation of air, standards of water, abatement of unsanitary conditions and enforcement of vaccination.

Federal and state governments often exercise public health powers together. However, by enacting the Supremacy Clause, the Federal government gets the final say. The Supremacy Clause states that the Federal government has the ultimate power (as granted in the United States constitution), even if the state is acting within its police powers.

58. B. State

Medical professionals may be credential through licensure, certification, or registration. These qualifications limit competition to qualified professionals, ensure quality of care and may serve as avenues to discipline healthcare providers.

Medical licensure is regulated by state laws. This includes licensure of practitioners, laboratories, and medical facilities. Each state may individually tailor requirements and scope of practice for licenses. For this reason, licensure varies state by state. For example, a state with a shortage of dentists may expand the scope of practice for dental hygienists to ensure access to specific dental services and encourage more dental hygienists to move to the state.

When a medical professional has become registered, it means that they have been approved to perform activities by a governing body.

The approval process may contingent on education, standardized testing, and experience.

Certification is a voluntary process that is regulated through professional organizations. Although voluntary, certified status may be necessary to enjoy benefits and incentives.

59. C. National Health Information Center (NHIC)

Of the programs listed, the only one operated under the Office of Disease Prevention and Health Promotion (ODPHP) is the NHIC. The ODPHP falls under the Assistant Secretary of Health within the DHHS. ODPHP sets national health goals and maintains programs that support these goals. This is accomplished through the NHIC, training programs for public health professionals (including a residency program) and management of the following websites: www.health.gov (resource for health information), www.healthypeople.gov (national Healthy People objectives), and www.healthfinder.gov (evidence-based information for health consumers).

The NHIC supports public health promotion. One main duty of the NHIC is to maintain a calendar of the National Health Observances to promote health awareness. It also updates the federal health information centers and clearinghouses NHIC connects health professionals to health consumers to relay accurate health information.

60. A. Strategic Planning

SWOT is a business tool used for strategic planning. It is an acronym for the following; strengths, weaknesses, opportunities, and threats. Proper use of SWOT will enable gap analysis, where an organization realizes the difference between its current resources and the resources it needs to fulfill its goals.

61. B. Technological

PEST is an acronym for political, environmental, sociological, and technological. PEST analysis is a strategic planning tool that allows an organization to identify where it stands in the marketplace, in terms of strengths and challenges.

62. A. Act

PDSA is an acronym that stands for plan, do, study, act. It represents a quality improvement tool. This acronym is sometimes identified as PDCA, where the "c" has been substituted for the "s" and stands for check.

To perform PDSA analysis, a specific opportunity for improvement is identified. Then each of the four steps are applied in a sequential basis.

63. C. Six Sigma

DMAIC is a component of Six Sigma. The name Six Sigma is derived from having the goal of being six standard deviations away from the mean, ≤3.4 defects per million. To accomplish this feat, six sigma uses the DMAIC acronym, which stands for the following:

- Define the opportunity for improvement
- Measure the most appropriate variables
- Analyze the data to determine the root problem
- Improve with solutions
- Control with continuous monitoring

64. D. Reliable

SMART is an acronym that outlines objectives to accomplish goals. It stands for specific, measurable, attainable, relevant, and time oriented. The 'r' in the S.M.A.R.T. stands for relevant, not reliable.

65. B. Hydrocodone Cough Syrup

The Secure and Responsible Drug Disposal Act (also known as Disposal Act) amended the CSA to regulate the way in which the Drug Enforcement Agency (DEA) is able to appropriately handle unused controlled prescription medications in the United States. This act authorizes collection of unused pharmaceuticals through three avenues: Depositing the unused medication in an approved collection receptacle, collection by law enforcement at a take-back event and use of appropriately designated mail-back packages. This law does not cover illicit drugs such as heroin or biohazardous agents such as used insulin needles. It also does not allow for collection of medications containing iodine or medical equipment containing mercury. The drugs may only be turned over by the prescription holder or a household member responsible for the prescription holder. In the case of death of the prescription holder, long-term care facilities and estate managers are delegated the right turn in controlled medications. The collecting entity may not ask for personally identifying information from the person turning in the medication.

The goal of the CSA is to appropriately reduce the amount of potentially hazardous controlled prescription medications in the United States. These medications are too commonly disposed of in waterways, where they lead to adverse environmental effects. Once collected appropriately, the most common form of ultimate disposal is through incineration.

66. D. Changing the Process Will Change the Outcome

There are three quality control stages to which healthcare organizations can focus on: Structure, process, and outcome. Structure involves the way the organization is composed, including the staff and the physical components of the facility. Process is the way the organization performs the work, including the skill of care and appropriateness of care. Outcome is the final results achieved by the organization. Changing the process typically produces a different outcome.

67. A. Healthcare Effectiveness Data and Information Set (HEDIS)

The HEDIS is a report card program that was developed by the NCQA, a nonprofit private organization that is dedicated to improving health quality. HEDIS measures can be used to directly compare performance measures from one plan to another. For example, once can use HEDIS to compare the amount of pneumococcal vaccine administered in one plan to another.

The National Guideline Clearinghouse is a resource hosted by the AHRQ that retains evidence-based clinical practice guidelines.

The National Quality Measures Clearinghouse is also hosted by the AHRQ. It retains a plethora of information on evidence-based quality measures and measure sets.

Departments of Vital Statistics do not distinguish between members of different healthcare plans. The information recorded by Departments of Vital Statistics are limited to data such as birth, death, marriage, and divorce.

The Hospital Quality Initiative is a voluntary reporting program in which hospitals publically report on their key quality measures, such as quality of care for heart attacks, pneumonia, and surgery. It is not further divided into patients belonging to individual health plans, as the goal of this initiative is for patients (consumers) to research and compare hospitals.

68. D. Determine the Root Cause of the Problem

Ishikawa diagrams are also commonly known as cause-and-effect diagrams and fishbone diagrams. An Ishikawa diagram reads from right to left. At the far right of the diagram is the problem to be addressed. Moving to the left, the diagram identifies root causes of the problem (s). These root causes are further broken into subcauses. Once the diagram has been drawn out, it takes the shape of fish bones.

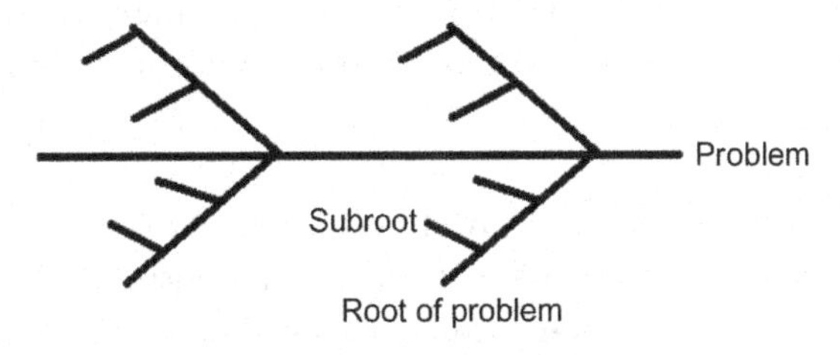

The incorrect options in this question each represent different management tools. Gantt charts are an administrative tool to help create a project schedule. On a Gantt chart, each member is assigned a task and a time period to complete that task. Often times, one task needs to be completed before another can be initiated. Funnel plots are tools to evaluate meta-analysis studies for possible publication bias. Authors of a meta-analysis study may advertently or inadvertently select studies which support the alternate hypothesis. Funnel plots are tools to prevent this bias from occurring. Flowcharts are the main tools used to map the decision-making process. Finally, pie charts are used to graphically show how a whole variable is divided.

69. D. Medical Paternalism

The physician in this case is most likely to state he used his best judgment in accordance with the bioethical principle of medical paternalism. In medical paternalism, a decision is made by a healthcare practitioner that the principle of beneficence trumps the principle of autonomy.

Autonomy refers to the patient's decision to make his/her own decisions. A decision is autonomous if the patient is competent to make decisions, the choice is voluntary and the patient is adequately informed of the risks and benefits of the procedure.

The principle of beneficence refers to helping a person achieve what is in their best interest. It encompasses preventing harm, removing harm, and promoting good. Medical paternalism is a more appropriate answer to this question, specifically because the patient's desires (and from her standpoint best interests) are outweighed by the beliefs of the physician.

Nonmaleficence is the bioethical principle stating that healthcare providers should refrain from causing harm.

Justice is the principle that states all patients in similar situations should be treated equally. Likewise, patients in different situations may be treated unequally. For example, a patient in the intensive care unit (ICU) should receive more healthcare resources than a patient in a clinic.

70. A. Beneficence

Refer to Answer #69 (directly above) for an explanation of the terms autonomy, beneficence, nonmaleficence, and justice.

71. D. A News Reporter Calls to Speak with the Treating Physician

A healthcare professional sharing personal health information with a member of the media is a clear violation of a patient's privacy and violates the HIPAA.

45 CFR 164.512(I) states that HIPAA privacy laws are different for workers' compensation insurers, workers' compensation administrative agencies and employers. Health information relating to a workers' compensation claim may be shared with need-to-know parties and must be the minimum necessary amount of information needed to work through the workers' compensation claim.

45 CFR 164.512(b) is legislation that allows entities to share personal health information with government entities for protection of the health of the public. It is an important tool for identifying risks to the general public and individual patients.

Legal and judicial enforcement are allowed access to a patient's address, date of birth, social security number, blood type, type of injury, date and type of treatment, date and time of death, and description of physical characteristics. However, under [45 CFR 164.512(f)(2)], a court may petition for personal health information in the form of DNA, dental records, and body fluid/tissue typing. Pertinent personal health information may also be shared to protect correctional officers in jails and prisons to from preventable injuries and infections.

Privacy Rule at 45 CFR § 164.512(j) states that a physician may contact a patient's family or law enforcement if they believe the patient to be at immediate risk to themselves or others.

72. B. Health Insurance Portability and Accountability Act (HIPAA)

The HIPAA was signed into law with the intentions to foster health insurance portability, fraud enforcement, and patient confidentiality.

73. B. Patient Perception of Care Provided

HCAHPS (pronounced H-Caps) is a nationally standardized survey of patient's perception of hospital care. Because HCAHPS is standardized, it can be used to compare hospitals to each other. CMS publishes these results on their webpage (www.hospitalcompare.hhs.gov) for consumers to compare hospitals.

74. B. Employee Retirement Income Security Act (ERISA)

The ERISA is federal legislation that dictates minimum requirements for health and retirement benefits for private industry. It is monumental legislation that has been expanded through both the COBRA and the HIPAA.

COBRA was passed to give workers and their families the option to continue their employee health benefits for a period of time after leaving a job.

Meanwhile, HIPAA was passed to increase protection of patient privacy and create safeguards of personal health information.

The Hill–Burton Act was passed in 1946 to update healthcare infrastructure in the United States.

The PSDA allows each [competent] individual to make healthcare decisions regarding their own medical treatment.

75. E. Parens Patriae

Parens patriae translates to "Parent of his country" and is a principle that mainly protects children and those that are intellectually disabled whom are unable to care for themselves.

An advanced directive is a statement completed by a person while they are competent, which directs their healthcare wishes in the case that they become incompetent.

Adverse selection is the tendency for those at high risk of illness to seek health insurance. If an insurance company does not insure lower-risk individuals to balance out the use of services, rates will increase.

Community rating refers to a method used by insurers to set premium rates based on demographic characteristics (gender and age), as opposed to rates based off health status or claims history (experience rating).

An entitlement program is a government program that provides resources to those that meet eligibility criteria. Examples include Medicaid and unemployment benefits.

76. C. ≥40 Years Old

The ADEA is a federal law that prohibits discrimination of workers age 40 and older. Some individual states provide similar protection to workers of other ages.

77. D. 36 Months

The COBRA guarantees continuation of employee (and their family) health insurance after leaving their job. Under COBRA, qualified individuals may be required to pay the entire premium for their coverage, without help from their former employer.

Qualified beneficiaries are typically entitled to 18 months of coverage after loss of benefits. However, in the case that the former employer is within 18 months of qualification for Medicare, 36 months of coverage is granted. Under no circumstance does COBRA last more than 36 months.

78. D. The ACA Allows Health Insurance Exclusion Due to Preexisting Conditions

The ACA passed in 2010 made revolutionary changes to the American healthcare marketplace. These changes amended the prior system rather than change the system altogether.

Prior to enactment of the ACA, health insurance companies enjoyed the ability to select the people they wanted to assure based off of their risk-profile and potential effect on their bottom line. The ACA forced regulation in the health insurance marketplace with several landmark amendments. First of all, it made it unlawful for insurance companies to continue their practice of not covering people with preexisting conditions. To help offset expenses from the health insurance companies [along with several other reasons], the ACA also called for an individual mandate, requiring everyone to obtain health insurance. The individual mandate allows individuals to be financially penalized if they fail to purchase health insurance. The ACA also created an employer mandate, where companies with more than 50 fulltime employees are penalized if they don't offer their [full time] employees an employer-sponsored health plan. It also required insurance companies to cover people up to age 26 on their parents' policy. In further regulation of the health insurance marketplace, the ACA limited the amount of money an insurance company may collect from the highest risk patients compared to the lowest and instituted a medical loss ratio, limiting the amount of premiums that may be kept by the insurance company in the form of profits.

Outside of the health insurance marketplace, the ACA also expanded Medicaid, established insurance exchanges, and made several other sweeping changes.

79. E. Social Security Act

Medicare and Medicaid were born through amendments to the SSA.

80. D. Only Hospitals That Receive Medicare Reimbursement

The EMTALA governs when and how patients may be refused treatment. It only applies to hospitals that accept Medicare reimbursement. Generally, it states that the hospitals and their on-call medical providers must treat or transfer (within reason) patients who present with emergency medical conditions. Doctors that violate EMTALA are subject to civil monetary penalties. Additionally, doctors that terminate their physician-patient relationship during emergency care are guilty of patient abandonment.

81. E. All of the Above Elements Must be Present for a Successful Malpractice Suit

A successful malpractice claim includes four main elements: Duty, breach, cause, and harm. All four of these elements are listed as potential answers to this question. In other words, a plaintiff must prove that a doctor–patient relationship existed, there was physician negligence, this negligence caused injury and this injury led to damages.

82. C. 4%

Roughly, 4% of all medical costs in the United States are spent on dental services. The largest payer of dental work is private insurance, followed closely by out-of-pocket expenditures, with the [federal and state] government coming in a distant third. The ACA of 2010 expanded Medicaid to many previously indigent patients. Children (up to age 21) receive benefits from Medicaid's EPSDT program, but Medicaid dental benefits are considered optional for adults, and many states limit this type of spending.

83. A. Clinical Laboratory Improvement Amendments (CLIA)

The CLIA is a set of regulatory standards that assures quality laboratory testing. It is regulated by the CMS. To certify CLIA compliance, laboratories review documentation and reporting of their results, scrutinize personnel credentialing, and partake in quality control efforts. To confirm accuracy, laboratories undergo proficiency testing, where the lab receives an unknown sample and are tasked with reaching the appropriate lab measurement/diagnosis. CLIA covers both private and public health laboratories. Public health laboratories are important to the medical and public health infrastructure in the United States.

The LRN is an integrated domestic and international network of laboratories that respond to public health problems through rapid testing and timely notification.

Outside of CLIA and LRN, the other answer options do not exist.

84. B. Food and Drug Administration (FDA)

Regulation for different types of GMOs in the United States depends on the type of GMO. The FDA is the agency responsible for approval of GMOs in food, drugs, and biological products. The FDA holds a stance that genetically modified foods must meet the same requirements as food derived from conventionally farmed plants. GMOs in plants (not consumed as food) are regulated by the US Department of Agriculture. Additionally, GMOs involving pesticides and microorganisms are regulated by the Environmental Protection Agency (EPA) under the Federal Insecticide, Fungicide, and Rodenticide Act (FIFRA) and the Toxic Substances Control Act (ToSCA).

GMOs are foods encrypted with foreign DNA to produce desired phenotypical properties that yield production increases, insect resistance, and increased nutritional content. This has great public health implications, as it has the potential to increase food safety and availability. In the future, GMOs may include vaccines and medications.

The United States is the world's leading producer of genetically modified crops. Proponents of GMOs include the National Research Council and the American Medical Association. These organizations have issued statements that there is no evidence GMOs present health safety risks. However, production of genetically modified foods is not without controversy. Due to fears to human and animal health (kidney and liver dysfunction are often cited), GMOs have been banned in many countries. There are parties on both side of the debate that feel there is still more information needed to evaluate the long-term effects of GMOs.

Because the federal government has created GMO legislation under national acts, state laws play a small role in GMO regulation due to the Federal Preemption Doctrine. Although state and local governments may not regulate GMO approval, they may take regulate GMOs through over avenues. For example, states may require labeling of GMO products and municipal governments may ban farming of GMO crops.

85. E. Urinary Tract Infections

Urinary tract infections (UTIs) are the most common type of healthcare-associated infection (HAI). Including catheter-associated urinary tract infections (CAUTIs), UTIs account for more than 30% of HAIs at acute care hospitals. With the highest use in intensive care units, roughly 20% of all hospitalized patients receive urinary catheters. Catheters are also frequently used in long-term care facilities.

Compared to other HAIs, morbidity and mortality from CAUTIs is relatively low. However, their burden on healthcare facilities is immense. CAUTIs are associated with increases in drug-resistant bacteria, hospital length of stay, hospital cost, morbidity, and mortality. They're the leading cause of secondary nosocomial bacteremia.

86. B. Hand Washing

Hand hygiene is regarded as the most important way to prevent nosocomial infections. Not only are pathogens transmitted from patient and equipment contact, but a healthcare worker's own flora may cause infection. Adherence to appropriate hand washing techniques remains as low as 40%. The most common reasons for lack of hand washing includes inconvenient sinks and sanitizing gel dispensers, busy work setting, cultural issues, lack of role models, and skin irritation from hand hygiene (most commonly contact dermatitis).

87. E. >100,000

Deaths from medical errors are underreported and difficult to classify. The number of annual deaths due to medical error is now estimated to top 250,000. Medical errors include unintended acts that do not achieve the planned outcome, error of execution, error of planning and error of diagnosis. In 1999, the Institute of Medicine (now known as the Academy of Medicine) brought attention to this issue by releasing a landmark report that estimated between 44,000–98,000 people die from medical errors annually. The numbers in this report are now thought to be an underestimate, even for that time. This report also estimated that 3% of hospitalized patients experience an adverse medical error.

As the science of medicine advances, there is more room for error. Nearly everyone will experience at least one diagnostic error in their lifetime and nearly 5% of the population will experience one in any given year. Postmortem studies indicate that diagnostic errors contribute to around 10% of all deaths. Human error is inevitable and continuous quality improvement programs should be set in place to better measure and design safer programs.

88. C. Pay-for-performance

Under the pay-for-performance model, financial incentives are provided for clinicians if predetermined patient care goals are met. This model is used to provide an incentive for clinicians to provide quality and appropriate care. Under the fee-for-service model, clinicians are reimbursed per individual action performed. Theoretically, if the clinician works harder and performs more services, he/she will be compensated for the extra work. The main drawback of fee-for-service is that the clinician has incentive to perform unnecessary tests and procedures. In a capitated model, the clinician is paid a set rate for each patient. The clinician assumes more financial risk and has the incentive to keep patients healthy to minimize the costs of care. The main drawback for the capitated model is that the clinician has an incentive to minimize the amount of care provided, which may come at the cost of patient wellbeing. A DRG is a type of hospital reimbursement system where the hospital receives a set payment per each patient's diagnosis during their hospital stay. Hospitals have incentive to minimize the cost of services used during the hospitalization.

89. B. The Patient Has a Demand for Treatment

In health economics, price and demand may differ from the free market. This is because healthcare may not always be a free market, as there are constraints such as government regulation and third-party payment. In healthcare, the need for a medical service exists when an individual has a condition for which he/she believes there to be a beneficial treatment. Meanwhile, demand for care exists when the individual thinks he/she has a need and is willing to spend resources to receive care. Both need and demand may coexist. Utilization of services occurs when services are rendered for a need or demand of healthcare services.

Needs may be broken down into four main categories:

1. Normative need- Services determined by experts to be essential. This category includes standard guidelines for both preventive and clinical care.
2. Felt need- Is the subjective view of the patient. It is not always based upon physiological necessity.
3. Expressed need- A Felt Need that is acted upon.
4. Compared need-A need arising out of comparison to a similar cohort. For example, community may be relatively in need for resources that a comparable community has.

90. E. World Health Organization (WHO)

The SAGE on Immunization is the primary advisory group to the WHO for immunization policies and strategies.

91. C. State Governments

Part of the Health Planning and Resources Development Act, Certificates of Need (CON) are required for healthcare providers to expand, acquire, or create new services. The CON process requires healthcare providers to submit their plans and a justification of need to their local review body. This review body then submits their recommendation to the state's review board, which makes the final determination. Each state maintains its own CON implementation laws.

CONs were developed to reduce exponentially growing healthcare costs. It has been shown that increased availability of medical resources is associated with increased utilization of medical resources.

92. C. Medicaid

There are two main types of public assistance: Entitlement and means-tested. In entitlement programs, the government's laws state that people within a group are entitled to a service. Meanwhile, means-tested programs encompass the government providing assistance to those that demonstrate need due to limited resources.

The two types of public assistance may be broken down into three categories of programs as shown in the table below:

Type of Public Assistance Program	Example of Program
Non-means-tested entitlement	Social Security, Medicare
Means-tested entitlement	Medicaid, SNAP
Means-tested nonentitlement	Temporary Assistance to Needy Families (TANF)

Medicaid is a government program operated by the states with assistance from the federal government to provide help to those that meet means-based criteria. The federal government sets requirements for Medicaid inclusion criteria and services. Individual states may choose to expand their own eligible Medicaid population and services.

93. A. Attorney General of the United States

The CSA is composed of three different sections to regulate the use of pharmaceutical agents. These three sections establish drug rehab programs, create control mechanisms for registration and distribution of controlled substances, and legislate regulations for importation and exportation of controlled substances.

Controlled substances are pharmaceutical agents deemed to have potential for harmful effects and abuse. They are categorized into five schedules (classifications). Schedule I agents carry extreme potential for harm and abuse, while Schedule IV agents have the least propensity for harm and abuse. The Attorney General of the United States has the authority to determine a medication's schedule. The Secretary of the DHHS may provide input to the Attorney General.

The DEA has been tasked with the responsibility to enforce the CSA on the federal level. Manufacturers, distributors, and dispensers of controlled substances must register with the DEA.

94. A. I

The United States Attorney General has the responsibility to classify controlled pharmaceuticals into five distinct classifications (schedules) that are controlled and enforcement by the DEA.

Schedule I agents are drugs without currently accepted medical use. These drugs have a high potential for abuse and harm. Examples of schedule I drugs include ecstasy and heroin.

Schedule II drugs also have a high potential for physical and psychological dependence and physical harm. However, if used appropriately, schedule II drugs have a medical purpose. Examples include methadone and morphine.

Schedule III drugs have intermediate abuse potential. Examples include anabolic steroids and codeine in combination with acetaminophen.

Schedule IV drugs have less abuse potential than schedule III agents. Most benzodiazepines are schedule IV drugs.

Schedule V drugs have the least potential for abuse of all the scheduled classes. Many drugs in this class are used for antidiarrheal, antitussive, and analgesic purposes. Specific examples include promethazine with codeine and pregabalin.

95. A. Allow Clinicians and Law Enforcement to View a Person's History of Controlled Prescription Medication

Prescription drug monitoring programs (PDMPs) are electronic databases operated by individual states to monitor prescribing, dispensing, and drug utilization patterns. They're intended to allow clinicians to see what medications their patients are receiving from other sources and to identify abusive or illegal activity. It is thought that this will help identify patients "doctor shopping" for controlled medications and potentially reduce prescription drug abuse.

All PDMPs monitor the use of scheduled medications, but some states have elected to include other prescription drugs. They usually work by having pharmacies directly reporting to the PDMP when a prescription is filled. States may also choose who gets to access these records. Common parties with access include prescribers, pharmacists, licensing boards, law enforcement, drug control agencies, addiction treatment programs, and public health agencies.

96. E. Pay-for-performance

The MIPS and Alternative Payment Models (APMs) were designed within the Medicare Access and Children's Health Insurance Program (CHIP) Reauthorization Act (MACRA) of 2015 to help address payment to providers from the Centers for Medicaid and Medicare Services (CMS)

MIPS is used to include a pay-for-performance element for physicians that bill under fee-for-service. Under MIPS, clinicians receive financial compensation based upon composite score of four differently weighed elements. These four elements include quality of care, use of resources, advancing care information (formerly known as meaningful use) and clinical practice improvement activities. Quality of care refers to information submitted to the PQRS, and holds the largest weight of the four categories.

The APMs offer alternative methods of paying health providers for care rendered to Medicare beneficiaries within accountable care organizations and PCMHs.

97. C. Center for Medicare and Medicaid Services

The PQRS is a reporting system operated by the CMS to assess the quality of care delivered to Medicare beneficiaries. This system has been linked with the Medicare Access and CHIP Reauthorization

Act (MACRA) to financially reimburse or penalize clinicians based upon performance measures. It represents a transition from a fee-for service payment method to one that is based upon pay-per-performance. By doing so, CMS believes it is making quality of care a higher priority.

CMS determines each and every quality measure included in PQRS. Many of these measures are suggested from outside parties such as the NCQA, specialty societies and the HRSA.

The AHRQ is a federal agency within the DHHS that is tasked with improving the quality, safety, efficiency, and effectiveness of healthcare in the United States. It houses the USPSTF, National Guideline Clearinghouse, National Quality Measures Clearinghouse, and much more.

The FSMB is a national organization that represents the state and territory medical (including Osteopathic) boards within the United States. It supports its member boards through public health, licensing, discipline, and clinician regulation processes.

The HEDIS is a report card program that was developed by the NCQA, a nonprofit private organization that is dedicated to improving health quality. HEDIS measures can be used to directly compare performance measures from one plan to another. For example, once can use HEDIS to compare the amount of pneumococcal vaccine administered in one plan to another.

BIBLIOGRAPHY

[1] Emanuel EJ. Reinventing American health care: how the affordable care act will improve our terribly complex, blatantly unjust, outrageously expensive, grossly inefficient, error prone system. New York: PublicAffairs, U.S; March 4 2014. p. 34–67.

[2] The Joint Commission. About The Joint Commission, <https://www.jointcommission.org/about_us/about_the_joint_commission_main.aspx>; [accessed October 6, 2016].

[3] Harrington MK. Health care finance and the mechanics of insurance and reimbursement. Burlington, MA: Jones and Bartlett Publishers; 2016. p. 279.

[4] IOM (Institute of Medicine). Graduate medical education that meets the nation's health needs. Washington, DC: The National Academies Press; 2014.

[5] Bodenheimer T, Grumbach K. Understanding health policy a clinical approach. 6th ed New York: MacGraw Hill; 2012. p. 146–51.

[6–7] Bodenheimer T, Grumbach K. Understanding health policy a clinical approach. 6th ed New York: MacGraw Hill; 2012. p. 6–11.

[8] EPSDT. A guide for states: coverage in the medicaid benefit for children and adolescents. CMS, <http://www.medicaid.gov/Medicaid-CHIP-Program-

Information/By-Topics/Benefits/Early-and-Periodic-Screening-Diagnostic-and-Treatment.html>; June 2014.

[9] Health Resources and Services Administration. Shortage designation: health professional shortage areas and medically underserved areas/populations, <http://www.hrsa.gov/shortage/>; [accessed 06.10.2016].

[10] Medicare Benefit Policy Manua. Rural Health Clinic (RHC) and Federally Qualified Health Center (FQHC) Services. Rev. 220, 15.01.2016, <https://cms.gov/regulations-and-guidance/guidance/manuals/downloans/bp102c13.pdf>; [accessed 09.10.2016].

[11] Medicare claims processing manual (Publication 100−04), <http://www.cms.gov/regulations-and-guidance/guidance/manuals/internet-only-manuals-ioms-items/cms018912.html>; [accessed 09.10.2016].

[12] Bodenheimer T, Grumbach K. Understanding health policy a clinical approach. 6th ed New York: MacGraw Hill; 2012. p. 161.

[13] NCQA. <http://www.ncqa.org/aboutncqa.aspx>; [accessed 09.10.2016].

[14] Bodenheimer T, Grumbach K. Understanding health policy a clinical approach. 6th ed New York: MacGraw Hill; 2012. p. 32.

[15] Bodenheimer T, Grumbach K. 6th ed Understanding health policy a clinical approach, 36. New York: MacGraw Hill; 2012. p. 65−6.

[16] Task Force 101 Resources. US Preventive Services Task Force, <http://www.uspreventiveservicestaskforce.org/page/name/task-force-101-resources>; [accessed 09.10.2016].

[17] National Vaccine Injury Compensation Program, <http://www.hrsa.gov/vaccinecompensation>; [accessed 09.10. 2016].

[18] Health Insurance Market Reforms. Rate Restrictions. Focus on Health Reform, <https://kaiserfamilyfoundation.files.wordpress.com/2013/01/8328.pdf>; June 2012 [accessed 09.10.2016].

[19] Emanuel EJ. Reinventing American health care: how the affordable care act will improve our terribly complex, blatantly unjust, outrageously expensive, grossly inefficient, error prone system. New York: PublicAffairs; March 4 2014. p. 34−67.

[20] Bodenheimer T, Grumbach K. Understanding health policy a clinical approach. 6th ed New York: MacGraw Hill; 2012. p. 109−10.

[21] Schulte MF. Healthcare delivery in the U.S.A.: an introduction. Boca Raton: CRC Press; 2009. p. 18.

[22−23] Bodenheimer T, Grumbach K. Understanding health policy a clinical approach. 6th ed New York: MacGraw Hill; 2012. p. 11−2.

[24] Federal Medical Assistance Percentages or Federal Financial Participation in State Assistance Expenditures. ASPE, <https://aspe.hhs.gov/federal-medical-assistance-percentages-or-federal-financial-participation-state-assistance-expenditures>; March 1, 2015 [accessed 09.10.2016].

[25] Elliott R, Payne K. Essentials of economic evaluation in healthcare. London: Pharmaceutical Press; January 1 2005. p. 56.

[26] Yih Y, editor. Handbook of healthcare delivery systems. Boca Raton, FL: Taylor & Francis; December 16 2010.

[27] Henderson J. Health economics and policy. 6th ed Stamford, CT: Cengage Learning; 2015. p. 114−5.

[28−29] McCarthy RL, Schafermeyer KW. Introduction to health care delivery: a primer for pharmacists. 5th ed Sudbury, MA: Jones and Bartlett Publishers; 2012. p. 480−1.

[30] Kongstvedt PR. Kongstvedt. Essentials of managed health care. 6th ed Sudbury, MA: Jones and Bartlett Learning; 2013. p. 571−2.

[31–32] Jonas S, Goldsteen RL, Goldsteen K. An introduction to the U.S. health care system. 6th ed New York: Springer Publishing Co; 2007. p. 168.

[33] Claxton G, DiJulio B, Whitmore H, et al. Health benefits in 2010: premiums rise modestly, workers pay more toward coverage. Health Affair 2010;29 (10):1942–50 doi:10.1377/hlthaff.2010.0725.

[34] Siu AL, Bibbins-Domingo K, Grossman D. Evidence-based clinical prevention in the era of the patient protection and affordable care act. JAMA 2015;314 (19):2021 doi:10.1001/jama.2015.13154.

[35] Health reform's independent payment advisory board. Caring for the ages. 2010;11(10):9. doi:10.1016/s1526-4114(10)60252-1.

[36–37] Bodenheimer T, Grumbach K. Understanding health policy a clinical approach. 6th ed New York: MacGraw Hill; 2012. p. 31–41.

[38] Bodenheimer T, Grumbach K. Understanding health policy a clinical approach. 6th ed New York: MacGraw Hill; 2012. p. 108–9.

[39] Pelletier K, Herman P, Metz RD, Nelson C. Health and medical economics: applications to integrative medicine, commissioned for the IOM summit on integrative medicine and the health of the public, <https://www.nationalacademies.org/hmd/~/media/Files/Activity%20Files/Quality/IntegrativeMed/Health%20and%20Medical%20Economics--Applications%20to%20Integrative%20Medicine.pdf>; 2009 [accessed 09.10.2016].

[40] Robinson R. Cost-utility analysis. BMJ 1993;307(6908):859–62 doi:10.1136/bmj.307.6908.859.

[41] CDC. National Healthcare Safety Network, <http://www.cdc.gov/nhsn/index.html>, [accessed 09.10.2016].

[42] National Practitioner Data Bank. Health Resources and Services Administration, <http://www.npdb.hrsa.gov/>; [accessed 09.10.2016].

[43] Agencies and Programs. Department of Labor, <http://www.dol.gov/general/dol-agencies>; [accessed 09.10.2016].

[44] Women, infants, and children (WIC). USDA, <http://www.fns.usda.gov/wic/>; [accessed 09.10.2016].

[45] Supplemental nutrition assistance program (SNAP). USDA, <http://www.fns.usda.gov/snap/supplemental-nutrition-assistance-program-snap>; [accessed 09.10.2016].

[46] State and Local Health Department Governance Classification System. ASTHO, <http://www.astho.org/Research/Data-and-Analysis/State-and-Local-Governance-Classification-Tree/>; [accessed 09.10.2016].

[47] National Association of County and City Health Officials. About NACCHO, <http://www.naccho.org/about>; [accessed 09.10.2016].

[48] Johnson JA. Introduction to public health management, organizations, and policy. Boston, MA: Delmar, Cengage Learning; 2013. p. 147–9.

[49] CDC. The National Institute for Occupational Safety and Health (NIOSH), <http://www.cdc.gov/niosh/about.html>; [accessed 09.10.2016].

[50] USDA—A Quick Reference Guide, <http://www.usda.gov/documents/about-usda-quick-reference-guide.pdf>; [accessed 09.10.2016].

[51] Assistant Secretary for Public Affairs (ASPA). HHS organization chart, <http://www.hhs.gov/about/orgchart/index.html>; May 25, 2016 [accessed 09.10.2016].

[52] HHS. About HHS, <http://www.hhs.gov/about/index.html#>; [accessed 09.10.2016].

[53] Office VWS. VA Organization, <http://www.va.gov/landing_organizations.htm>; [accessed 09.10.2016].

[54] DOJ Agencies. <https://www.justice.gov/agencies>; [accessed 09.10.2016].

[55–56] Social security act table of contents. Social Secutiry Administration, <https://www.ssa.gov/OP_Home/ssact/ssact-toc.htm>; [accessed 09.10.2016].

[57] Gostin LO. Public health law: power, duty, restraint. 2nd ed Berkeley: University of California Press; 2008. p. 77–95.

[58] Durham LS. Study guide to accompany Lippincott Williams and Wilkins' administrative medical assisting, second edition. 2nd ed Philadelphia, PA: Lippincott Williams and Wilkins; March 1 2008. p. 30–1.

[59] ODPHP. About ODPHP, <http://health.gov/about-us/>; [accessed 09.10.2016].

[60–61] Bialek RG, Duffy GL, Moran JW. The public health quality improvement handbook. Milwaukee, WI: ASQ Quality Press; 2009. p. 25–6.

[62–63] Begun JW, Malcolm JK. Leading public health: a competency framework. New York: Springer Publishing Co; 2014. p. 222–3.

[64] Yemm G. The financial times essential guide to leading your team how to set goals, measure performance and reward talent. Harlow, England: FT Publishing International 2012;37–9.

[65] Drug disposal information. Drug Enforcement Agency, <http://www.deadiversion.usdoj.gov/drug_disposal/index.html>; [accessed 09.10.2016].

[66] Mullner RM. Health and medicine. Thousand Oaks, CA: Sage; 2011. p. 62–4.

[67] Bodenheimer T, Grumbach K. Understanding health policy a clinical approach. 6th ed New York: MacGraw Hill; 2012. p. 126.

[68] Scutchfield FD, Keck W. Scutchfield and Keck's principles of public health practice. 4th ed Boston, MA: Cengage Learning; 2017. p. 280–1.

[69–70] Paola FA, Walker R, Nixon LL. Medical ethics and humanities. Sudbury, MA: Jones and Bartlett Publishers; 2010. p. 61.

[71a] When does the privacy rule allow covered entities to disclose information to law enforcement. Department of Health and Human Services, <http://www.hhs.gov/ocr/privacy/hipaa/faq/disclosures_for_law_enforcement_purposes/505.html>; [accessed 09.10.2016].

[71b] HIPAA. Privacy Rule and Sharing Information Related to Mental Health. Department of Health and Human Services, <http://www.hhs.gov/hipaa/for-professionals/special-topics/mental-health/index.html>; [accessed 09.10.2016].

[72] Herold R, Beaver K. The practical guide to HIPAA privacy and security compliance. 2nd ed Boca Raton, FL: Taylor & Francis; 2015. p. 3.

[73] HCAHPS: Patients' Perspectives of Care Survey. Centers for Medicare and Medicaid Services, <http://www.cms.gov/Medicare/Quality-Initiatives-Patient-Assessment-Instruments/HospitalQualityInits/HospitalHCAHPS.html>; September 25, 2014 [accessed 09.10.2016].

[74] Beracochea E, Weinstein C, Evans D, editors. Rights-based approaches to public health. New York: Springer Publishing Co; 2011.

[75] Slee DA, Slee VN, Schmidt JH. Slee's health care terms. 5th ed Sudbury, MA: Jones & Bartlett Publishers; 2008. p. 427.

[76] Age discrimination. U.S. Equal Employment Opportunity Commission, <http://www.eeoc.gov/laws/types/age.cfm>; [accessed 09.10.2016].

[77] An Employees Guide to Health Benefits Under COBRA. Employee Benefits Security Administration, <https://www.dol.gov/sites/default/files/ebsa/laws-and-regulations/laws/cobra/COBRAemployee.pdf>; [accessed 09.10.2016].

[78] Bodenheimer T, Grumbach K. Understanding health policy a clinical approach. 6th ed New York: MacGraw Hill; 2012. p. 193–7.

[79] Rogers K, editor. Battling and managing disease. New York: Britannica Educational Pub. in association with Rosen Educational Services; 2011. p. 42.

[80–81] O'Connor M. The physician-patient relationship and the professional standard of care reevaluating medical negligence principles to achieve the goals of tort reform. Tort Trial Insur Pract Law J 2010;46.1(109):33.

[82] Wall T, Nasseh K, Vujicic MUS. Dental Spending Remains Flat through 2012. Health Policy Institute Research Brief. American Dental Association, <http://www.ada.org/sections/professionalResources/pdfs/HPRCBrief_0114_1.pdf>; January, 2014 [accessed 09.10.2016].

[83] Ball JR, Balogh E. Improving diagnosis in health care: highlights of a report from the national academies of sciences, engineering, and medicine. Ann Intern Med 2015;164(1):59 doi:10.7326/m15-2256.

[84] Acosta L. Restrictions on genetically modified organisms: United States. Library of Congress, <https://www.loc.gov/law/help/restrictions-on-gmos/usa.php>; [accessed 09.10.2016].

[85] Guideline for Prevention of Catheter-Associated Urinary Tract Infections 2009. Healthcare Infection Control Practices Advisory Committee, <http://www.cdc.gov/hicpac/pdf/CAUTI/CAUTIguideline2009final.pdf>; [accessed 09.10.2016].

[86a] Landers T, Abusalem S, Coty M-B, Bingham J. Patient-centered hand hygiene: the next step in infection prevention. Am J Infect Control 2012;40(4):S11–7 doi:10.1016/j.ajic.2012.02.006.

[86b] Ellingson K, Haas JP, Aiello AE, et al. Strategies to prevent healthcare-associated infections through hand hygiene. Infect Cont Hosp Ep 2014;35 (08):937–60 doi:10.1086/651677.

[87] Makary M, Daniel M. Medical error—the third leading cause of death in the US. BMJ 2016;353:2139 doi:10.1136/bmj.i2139.

[88] Bodenheimer T, Grumbach K. Understanding health policy a clinical approach. 6th ed New York: MacGraw Hill; 2012. p. 30.

[89] Tulchinsky TH, Varavikova EA. The new public health. 3rd ed San Diego, CA: Elsevier Science; 2009. p. 408–11.

[90] WHO. Strategic advisory group of experts (SAGE) on immunization. World Health Organization, <http://www.who.int/immunization/policy/sage/en/>; [accessed 09.10.2016].

[91] Schulte MF. Healthcare delivery in the U.S.A.: an Introduction. Boca Raton: CRC Press; 2009. p. 112.

[92] Harris R. Long-term government funded programs: a study of their impact on poverty in the United States. Boca Raton, FL: Universal-Publishers; 2006. p. 112–4.

[93–94] Gabay M. The federal controlled substances act: schedules and pharmacy registration. Hosp Pharm 2013;48:6.

[95] Rutkow L, Turner L, Lucas E, Hwang C, Alexander GC. Most primary care physicians are aware of prescription drug monitoring programs, but many find the data difficult to access. Health Affair 2015;34(3)):484–92 doi: 10.1377/hlthaff.2014.1085.

[96] Hirsch J, Leslie-Mazwi T, Nicola G, et al. PQRS and the MACRA: value-based payments have moved from concept to reality. Am J Neuroradiol 2016;37(12):2195–200 <http://www.ajnr.org/content/37/12/2195.full.pdf+html>; [accessed 08.02.2017].

[97] Frankel BA, Bishop TF. A Cross-Sectional Assessment of the Quality of Physician Quality Reporting System Measures. J Gen Intern Med 2016;31:840 doi:10.1007/s11606-016-3693-3.

CHAPTER THREE

Epidemiology and Biostatistics

3.1 EPIDEMIOLOGY AND BIOSTATISTICS QUESTIONS

1. Which of the following types of studies is susceptible to ecological fallacy?
   A. Cross-sectional
   B. Prospective cohort
   C. Retrospective cohort
   D. Randomized control trials
   E. Meta-analysis
2. In researching preventable colon cancer mortality, which study design would be scientifically and ethically most appropriate to compare the effectiveness of utilizing a screening colonoscopy compared to not performing any screening technique?
   A. Randomized control trial
   B. Cohort
   C. Case-control
   D. Ecological
   E. Survival analysis
3. Which of the following life events is typically not recorded by a state health department's Department of Vital Statistics?
   A. Birth
   B. Death
   C. Marriage
   D. Divorce
   E. Cancer
4. Which organization is responsible for compiling the nation's health statistics? Example, statistics include injuries, nutrition, vital statistics, and health insurance coverage.
   A. Federal Bureau of Investigation
   B. Federal Health Statistics Center

Board Review in Preventive Medicine and Public Health.
DOI: http://dx.doi.org/10.1016/B978-0-12-813778-9.00003-7

C. National Center for Health Statistics
D. National Library of Medicine
E. There is no centralized organization that records these statistics

5. A study to determine the effects of a new skin cream to reduce wrinkles is undergoing FDA approval. The investigators propose using a double-blinded study. Understanding the subjective nature of this experimental compound, the FDA Dermatologic and Ophthalmic Drug Advisory Committee recommends that the experimental wrinkle cream and the placebo cream undergo a triple-blinded study.

Which of the following parties is included in the triple-blinded study, but not the double-blinded study?
A. Institutional Review Board
B. Shareholders
C. Statisticians that analyze the data
D. Study participants
E. Study investigators

6. Which of the following study designs is not observational?
A. Case-control
B. Prospective cohort
C. Retrospective cohort
D. Randomized controlled
E. Cross-sectional

7. A young physician is surveying the incidences of disease B and disease C. He concludes that disease C has a higher incidence than disease B. It is later found out that the two diseases have the exact same incidence and that disease B has a much faster disease onset leading to mortality. What is this phenomenon called?
A. Length bias
B. Lead-time bias
C. Incidence bias
D. Mortality bias
E. Morbidity bias

8. Which of the following is an example of nondifferential bias?
A. Only selecting men to participate in a study about marital happiness of spouses
B. Performing a survey of hospital experience of congestive heart failure (CHF) patients one week after admission for CHF, in the patients that have not yet been discharged

C. An investigator is trying to prove that cabbage consumption in pregnancy causes autism, so he interrogates mothers of autistic children about cabbage consumption, but not mothers of children without autism.
D. A scale that records all of the patients 2500 kg more than they actually weigh
E. Randomly assigning 1000 study participants to separate arms in a study

9. A new screening test is applying for FDA approval. The owners of the test say that they can increase the length of survival in patients with pancreatic cancer. Their studies show that by diagnosing it 6 months earlier than any other test, they can increase the length of survival with the disease.

The FDA approval panel responds that this test has what type of bias?

A. Confounding
B. Hawthorne effect
C. Misclassification
D. Lead-time
E. Length

10. A hospital administrator wants to conduct a study to investigate the benefits of a hand soap that promises to reduce iatrogenic hospital infections. She decides to enroll 400 nurses in the study. The study is designed so that half of the nurses are instructed to use the new soap, while the other half are instructed to use the old soap. To ensure that the nurses are using soap, the preventive medicine physician on staff is to randomly give the nurses a "sniff test" to make sure their hands smell like soap. Because the preventive medicine physician only works on the first floor, he is only able to sniff nurses that work on the first floor. The nurses on the second floor are never subjected to the sniff test.

When the study results are tabulated, it is found that there is no difference between iatrogenic infections between the new soap and old soap. However, it is found that there were more iatrogenic infections on the second floor than the first.

Which answer best explains why there were more infections in patients on the second floor?

A. Regression towards the mean
B. Healthy worker effect

C. Placebo effect
D. Hawthorne effect
E. Random error

11. A biochemist at a pharmaceutical company created a new drug that lowers blood pressure. He instructs the trial investigators to only include study participants that are not taking an antihypertensive and have a systolic blood pressure greater than 180. After screening by the investigator, those that met the criteria were told to return the following month for the first dose of the investigational drug. On the day that the participants are set to start receiving the new drug, the study investigator has to disqualify many of the study participants because they no longer meet the systolic blood pressure requirements.

What is the best explanation for the participants no longer meeting study criteria?

A. Hawthorne effect
B. Information bias
C. Neyman bias
D. Recall bias
E. Regression towards the mean

12. Which of the following is a control for confounding primarily used in the analysis stage?

A. Randomization
B. Restriction
C. Matching
D. Stratification
E. Hypothesis testing

13. An epidemiology student conducted a study to find the strength of association between smoking and cardiomyopathy in his community. For ease of conducting his research, the student surveyed subjects at his local bar. To his surprise, he found a higher correlation between smoking and cardiomyopathy than other published studies.

Which factor is a cofounder?

A. Alcohol
B. Lung cancer
C. Smoking
D. Bias between those that took the survey and refused the survey
E. There are no cofounders

14. Bayes' theorem requires which of the following to calculate?
A. Sensitivity, specificity, and negative predictive value
B. Specificity, false-positive rate, and incidence
C. Sensitivity, specificity, and prevalence
D. NPV, incidence, and prevalence
E. NPV, false-positive rate, and sensitivity

15. Which of the following is the denominator of incidence density?
A. Geographic region
B. Person-time
C. Population at risk
D. Prevalence
E. None of the above

16. A group of scientists believe they have found a single dose universal influenza vaccine that is resistant to antigenic drift and antigenic shift. The FDA grants them a phase 1 trial to test the safety of this new vaccination. The incidence of influenza is shown in the following table:

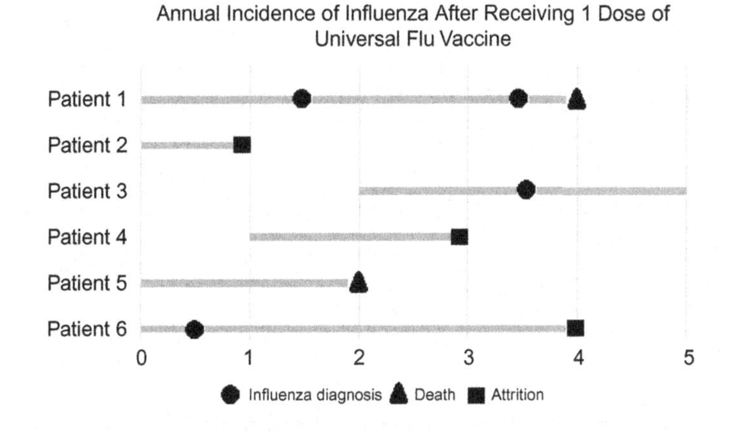

What is the incidence density of Influenza after receiving the experimental influenza vaccine?
A. 0.13 cases/person-year
B. 0.25 cases/person-year
C. 0.5
D. 3
E. 4

17. A large hospital chain plans to assess the pain scales of 100 patients in the orthopedic units of their 40 hospitals. The same 1–10 scale was used for all of the patients on the day after surgery. Each patient's score was only recorded one time and there were no repeat patients that were recorded on separate visits.

Which of the following hypotheses states that the hospital chain may assume their data approaches normal distribution?

A. Hawthorne effect
B. Central limit theorem
C. Inferential statistics
D. Binomial distribution
E. Kaplan–Meier function

18. A mother is contacted by her son's teacher to discuss his poor academic performance. Because of this discussion, the mother hires a professional to evaluate her son's intelligence quotient (IQ). His IQ is found to be two standard deviations below the mean.

What is his IQ?

A. 50
B. 70
C. 90
D. 110
E. 130

19. A father believes that his daughter is intellectually advanced and wants to prove it to his friends with an IQ test. He learns that his daughter's IQ is 105.

What is the *Z-score* for an IQ of 105?

A. − 0.66
B. 0
C. 0.33
D. 0.66
E. 1.33

20. A sample was drawn of the following numbers: 5, 6, 8, 10, 11, 11, 11, 12, 13, 13, 3000.

Which of the measures of central tendency is heavily influenced by the outlier?

A. Mean
B. Mode
C. Median
D. Geometric mean
E. All are appropriate to accurately depict this dataset

21. The risk manager at a textile factory has noticed a 30% increase in workers' compensation claims over the last 20 years. She attributes the change the national obesity epidemic. The BMI of the last 10 workers to file workers' compensation claims have been as follows: 33, 30, 18, 38, 34, 27, 41, 29, 25, 18.

Of the last 10 claims, what percentage of the workers can be classified as obese?

A. 20%
B. 30%
C. 40%
D. 50%
E. 60%

22. If a distribution is skewed to the right, what can be said about the mean, median, and mode?

A. Mean = median = mode
B. Mean = median < mode
C. Mean < median < mode
D. Mean > median > mode
E. None of the above

23. You are the director of four local for-profit urgent care clinics, currently employing 11 salaried physicians. At the end of the month, you decide that you would like to recognize the physicians that have seen the most patients. Arbitrarily, you decide that the top 50th percentile of physicians deserve your acknowledgment.

The physician counts are as follows:

Center 1:	Physician A-	300 patients
	Physician B-	250 patients
	Physician C-	360 patients
Center 2:	Physician D-	250 patients
	Physician E-	410 patients
	Physician F-	220 patients
Center 3:	Physician G-	315 patients
	Physician H-	335 patients
	Physician I-	390 patients
Center 4:	Physician J-	400 patients
	Physician K-	150 patients

Which physician represents the median number of the patients seen?

A. B
B. G
C. A
D. K
E. I

24. As the number of subjects in a trial increases, which of the following is always true?

A. The number needed to treat increases
B. The confidence interval becomes smaller
C. The power of the study decreases
D. Clinical relevance increases
E. Positive predictive value decreases

25. You are reading a study about an experimental new weight loss drug given to 500 obese subjects. The drug yielded a weight loss of 20 lbs, with study participants on the experimental medication dropping from an average of 270–250 lbs in 3 months (95% CI = 1–39 lbs). Meanwhile, similarly matched study participants taking a placebo did not lose any weight.

Using the information available, what does this mean?

A. If the drug is tested on 1000 similar sample populations, 950 of these sample populations would likely lose a mean between 1 and 39 lbs.
B. If the drug is tested on 1000 similar patients, 95% of these patients would likely lose between 1 and 39 lbs.
C. The findings are not clinically significant
D. The findings are not statistically significant
E. The risk ratio shows a mild statistical benefit

26. A researcher is questioning whether he should use *p*-values or confidence intervals to publish the results of his fluoridation prospective cohort study. Which of the following is true regarding confidence intervals?

A. Both *p*-values and confidence intervals are without units of measurement
B. Confidence intervals that include 1 are always statistically significant
C. Larger samples produce smaller confidence intervals

D. Numbers within confidence intervals all have equal clinical importance
E. *p*-Values are not comparable to confidence intervals

27. 300 Medical Students are forced to take an art history course because the school administration believes that it will lead to well-rounded physicians. The final grade is based off of one test. All of those scoring higher than two deviations below the mean receive a passing score.

The scores were normally distributed. The average test score for the class was 50%. The standard deviation was 10% and the highest score was 90%. Approximately what percentage of students passed the exam?

A. 99%
B. 98%
C. 95%
D. 90%
E. 68%

28. A psychiatrist at a drug rehabilitation facility has noticed that 60% of her patients are recovering from alcohol abuse and 50% are recovering from opiate abuse. Some patients are recovering from both. The rest of her patients abuse a plethora of other recreational drugs.

What is the probability that her next patient will be recovering from either alcohol or drug abuse?

A. 0.3
B. 0.7
C. 0.8
D. 1.0
E. 1.10

29. Which of the following standardized mortality ratios (SMR) is most likely to be found in a population experiencing an unexpected lethal epidemic?

A. 0.1
B. 0.5
C. 0.8
D. 1.0
E. 1.4

30. A young epidemiologist wishes to calculate the standardized mortality ratio (SMR) of drowning deaths in his landlocked state. Which of the following is the best way to calculate this number?

A. Henry's law
B. Interquartile range
C. Stratification
D. Direct adjustment
E. Indirect adjustment

31. A preventive medicine resident contemplated reading a book full of practice questions and answers to help him prepare for the board exam. He hypothesized that reading the book would help increase his test score. He asked former residents that have taken the test about whether or not they had read the book and how well they scored.

Which of the following statements represents his null hypothesis?

A. Reading the book would increase his test score
B. Reading the book is not associated with his test score
C. Reading the book would decrease his test score
D. The null hypothesis may not be developed until the *p*-value is available
E. None of the above

32. Which of the following statements is true regarding a null hypothesis, alternative hypothesis and *p*-value?

A. Accept the alternative hypothesis when there is a high *p*-value
B. Accept the null hypothesis when there is a high *p*-value
C. Accept the alternative hypothesis when there is a low *p*-value
D. Accept the null hypothesis when there is a low *p*-value
E. There is no relationship between the null hypothesis, alternative hypothesis and *p*-value

33. Which of the following accurately describes a Type I error?

A. False-positive
B. False null is rejected
C. True null is accepted
D. True null is rejected
E. More than one of the above

34. Which is true of a purified protein derivative (PPD) tuberculosis test in a patient that has tuberculosis and poorly controlled HIV?

A. This is a type I error because the result is likely positive when the disease is not present

B. This is a type I error because the result is likely negative when the disease is present
C. This is a type II error because the result is likely positive when the disease is not present
D. This is a type II error because the result is likely negative when the disease is present
E. There is no error, as the test will correctly identify when the disease is present

35. A professor wants to analyze test scores of his 225 students taking a biostatistics exam. If the average score of the exam is 80 and the standard deviation of the scores is 15, what is the 95% confidence interval?

A. 65, 95
B. 66, 94
C. 76, 82
D. 78, 82
E. 80, 84

36. What happens to the PPV of a test if there is an increase in the prevalence of the disease being tested?

A. Increases
B. Decreases
C. Remains the same
D. No relationship
E. Need more information

37. A researcher wishes to find out the risk ratio of clinical depression as a result of poor standardized test performance. He hypothesizes that those scoring greater than the 65th percentile suffer less depression. What can he do to increase the power of his study?

A. Decrease the number of test-takers included in the study
B. Decrease the alpha level of the study
C. Increase the beta of the study
D. Increasing the null from 65th to 75th percentile
E. None of the above

38. Which of the following variables is not considered when calculating the necessary sample size of a clinical study?

A. Degree of precision desired

B. Expected attrition rates
C. Size of the population under investigation
D. Type of study being conducted
E. All of the above are considerations when calculating sample size

39. What does the y represent in the following equation?

$$y = a + b_1x_1 + b_2x_2 + e$$

A. Adjustment coefficient
B. Dependent variable
C. Error
D. Independent variable
E. Regression constant

40. If the Pearson product correlation coefficient (r) between the time of sunset and amount of traffic on the highway is 0.75, what can be inferred without any more detail about the relationship between sunset and traffic?

A. There is a negative correlation between sunset and traffic
B. There is no correlation between sunset and traffic
C. There is a positive correlation between sunset and traffic
D. Either sunset or traffic is causative of the other
E. None of the above

41. A medical student would like to learn the relationship between heart rate and blood pressure. While in a clinic, he creates the following scatterplot of the heart rate and blood pressure of his 10 patients.

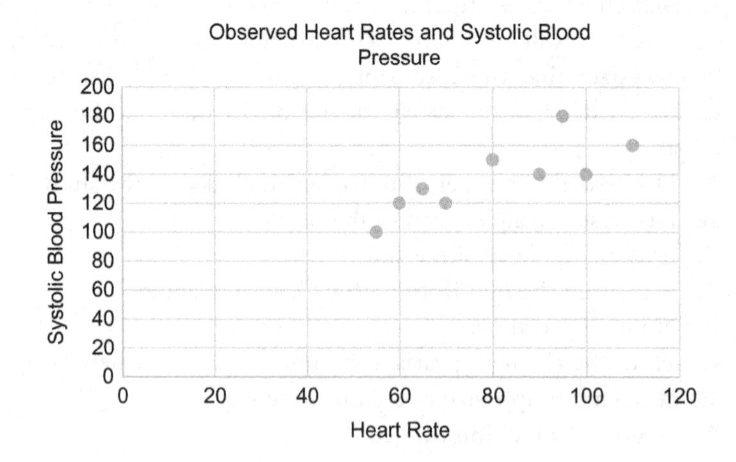

According to this scatterplot, what is the relationship between systolic blood pressure and heart rate?

A. Negative exponential
B. Negative linear
C. No correlation
D. Positive exponential
E. Positive linear

42. A teenager believes that the severity of his acne vulgaris is related to the number of hours he works at a fast food restaurant. Every Sunday for 5 months, he meticulously counts the number of comedones on his face and records the number of hours he worked the week prior. His older sister, an epidemiology student, helps him calculate the Pearson correlation coefficient of 0.5 for the correlation of hours worked and number of comedones.

Which answer best describes the relation of this teenager's number of hours worked and severity of acne?

A. Strongly negative
B. Mildly negative
C. No relation
D. Mildly positive
E. Strongly positive

43. Suppose that developers of a new water coliform test are trying to identify the time of year when water coliform levels are the highest. To find this time, biostatisticians analyze the coliform levels while independently adjusting for water temperature, air temperature, level of sun exposure, water pH and several other factors.

Which statistical test is most appropriate?

A. Chi-square
B. Kappa
C. Multiple regression
D. t-Test
E. Wilcoxon signed rank test

44. When is the actuarial method of survival analysis preferred over the Kaplan–Meier method of survival analysis?

A. Never
B. Small sample size
C. Studying survival rates in fixed periods
D. There is high loss to follow-up
E. The two methods are not comparable

45. Which type of study is a funnel plot most likely to be used in?
A. Cohort
B. Cross-sectional
C. Ecological
D. Meta-analysis
E. Randomized controlled trial

46. If a new medication reduces the risk of a disease by 5%, how many people would needed to be treated to benefit one person?
A. 5
B. 10
C. 15
D. 20
E. 25

47. When determining whether there is a causal association between two variables, which of the following is not a consideration?
A. Consistency of association
B. Strength of association
C. Evidence of association
D. Analogy of other similar associations
E. All of the above are considerations

48. Which variables are needed to calculate the standard error?
A. *p*-value, mean
B. Standard deviation, number of samples
C. Incidence, prevalence
D. Expected outcome, observed outcome
E. Alpha, beta

49. A physician is skeptical of a new weight loss medication that recently became available over the counter. He is unable to find any studies demonstrating the drug's efficacy, so he decides to conduct his own study to see how well it works. He follows 60 of his weight loss patients for one year. 30 are placed on the medication and 30 are placed on a placebo. After one year of collecting data, he conducts a t-test, which confirms his null hypothesis and finds that the drug does not work.

Without more information, which of the following can be stated as true?
A. The results of this study are not internally valid
B. The results of this study are externally valid
C. The physician's conclusion applies to all patients taking this drug

D. The results are not clinically significant to the 60 patients screened
E. None of the above are true

50. A math professor wishes to throw a party to celebrate the end of the year with his students. To help him decide what kind of ice cream to get, he decides to poll his students. The results are as follows:

	Favorite ice cream			
	Chocolate	**Strawberry**	**Vanilla**	
Male	30	30	60	**120**
Female	70	20	50	**140**
	100	**50**	**110**	

In looking at the favorite ice cream by gender, the professor feels that there are distinct differences between men and women in his class. He decides to conduct a chi-squared (χ^2) analysis to test his hypothesis. How many degrees of freedom (*df*) are there in this χ^2 analysis?
A. 1
B. 2
C. 3
D. 4
E. 6

51. A county health department has recently begun a new fitness initiative by installing outdoor gyms complete with public exercise equipment in local parks. Several months after constructing the newest outdoor gym, an employee of the health department returned to the site to reevaluate the project and research improvements for the next project. To create objective data, the health department employee kept records of which genders exclusively used one of the two types of exercise equipment: Aerobic (outdoor treadmills and exercise bicycles) and anaerobic (weight resistance). People that used both aerobic and anaerobic equipment were not recorded.

The results are as follows:

Exclusive use of outdoor exercise equipment by gender

	Aerobic	**Anaerobic**	
Men	75	50	**125**
Women	25	125	**150**
	100	**175**	**275**

After looking at the data, the health department employee hypothesizes that different genders prefer to use different types of gym equipment. He decided to use chi-squared (χ^2) analysis to determine if this difference is due to chance alone. What is the approximate test statistic that the employee will use to compare to the critical value to accept or reject the null hypothesis?

A. 2
B. 30
C. 43
D. 57
E. 275

52. Two preventive medicine residents are studying for their board exams. They tally their answers after the true or false portion of their practice tests. They agree on the true answers for 40 questions and agree on the false answers for 30 questions. However, 20 of the questions were answered to be true by Resident A and false by Resident B. On the other hand, 10 questions were answered true by Resident B and false by Resident A.

What is the Kappa ratio?

A. 30%
B. 40%
C. 50%
D. 60%
E. 70%

53. If you would like to compare the blood pressure of patients before and after drinking three cups of coffee, which test would be most appropriate?

A. Student's t-test
B. Paired t-test
C. Chi-squared
D. Pearson's correlation
E. Kaplan–Meier analysis

54. Which of the following tests does ANOVA use?

A. Paired t-test
B. Wilcoxon signed rank test
C. F-test
D. χ^2
E. Regression analysis

55. A truck driver working towards a degree in public health notices that the all of the different coffee brands at different gas stations taste the same. He decides to hold a study see if other people share his opinion. He gathers the opinions of six of his fellow truck drivers to see how they like 10 competing gas station coffees.

Which test is best for the truck driver to test his hypothesis that all the brands taste the same?

A. ANOVA
B. Chi-squared
C. Mann–Whitney U test
D. Multiple regression
E. Odds ratio

56. After having a newborn child, a professional basketball player takes criticism because he is not playing well. He believes that his lack of sleep is the main problem. To test this theory, he records his nightly amount of sleep (in hours) and the amount of points he scores in the game the next day. After 25 nights and games, he charted his findings on a scatterplot. The results appear linear and have a Pearson correlation coefficient of 0.8.

What percent of his point production is tied to his nightly sleep?

A. 16%
B. 20%
C. 40%
D. 64%
E. 80%

57. One month after initiating a vector control program to control the mosquito population with insecticides, the state health department would like to check the incidence of mosquito-borne illness in the 15 counties within the program. The insecticide used in each county was different and the state health department would like to know if there is a difference in efficacy between the different insecticides. Because the three counties have different demographics and different amounts of outdoor attractions, the epidemiologist would like to control for confounding variables.

Which test would be best to test whether or not one insecticide reduces mosquito-borne illness more than others, while controlling for confounding from demographics and differing risk of exposure (using outdoor attractions as a surrogate)?

A. ANOVA
B. ANCOVA
C. Chi-squared

D. Kappa
E. t-Test

58. What do each of the following three epidemiologic tools have in common?

1. Cohort study
2. Epidemic curve
3. Longitudinal data

A. Each demonstrates secular patterns
B. Each is a type of time-series analysis
C. Each is analyzed through multiple regression
D. Each will always show causation of correlation
E. None of the above

59. If you are recording temperatures on the Fahrenheit or Celsius Scale, which type of data are you using?

A. Nominal
B. Ordinal
C. Interval
D. Ratio
E. Combination of two or more of the above

60. The risk manager at a textile factory has noticed a 30% increase in workers' compensation claims over the last 20 years. She attributes the change the national obesity pandemic. The BMI of the last 10 workers to file workers' compensation claims have been as follows: 33, 30, 18, 38, 34, 27, 41, 29, 25, 18.

When stratified into groups according to BMI classification (underweight, obese, etc.), what type of variable is represented?

A. Interval
B. Nominal
C. Ordinal
D. Ratio
E. Numerical

61. The Kelvin scale is an example of which type of variable?

A. Nominal
B. Ordinal
C. Interval
D. Ratio
E. None of the above

62. Which of the following is true of binomial distribution?

A. Lower n (number of observations) has no effect on a binomial distribution

B. Lower p (success) rate skews the distribution to the right
C. Normal distribution is never a viable substitute for binomial distribution
D. There is one universal binomial distribution curve to be used in every scenario
E. Variables in the binomial distribution are continuous

63. It is appropriate to use Fisher's exact test when which of the following tests do not meet necessary criteria?
A. ANCOVA
B. Chi-square
C. Likelihood ratio
D. Paired t-test
E. Wilcoxon test

64. Which type of variables are used in McNemar's test?
A. Continuous
B. Dichotomous
C. Interval
D. Ratio
E. None of the above

65. An epidemiologist would like to compare the heart rate of 10 men and 10 women with the Zika virus. Which of the following tests is best suited to compare the nonparametric medians of two groups?
A. ANOVA
B. Chi-squared
C. Mann–Whitney U test
D. Multiple regression
E. t-Test

66. After initiating a new workplace safety program, the risk manager would like to know the how well the employees know the safety protocol. He administers a test to examine their knowledge.
The test scores are as follows:
66, 62, 73, 70, 68, 59, 74, 76, 65, 65, 71, 62, 67, 69, 70, 62, 71, 72, 64
What is the interquartile range?
A. 5
B. 6
C. 7
D. 8
E. 9

67. Unable to relax on her vacation, an epidemiology professor reflects on her teaching career and feels that her Monday students score higher test results than her Friday students. Although she cannot remember the exact average scores of her 100 + size student classroom over the semester, she can remember how her two classes performed comparatively on the last nine exams.

She decides to plot these last nine exam scores into the following table.

Comparative Monday and Friday classroom test performance during the semester

	Week 1	Week 2	Week 3	Week 4	Week 5	Week 6	Week 7	Week 8	Week 9
Monday	Better	Worse	Better	Better	Better	Better	Worse	Better	Better
Friday	Worse	Better	Worse	Worse	Worse	Worse	Better	Worse	Worse

Assuming that the two classes are identical, outside of the date they go to class and take tests, which test would be most appropriate to analyze this data?

A. Chi-squared test
B. Paired t-test
C. Student's t-test
D. Sign test
E. Wilcoxon signed rank test

68. The Wilcoxon signed rank test is the nonparametric equivalent of which of the following tests?

A. Student's t-test
B. Paired t-test
C. Chi-squared
D. ANOVA
E. Pearson correlation

69. While calculating a Pearson correlation coefficient, a statistician realizes that the variables he is working with are not normally distributed. What is the best alternative to identify the association between the variables he is working on?

A. ANOVA
B. Chi-square (χ^2)
C. Mann–Whitney U test
D. Sign test
E. Spearman rank correlation

70. Which of the following tests is the nonparametric counterpart of ANOVA?

A. Kruskal–Wallis one-way test
B. Mann–Whitney U test
C. Spearman rank correlation coefficient
D. Wilcoxon test
E. None of the above

71. After the discovery of localized Cancer Cluster X, researchers and physicians are unable to develop a treatment. After the first year of discovery, 80% of those originally identified had survived. Of those that survived the first year, the survival rate at the end of the second year was 50%. Of those that survived the second year, the survival rate at the end of the third year was 50%. Of those that survived the third year, the survival rate at the end of the fourth year was 40%. The five year survival rate was 5%.

What is the 3 year survival rate?

A. 50%
B. 40%
C. 25%
D. 20%
E. 8%

72. Logistic regression is used when the dependent variable has how many potential outcomes?

A. 0
B. 1
C. 2
D. >2
E. All of the above

73. A physician's group recently began seeing patients in a new office. There is mixed feedback about the office from both physicians and patients. The office administrator is worried that it is affecting the physician-patient relationship, so she asks a biostatistician to evaluate the contentment of the physician–patient relationship from both the physician and the patient viewpoints. Which test would be the best tool to evaluate this measurement?

A. Analysis of covariance
B. Chi-square
C. Linear regression
D. Multivariate analysis of variance
E. Spearman rank correlation coefficient

74. Which of the following types of bias is most common in meta-analysis studies?
A. Recall bias
B. Observer bias
C. Response bias
D. Attrition bias
E. Publication bias

75. According to international agreement, a fetal death between which gestational ages is considered an intermediate fetal death?
A. 20–24 weeks gestation
B. 20–26 weeks gestation
C. 20–28 weeks gestation
D. 20–30 weeks gestation
E. 20–32 weeks gestation

76. What is the denominator used in the maternal mortality rate?
A. Number of pregnancies
B. Number of live births
C. Number of women aged 15–44
D. Number of pregnant women that die due to complications of pregnancy
E. Number of women having a high risk pregnancy

77. Which best describes the demographic gap?
A. Differences between birth and death rates
B. Differences of income
C. Differences between immigration and emigration
D. Differences of life expectancy
E. Differences of income

78. Which of the following groups has the highest life expectancy at birth?
A. Black Hispanic men
B. Black non-Hispanic women
C. White Hispanic women
D. White non-Hispanic men
E. White non-Hispanic women

79. Which of the following is an example of active surveillance?
A. Healthcare facilities reporting influenza like illness (ILI)
B. Public health laboratory reporting positive chlamydia case
C. A clinician reporting a bite from a strange and aggressive dog
D. A news station calling a local health department about a salmonella outbreak

E. A health department epidemiologist calling the local hospital to ask about confirmed HIV cases

80. Which organization supports the National Notifiable Diseases Surveillance System (NNDSS)?
A. Agency for Healthcare Research and Quality
B. Centers for Disease Control and Prevention
C. Council of State and Territorial Epidemiologists
D. National Academy of Medicine
E. National Association of County and City Health Officials

81. Which entity works with individual states to determine which diseases are placed on a state's mandatory reportable disease list?
A. Agency for Healthcare Research and Quality
B. Centers for Disease Control and Prevention
C. Congressional Health Committee
D. Council of State and Territorial Epidemiologists
E. National Association of County and City Health Officials

82. When must health practitioners report mandatory reportable diseases?
A. When the condition is designated as "notifiable" by the state
B. When the condition is designated as "notifiable" by the Surgeon General
C. When the condition has a high case fatality rate
D. When the condition is highly infectious
E. When the CDC determines that the disease is beyond the scope of a local health department

83. In FDA Phase 4 drug testing, a pharmaceutical company has become aware of multiple fatal adverse events in patients taking their new drug. Fortunately, the company has developed a simple screening test to identify those at risk of complications. If this test does not work correctly, the company will be forced to remove their potential blockbuster drug from the market. This screening test should have a high _____ in order to appropriately identify those at increased risk of complications.
A. Specificity
B. Sensitivity
C. Positive predictive value
D. Negative predictive value
E. Prevalence

84. Which of the following is not a consideration to make when creating screening recommendations?
A. Ethical consequences of performing test
B. Predictive value of results
C. Stress related to testing
D. Validity of test results
E. All of the above are considerations to make when creating screening recommendations

85. In July 2014, there was a norovirus outbreak in Tulsa, Oklahoma. If one were to create an epidemic curve, which of the following choices would be the best title?
A. Outbreak
B. Norovirus Outbreak
C. Norovirus Outbreak in Tulsa, Oklahoma
D. Cases of Norovirus in Tulsa, Oklahoma, July, 2014
E. Cases of Norovirus in Tulsa by Date of Onset, Tulsa, Oklahoma, July, 2014

86. When the annual incidence of an illness is plotted over one year, what is the appropriate way to place the months with the highest incidence of disease?
A. The beginning
B. The middle
C. The end
D. Depends on disease
E. The graph should always start in January

87. What are the two variables plotted on an epidemic curve?
A. Number of cases and time
B. Number of ill and number of dead
C. Sensitivity and specificity
D. Suspected cause of illness 1 and suspected cause of illness 2
E. Type of outbreak and time

88. Which type of epidemiologic outbreak occurs entirely within one incubation period?
A. Common source
B. Mixed
C. Continuous common source
D. Propagated
E. All of the above outbreak patterns are more than one incubation period

89. An epidemiologist sets out to create an epidemic curve for a novel coronavirus traveling through his community. As expected, she discovers that the only route of transmission is person to person through respiratory droplets. What type of pattern will be seen on the epidemic curve?

A. Common source
B. Mixed
C. Continuous common source
D. Propagated
E. None of the above

90. Right before the beginning of rental season, an apartment complex decided to increase community spirit by hosting a potluck pool party. The party was a wild success, with five people bringing food for the 70 people in attendance. After the party, the apartment complex management began to hear complaints about diarrheal illness attributed to the party. To find the culprit of the illness, management surveyed the 70 people that went to the party about their health status. The results were tallied into the following table.

	Ate		Did not eat	
Food	**Ill**	**Well**	**Ill**	**Well**
A. Chicken	20	18	22	10
B. Burger	31	11	6	22
C. Hotdog	12	15	15	28
D. Oranges	25	30	3	12
E. Egg roll	1	5	30	34

Which food is most likely the cause of the diarrheal illness?

A. Chicken tender
B. Burger
C. Hotdog
D. Watermelon
E. Egg roll

91. Jane is a registered nurse that traveled to Africa to assist in the containment of a suspected smallpox outbreak. At the end of her second day in the field, Jane noticed a gaping hole in her personal protective equipment. Earlier that day, she was directly exposed to a young man with symptoms that appeared to be due to smallpox. When she reports the exposure to another investigator the following morning, she is asymptomatic. Through an administrative error, she never received the smallpox vaccine. What should Jane do to prevent the further spread of this suspected smallpox outbreak?

A. Fly back home to the United States
B. Isolation
C. Quarantine
D. Antibiotic therapy
E. Nothing, she is asymptomatic

92. A person flies home from Nigeria to Tampa, Florida. Through Nigerian surveillance and investigation, this person is identified to American authorities as possibly being exposed to Ebola. Which government authority has the right to detain this passenger at the airport and quarantine him until he is declared free of the disease?

A. Tampa
B. Nigeria
C. United States
D. Nobody has the authority to detain this passenger
E. World Health Organization

93. Which of the following tools uses maps to identify populations at risk of health disorders, disease risk factors, health interventions and health outcomes?

A. Epidemic curve
B. Gantt chart
C. Geographic information systems
D. Ishikawa diagram
E. All of the above

94. An insurance company would like to know if members enrolled in their plans are honest when they say they do not use tobacco products. To test tobacco abuse, the company hired a biochemist to create a test for tobacco metabolites using a saliva sample. To evaluate the test, a biostatistician randomly tested a group of 1,000 subjects. 17% of these subjects admit to regular tobacco use. The test has 90% sensitivity and 90% specificity.

What is the PPV of this test?

A. 0.17
B. 0.53
C. 0.65
D. 0.90
E. 0.98

95. What happens to the NPV if the prevalence of a disease in a population increases?

A. It stays the same
B. It increases
C. It decreases
D. There is no effect
E. Depends on other variables

96. What happens to the sensitivity as the prevalence of a disease in a population increases?

A. It stays the same
B. It increases
C. It decreases
D. Depends on other variables
E. None of the above

97. Complete the following equation:

Specificity + __________ = 100%

A. Sensitivity
B. True-positive
C. True-negative
D. False-positive
E. False-negative

98. Complete the following equation;

Sensitivity + ________ = 1

A. False-negative
B. False-positive
C. Specificity
D. True-negative
E. True-positive

99. A news reporter is writing an article about a new revolutionary test for penicillin allergy. The test will allow clinicians to perform a point of contact finger prick blood analysis test to see if patients are indeed allergic to penicillin. She reports that the test will identify nine out of every ten people that actually have the allergy. What is the name for this metric?

A. False-negative rate
B. Negative predictive value
C. Positive predictive value
D. Sensitivity
E. Specificity

100. In the receiver operating characteristic (ROC) Curve, the y-axis is the sensitivity and the x-axis is the __________.

A. Specificity
B. Positive predictive value
C. Negative predictive value
D. False-positive error rate
E. False-negative error rate

101. An epidemiologist at the CDC wishes to analyze vital statistics to follow trends in birth records. Which national source could she look to that will provide her with collected information from health departments?

A. Federal Department of Vital Statistics
B. Federation of State Medical Boards
C. Health Resources and Services Administration
D. National Vital Statistics System
E. There is no national collaboration; individual states maintain their own vital statistics

102. The US Outpatient Influenza-like Illness Surveillance Network (ILINet) records the number of cases of influenza-like illness (ILI). These ILI cases are compared to other time periods to determine changes in trends.

Which of the following answers does not meet criteria for influenza-like illness (ILI)?

A. Temperature greater than 100°F (37.8°C)
B. Cough
C. Sore throat
D. Myalgias
E. Positive influenza type B

103. How is data for the National Health and Nutrition Examination Survey (NHANES) gathered?

A. Home interview
B. Phone interview
C. Physical in-person exam
D. Options A and C
E. Options B and C

104. An epidemiology student is trying to figure out how to research the risk of adverse events linked to exposure to pharmaceuticals during pregnancy. He is on a budget with limited time and resources, but would like to obtain widespread accurate data. Which is the best way for this student to obtain the data?

A. Access data from his state's Department of Vital Statistics
B. Access a pregnancy medication exposure registry
C. Conduct case-control study
D. Conduct ecological study
E. Conduct randomized control trial

105. Which organization is responsible for publishing the International Classification of Diseases (ICD)?

A. Department of Health and Human Services (DHHS)
B. Health Resources and Services Administration (HRSA)
C. International Classification of Diseases Executive Committee (ICDEC)
D. International Medical Billers Coalition (IMBC)
E. World Health Organization (WHO)

106. Which group in the United States experiences the highest infant mortality rate (IMR)?

A. American-Indian and Alaska Native
B. Asian and Pacific islander
C. Black
D. Latino
E. White

107. The neonatal mortality rate includes death up to which age?

A. 0 days
B. 7 days
C. 28 days
D. 6 months
E. 1 year

108. The IMR of the United States falls within which range?

A. <3 per 100,000
B. 3–5 per 100,000
C. 3–5 per 1000
D. 5–7 per 100,000
E. 5–7 per 1000

109. How frequently is the population census of the United States held?

A. Every year
B. Every 2 years
C. Every 3 years

D. Every 5 years
E. Every 10 years

110. Which group is the primary target of the Youth Risk Behavior Surveillance System (YRBSS)?
A. Those 15 years of age and below
B. Parents of those 18 years of age and below
C. Students in school
D. Random telephone survey
E. Both A and B

111. How is data obtained for the Behavioral Risk Factor Surveillance System (BRFSS)?
A. In person interview
B. Physical examination
C. Telephone interview
D. Internet survey
E. Combination of the above choices

112. Which of the following is an experimental study?
A. Evaluating the number of malaria cases in a community during summer and winter
B. Following the number of newly hired coal workers that develop pneumoconiosis
C. Asking mothers of children with neural tube defects about their use of folic acid
D. Administering a new type of drug to compare it to the safety of the old drug
E. Comparing the number of deaths observed deaths in a population to the number of expected deaths

113. Which of the following differentiates an experimental study from a quasi-experimental study?
A. The alpha level
B. The beta level
C. The exposure
D. The outcome
E. The hypothesis

114. A drug researcher is testing two experimental drugs, Drug A and Drug B. Both drugs kill 850 out of 1000 lab rats during the first 10 days of taking the medication. 90% of the Drug A rats that died succumbed during the first two days. Meanwhile, 90% of the Drug B rates that died lived until days #9 and #10.

Which of the following is true?

A. Drug A has a higher risk of death
B. Drug B has a higher risk of death
C. Drug A has a higher rate of death
D. Drug B has a higher rate of death
E. The death rate and risk of both drugs are identical

115. Which of the following is considered to be a crude rate?

A. Number of bicycle injuries in Florida per year
B. Number of bicycle injuries in Florida per the state population in a year
C. Number of bicycle injuries, aged 24–35, in Florida per the state population in a year
D. Number of bicycle injuries, with black hair aged 24–35, in Florida per the state population in a year
E. Cannot be determined from this information

116. Which of the following is true when comparing the life expectancy of a person at birth to the life expectancy of a person at age 60?

A. Life expectancy is higher at birth
B. Life expectancy is higher at age 60
C. There is no relationship between the two
D. It is the same
E. Impossible to tell with the data given

117. In the United States, what is the leading cause of death in those age 1–44?

A. Suicide
B. Homicide
C. Unintentional injuries
D. Cancer
E. Heart disease

118. In the United States, what is the highest cause of years of potential life lost?

A. Cancer
B. Heart disease
C. Homicide
D. Motor vehicle accident
E. Obesity

119. Using 75 years as the endpoint, which of the following scenarios contributes to the highest number of years of potential life lost (YPLL)?

A. Contaminated equipment at a nursing home leading to deaths of 10 patients, all aged older than 75
B. Drowning of a 10-year-old
C. Homicide of a 50-year-old man and his 48-year-old wife
D. Intentional drug overdose of a 35-year-old and his 55-year-old companion
E. Motor vehicle accident killing four people aged 68

120. A gas leak in an underground subway kills 50 people. The number of people killed have been arranged into the following table based upon age:

People killed in gas leak in underground subway

Age group	<1	1–14	15–24	25–34	35–44	45–54	55–64	65–74	>75
Number of deaths	5	5	5	5	5	5	5	5	10

Assuming an endpoint of 75 years of age, what is the approximate overall number of years of potential life lost (YPLL)?
A. 500
B. 975
C. 1625
D. 1875
E. Need more information

121. Assuming the sum of immigration and emigration is zero, what type of population growth is represented in the following population pyramid?

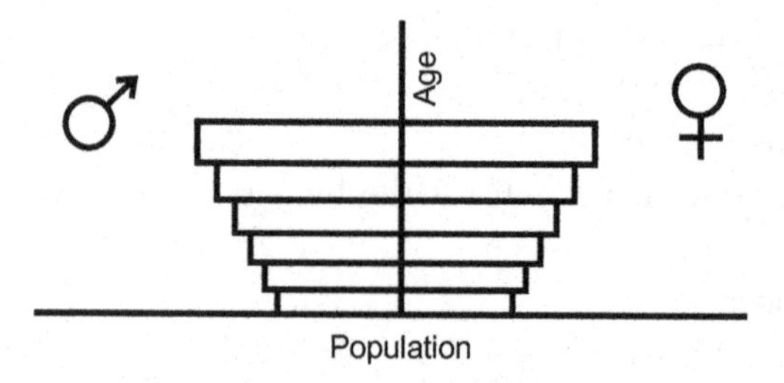

A. Growing
B. Stable
C. Declining
D. No specific pattern
E. Impossible to tell

122. A startup air conditioning company states that their revolutionary product can eliminate sick building syndrome (SBS). After a local office building of 1000 workers upgrades to their product, they notice that the prevalence of SBS reduces from 225 employees to 125 employees. What is the risk difference attributed to the new air conditioning system?

A. −0.1

B. −0.13

C. −0.2

D. −0.23

E. 0.3

123. After the completion of a stage 2 study in 300 teenage girls for a new mononucleosis vaccination, a biotech company has found that 80 of the 200 study participants exposed to the experimental immunization developed mononucleosis. Meanwhile, 80 of the 100 girls given the placebo developed mononucleosis. What is the vaccine efficacy percentage?

A. 26.7%

B. 40%

C. 50%

D. 53.3%

E. 80%

124. A nursing student is studying the relationship between red meat consumption and colon cancer. He wants to figure out the percent of colon cancer cases caused by elevated red meat consumption amongst red meat eaters. Which equation is the most appropriate way to calculate this relationship?

A. Attributable risk

B. Attributable risk percent

C. Population attributable risk

D. Population attributable risk percent

E. Risk ratio

125. 20 years after accidental exposure to Chemical X at a texture mill, the National Institute of Occupational Safety and Health (NIOSH) wishes to conduct an analysis between the link of Chemical X and brain cancer. A retrospective cohort study is conducted in 300 former employees. 100 of them were exposed to Chemical X. 15% of those exposed to Chemical X had received a diagnosis of brain cancer. Meanwhile, 10% of those that were not exposed to Chemical X had received a diagnosis of brain cancer.

What is the relative risk (*RR*) of diagnosis of brain cancer in those exposed to Chemical X, compared to those unexposed to Chemical X?

A. 0.667
B. 1.0
C. 1.5
D. 1.8
E. 2.5

126. A case-control study is analyzed with which of the following tools?

A. Attack rate
B. Cox regression analysis
C. Odds ratio
D. Meta-analysis
E. Risk ratio

127. What is the name of the program conducted by the CDC in conjunction with state health departments to monitor trends in women and infant care by surveying pregnant women and women that have just given birth?

A. Federal Obstetrics Monitoring Program (FOMP)
B. Obstetrics Surveillance Network (OSN)
C. Pregnancy Risk Assessment Monitoring System (PRAMS)
D. Survey of Maternal Care (SMC)
E. Women and Infant Survey (WIS)

128. In which case is the odds ratio a poor estimate of the risk ratio?

A. A case-control study is used to yield the odds ratio
B. If the prevalence is $>10\%$
C. The total number of subjects is >100
D. The outcome is rare
E. When causation has been established

129. A body builder believes that the trendy unregulated supplement he has been taking is responsible for his constipation. He asks his friend, a biostatistician, to look into this claim. The biostatistician decides to hold a study to quantify the relationship between this supplement and constipation. He randomly samples 120 gym members with similar demographics. Fifty-five of those surveyed were also on the supplement. Of those on the supplement, seven had constipation. Of the gym members not on the supplement, 13 had constipation.

The data can be filled into the following 2 × 2 table:

Relationship of supplement to constipation over 1-month observation

		Disease	
		Yes	No
Exposure (risk factor)	Yes		
	No		X

Which value goes in box X?

A. 48
B. 52
C. 55
D. 65
E. 100

130. What alcohol's relationship to cirrhosis?

A. Necessary and sufficient
B. Necessary and not sufficient
C. Not necessary and not sufficient
D. Not necessary and sufficient
E. None of the above

131. Healthcare is typically _____ for improved health status when one has an infectious health ailment such as HIV.

Which of the following options best answers this question?

A. Efficient
B. Sufficient
C. Necessary
D. Sufficient and necessary
E. None of the above

132. Which of the following conditions is classified as nondifferential error?

A. A case-control study where exposed subjects are misclassified as unexposed and a similar number of unexposed are classified as exposed
B. A case-control study where exposed subjects are misclassified as unexposed, but no unexposed subjects are misclassified

C. Interviewing mothers of children with birth defects about chemical exposures in pregnancy
D. Study participants receiving the experimental drug dropping out of a study due to adverse effects, while subjects in the placebo group remain in the study
E. None of the above

133. A study selects 1800 men aged 57–60 for a trial of a new type of cholesterol drug. 89% of the subjects on the drug are found to have decreased cholesterol with an alpha value of 0.01, averaging a 30 mg/dL decrease from those on the placebo. Over time, the risk of cardiovascular events was reduced 15% compared to those in the placebo group. When considering whether to approve this drug for all people at risk of dyslipidemia, which of the following might the US Food and Drug Administration (FDA) say about this study?
A. The drug has poor clinical significance
B. The trial lacks internal validity
C. The trial lacks external validity
D. The trial is not statistically significant
E. There is no need for new cholesterol medications

134. Which of the following contributes to the difference between vaccine effectiveness and vaccine efficacy?
A. Antivaccination public sentiment
B. Limited access to vaccinations
C. Improper storage of a vaccine
D. Prohibitive cost of the vaccine
E. All of the above

135. A researcher has proposed a formula to predict the number of soldiers that develop post-traumatic stress disorder (PTSD) based on their wartime duties.

Prior to initiating the study, the researcher wishes to see how accurate the formula is if the expected parameters were to vary. What is the name of the process used to accomplish this?
A. Attributable risk
B. Data organization
C. Sensitivity analysis
D. Standard deviation
E. None of the above

136. Which term best describes the severity of an infection once the disease occurs?

A. Infectivity
B. Immunogenicity
C. Pathogenicity
D. Secondary attack rate
E. Virulence

137. As the director of a local health department, the director of the emergency preparedness division submits a report to you about an abrupt increase in the number of influenza-like-illness (ILI) cases reported by local clinics and health departments. Additionally, the report cites increased purchasing of over the counter cold remedy medications from local pharmacies. Local clinicians report an increase in the number of confirmed influenza cases.

After speaking with a representative at the CDC, it appears that the influenza virus has mutated to a form that is not covered by the annual vaccination. Furthermore, it has turned into a strain that only those older than 45 years old demonstrate any immunity.

What is the best explanation for this influenza epidemic?

A. Antigenic drift
B. Antigenic shift
C. Malaria coinfection
D. Recall bias
E. Resistance to neuraminidase inhibitor

138. A geographic territory experiences an outbreak of laboratory confirmed influenza in chickens. The prevalence of humans diagnosed with influenza is unchanged. Prevalence of influenza in chickens had been stable for the prior two decades. The surrounding region has not noticed a change in influenza incidence in humans or animals.

How is this outbreak in chickens best categorized?

A. Endemic
B. Enzootic
C. Epidemic
D. Epizootic
E. Pandemic

139. What is the first step in an epidemiologic investigation?

A. Characterize the epidemic (time, place)
B. Develop a hypothesis
C. Establish a case definition

D. Establish a diagnosis
E. Determine whether or not an epidemic is occurring

140. In clinical trial testing for a new drug, what does it mean when the ED_{50} and TD_{50} are close to each other?
A. Depends on the type of drug
B. The drug is widely considered to be safe
C. The drug has little margin for safety
D. The LD_{50} is likely close to the ED_{50} and TD_{50}
E. None of the above

141. What is the name of the joint venture between the CDC and the Food and Drug Administration (FDA) to monitor post-market vaccine safety for immunizations licensed in the United States?
A. American Inoculation Safety Network
B. Immunization Complication Monitoring Organization
C. Phase 4 Study Observation Program
D. US Shot Surveillance Alignment
E. Vaccine Adverse Event Reporting System

142. Which of the following best describes the purpose of an IRB?
A. To recommend physicians for privileges at a hospital
B. Analyze results of utilization review
C. Review the ethics of a research study
D. Review cases of clinician misconduct for the state licensing board
E. Determine accreditation status for educational institutions.

143. Which of the following is not necessary in order to obtain informed consent?
A. IRB approval
B. Legal decision making capacity
C. Presumption of competence
D. Understanding of risks and benefits
E. Voluntary decision making, without coercion

144. A researcher believes that the current gold standard of screening for a lethal disease yields more harm than benefit. This disease is currently curable with modern medicine and there are no other screening techniques. He proposes to the IRB that he conduct a cohort study where he screens some subjects at risk with the current test and allows the others to go unscreened.

What is a potential concern of the IRB?

A. The test may significantly alter the way that clinicians screen for this disease
B. The company that manufactures the treatment may suffer financial loss
C. It would be unethical to not screen patients for a curable disease
D. Harms exceeding benefits is not a legitimate concern to investigate
E. Disproving the benefit profile of the screening test would alter the test's indications

145. Which of the following groups are not included in interviews used in the National Survey on Drug Use and Health (NSDUH)?
A. Adults in apartments
B. College students in dormitories
C. Homeless people in shelters
D. Inmates in prison
E. All of the above are included

146. The Drug Abuse Warning Network (DAWN) is part of which organization?
A. Behavioral Risk Factor Surveillance System
B. Centers for Disease Control and Prevention
C. Department of Justice
D. Office of National Drug Control Policy
E. Substance Abuse and Mental Health Services Administration

147. Which of the following is the largest contributor to the IMR in the United States?
A. Congenital malformations
B. Low birth weight
C. Maternal complications
D. Sudden infant death syndrome
E. Respiratory distress of newborn

3.2 EPIDEMIOLOGY AND BIOSTATISTICS ANSWERS

1. **A. Cross-sectional**

   Cross-sectional studies look at a snapshot of the population being studied. Extrapolating the population findings to an individual level may lead to ecological fallacy, in which an association at the population level is not necessarily true at the individual level. This is especially true when there is a larger population (constituting a cross-sectional ecological study). For example, a cross-sectional ecological study showing that City B has a higher rate of mesothelioma than City C may falsely lead someone to believe that all residents of City B are more likely than residents of City C to get mesothelioma, regardless of asbestos exposure.

2. **B. Cohort**

   A RCT allows the investigator control over the exposure. Although an RCT would yield the most robust results, it would also be considered unethical to withhold a screening tool that has been shown by multiple previous studies to be effective.

   A cohort study would involve enrolling study participants based off their exposure, in this case whether or not he/she had a screening colonoscopy. Information gained from a cohort study can be used to compare case fatality ratios and determine how effective an intervention is.

   A case-control study categorizes study participants based off of their disease status rather than their exposure status and is thus less appropriate for this example.

3. **E. Cancer**

   The vital statistics sector of each state's department of health records birth, death, marriage, and divorce. Cancer is typically reported through registries recorded at health facilities. Hospital cancer registries send their records to the central cancer registry in its state. The state cancer registry will then submit it to the CDC's National Cancer Institute's Surveillance, Epidemiology, and End Results (SEER) Program.

4. **C. National Center for Health Statistics (NCHS)**

   The NCHS compiles statistical information in numerous categories from numerous sources (states, municipalities, private organizations, etc.). These statistics are used to guide public health decision making and create goals, such as the Healthy People program. In addition to storing health statistics, NCHS also collects health data. The NHANES is hosted within NCHS.

5. **C. Statisticians that analyze the data**

When a study has subjective outcomes, such as wrinkle reduction, blinding the parties is used to eliminate bias. Single-blinded studies blind the study participants. Double-blinded studies blind the study participants and the study investigators. Triple-blinded studies blind the study participants, study investigators, and the statisticians. Blinding is less common in objective studies, such as those recording lab results, but it still may be beneficial.

The IRB is the group that approves studies, mainly based on how ethical and feasible they are. There is no need to blind this group. Shareholders should not have influence over the internal workings of the study, so there is no need to blind this group either.

6. **D. Randomized controlled**

The main difference between an observational study and a controlled study is that controlled studies will manipulate the risk factor. In a randomized controlled trial (RCT), exposure to the risk factor is determined by those conducting the study; thus, it is an experimental study and not an observational one.

7. **A. Length bias**

Length bias occurs when a less aggressive disease appears to have a higher incidence. This is because slower moving diseases are more likely to be detected since the subject is alive for longer. To the contrary, diseases that cause mortality sooner are less likely to be detected.

Length bias is often confused with lead-time bias. Lead-time bias occurs when the diagnosis is made earlier and creates the illusion that the subject lived longer than if the diagnosis were made later. If a subject with a terminal disease is diagnosed 1 month earlier, he will still die at the same time. However, the records will indicate that he lived one month later with the earlier diagnosis.

8. **E. Randomly assigning 1000 study participants to separate arms in a study**

Nondifferential error/bias is also called random error, or chance error. If a sample is has equal amounts of error on both sides of the true value, the error will cancel out and the overall value will closely approximate the true value. Differential error/bias produces deviation in one direction from the true value, either above or below.

Answers A, B, C, and D are all examples of differential bias. All of these answers target only one of the two populations that should be interrogated equally. In answer "A," only the men and not the women are questioned about marital happiness. In answer "B," the CHF patients that have been discharged will not be questioned regarding their experience. The patients that have been discharged sooner may have a different perception of their hospital stay. In answer "C," the researcher contributes to recall bias by pressuring mothers of autistic children, but not the nonautistic. Mothers of children with mental or physical disabilities are more likely to reflect more heavily on their exposures and activities during pregnancy. In answer "D," the scale produces a differential bias by pushing the observed value in one direction away from the true value.

Only answer E is an example of nondifferential bias. Participants are randomized equally, without bias.

9. D. Lead-time bias

Lead-time bias is the appearance that early diagnosis of a disease prolongs survival with that disease. In this case, the FDA panel should be concerned that the early diagnosis of pancreatic cancer is not actually increasing the length of time that the patients live.

Confounding is not a type of bias. Confounding occurs when there is a variable interacting with the independent variable (exposure) and dependent variable (outcome).

The Hawthorne effect is also known as observer bias. It is the theory that people (including study participants) will change their behavior if they believe that they are being observed.

Misclassification bias occurs where there are errors in recording disease or exposure.

Length bias occurs when the prevalence of a disease with a longer lasting disease appears higher than the prevalence of shorter lasting diseases. Consider "Disease X," which lasts 1 month and "Disease Y," which lasts for 6 months. There are 5 months of extra opportunity for "Disease Y" to be discovered, leading to the appearance of a higher prevalence compared to "Disease X."

10. D. Hawthorne effect

The Hawthorne effect states that individual behavior is changes when a person is aware they are being observed. In this case, the nurses are more likely to use the soap because they are being observed. However, the nurses on the first floor are even more likely to use soap because there is a higher risk of being discovered to not be in compliance with study parameters.

Regression towards the mean states that the further a value is from the mean, the more likely future recordings are closer to the mean. For example, if an otherwise healthy patient presents to your clinic and is found to have high blood pressure, future blood pressure readings are expected to be closer to the true blood pressure.

The healthy worker effect states that workers are typically healthier than the general population because they are different from the general population, as ill and disabled people are typically unemployed.

The placebo effect, occurs when a person believes they are healthier because they are receiving treatment, even if the treatment is not scientifically effective.

Random error is an accepted discrepancy in clinical studies. It may be controlled for in all phases of study design. There is no reason to believe that random error is the source of the findings in this vignette.

11. E. Regression towards the mean

Regression towards the mean states that the further a value is from the mean, the more likely future recordings are closer to the mean. Systolic blood pressure is not a static measurement: It varies daily, and those recorded at outlier measures are more likely to be at the extreme for their norm. Over time, an individuals recorded blood pressures will average out to more accurately show their mean.

The Hawthorne effect is the observation that individual behavior changes once that individual is aware they are being observed.

Information bias is the use of erroneous study data and may result from imprecise and invalid study measures.

Neyman bias (also known as selective survival bias), occurs when cases in a study that survive have different exposures than those that die.

Recall bias occurs when those that suffer an adverse event recall their exposure history differently than those that did not suffer an event. A common example is that mothers of children with birth anomalies may recall their pregnancy differently than mothers of healthy children.

12. D. Stratification

Stratification is the only technique listed that reduces confounding during the analysis stage. It involves breaking the data into strata that can be more descriptive. For example, stratification of elementary school students would reveal that 3rd graders have higher understanding of math than 1st graders, but less than 5th graders. Randomization, restriction, and matching are all techniques to reduce confounding during the design stage of a study. Hypothesis testing is not a method used to control for confounding.

13. A. Alcohol

Confounding is a type of bias that occurs when a third variable influences (confounds with) the factor of interest and skews the observed result between exposure and disease. A confounding variable by definition must be associated with both the outcome and exposure. In this example, alcohol use is a confounder, as it is a risk factor for cardiovascular disease and is associated with cigarette smoking.

Confounding may be accounted for and controlled in studies. Controls to reduce confounding may be built into the design stage, or analysis stage of a study. Ways to control for confounding in the design stage include randomization, restriction, and matching. Ways to control for confounding in the analysis stage include standardization, stratification, and statistical modeling.

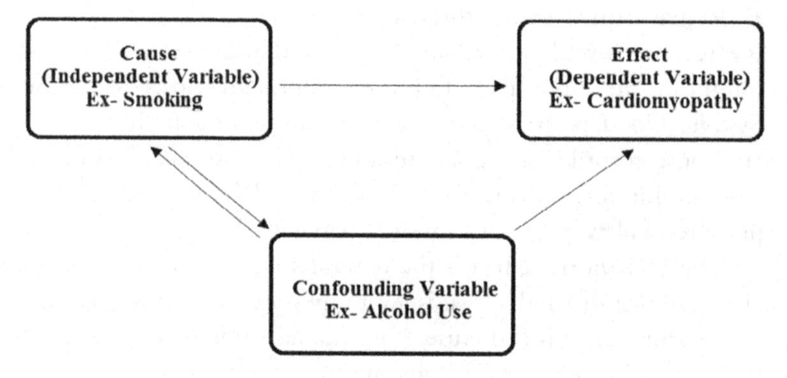

14. C. Sensitivity, specificity, and prevalence

Bayes' theorem is a mathematical tool for figuring out the probability of an event. It can be calculated every time new information is received that may alter the probability of the event.

Once a prior probability is known the next sequential event will yield a posterior probability. When more information is learned, the former posterior probability becomes the prior probability and a new posterior probability is calculated. Prevalence is equivalent to a prior probability. Bayes' theorem can be considered an alternative method of calculating the PPV.

The following formula may be used in place of Bayes' theorem:

$$PosteriorProbability = \frac{(Prevalence)(Sensitivity)}{(Prevalence)(Sensitivity) + [(1 - Prevalence)(1 - Specificity)]}$$

This formula may be difficult to remember, but the prevalence, specificity, and sensitivity can be used to fill out the cells of a 2 × 2 table, which will yield the PPV.

15. B. Person-time

Incidence density is a tool used to describe the number of new cases of a disease (incidence) per summation of time that each person is at risk of disease in a specified time and place. It is useful for observing dynamic populations (including clinical trials), where people are entering the leaving the risk pool. Additionally, it allows for each subject to be counted in the numerator more than once. This is important in cases where a subject experiences the disease of interest more than once.

For example, a researcher may be interested in observing a daycare to see how frequently children develop conjunctivitis over a 3-year period. In this setting, not all children are in the daycare for the same amount of time. Moreover, some children contract conjunctivitis more than once. If one child attends the daycare daily for 36 months and another is only there for 6 months, the overall person-time would be 42 months.

The downside of person-years is that a few amount of subjects may substantially influence the incidence density. This happens when a small number of people are observed over a long period of time are calculated along with a large number of people observed for a short duration.

16. B. 0.25 cases/person-year

Refer to Answer #15 (directly above) for an explanation of incidence density.

The time period of observation in this vignette is measured in years. Altogether, the six patients combined for 16 years of observation. During that time period, there were four cases of influenza.

$$IncidenceDensity = \frac{\#\ NewCases}{Summation\ of\ Person - Time}$$

$$= \frac{4}{16\ Person - Year} = \frac{0.25\,Cases}{Person - Year}$$

17. B. Central limit theorem

The central limit theorem states that when there are a large amount of mutually independent random variables, the mean population will approach normal distribution. A general rule of thumb is that for the central limit theorem to hold true, $N \geq 30$.

The Hawthorne effect is the phenomenon where subjects change their behavior because they are aware that they are being observed.

Inferential statistics methods allow one to make a statement about the general population by studying a smaller part of that population.

Binomial distribution describes data that has two discrete outcomes, typically success or failure.

Kaplan–Meier function is a tool used to examine survival analysis.

18. B. 70

The IQ is a tool for gauging intellectual abilities. It is determined through standardized testing to calculate the mental age, which is then divided by the chronological age and multiplied by 100. IQ was designed to follow a normal distribution with the central mean at 100 and the standard deviation of 15. As always, 68% of people fall within one standard deviation of the mean (85–115), 95% fall within two standard deviations of the norm (70–130), and 99% fall within three standard deviations of the norm (55–155). Observations more than two standard deviations from the mean are typically considered to be abnormal.

Two standard deviations below 100 is 70. An IQ of 70 nearly meets criteria for intellectual disability from the International Classification of Diseases (ICD) and the Diagnostic and Statistical Manual of Mental Disorders (DSM) categorize intellectual disability as follows: Mild (50–69), moderate (35–49), severe (20–34), and profound (>20). Individuals with intellectual disability experience reduced intellect and impaired adaptive functioning.

19. C. 0.33

The *Z-score* describes how many standard deviations are between an observed value and the mean. Observed values that are larger than the mean will have positive *Z-scores*, while observed values that are less than the mean will have negative *Z-scores*. It can be calculated through the following formula:

$$Z\text{-}score = \frac{x - \mu}{\sigma}$$

x is the observed value

μ is the mean of the distribution

σ is the standard deviation

To solve this problem, it is important to understand that the IQ scale was designed to follow a normal distribution with the central mean at 100 and the standard deviation of 15.

The *Z-score* for an IQ of 105 is calculated as follows:

$$Z\text{-}score\ IQ\ 105 = \frac{105 - 100}{15} = \frac{5}{15} = 0.334$$

20. A. Mean

In this set, 3000 is the outlier. This outlier would greatly skew the mean. With the full set, the mean is 281.8. Without the value of 3000 contributing to the mean, the overall mean drops to 10.

The geometric mean is typically used for extremely large numbers. It is most often used in logarithmic fashion. The geometric mean is found by multiplying all of the numbers (*n*) in the set and then taking the nth root of the product. Another method of calculating the geometric mean involves converting all of the numbers in a set into a logarithmic scale. The geometric mean cannot be used on negative numbers, or the number zero.

21. D. 50%

To answer this question, it is important to understand the definition of obesity. Obesity is determined based on the body mass index (BMI), which is calculated as a person's weight in kilograms (kg) divided by the person's height in meters squared (m^2).

Category	BMI (kg/m^2)
Underweight	<18.5
Healthy	18.5–24.9
Overweight	25.0–29.9
Obese	≥ 30.0

Five of the ten (50%) people filing claims are obese.

22. D. Mean > median > mode

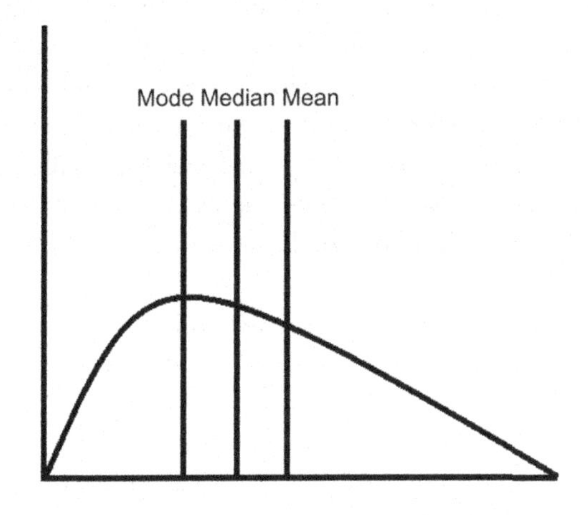

If a distribution is skewed to one side, there are fewer events on that side of the curve. This means that there are more events on the other side of the curve. An easy way to remember this is that skew rhymes with few.

In a skewed distribution, relative positions of the measures of central tendency is constant. From the tail (side with few) to the peak, the order is mean, median, and mode, as is shown in the diagram.

23. B. G

To solve this problem, you must first put the physicians in order of the number of patients they saw:

150, 220, 250, 250, 300, 315, 335, 360, 390, 400, 410

After placing the numbers in ascending order, you must then multiply the number of subjects by the percentage sought. This leads to $0.50 \times 11 = 5.5$. After rounding this number up, you get the 6th number. The 6th number in the ascending series is 315. Therefore, 315 represents the 50th percentile. The 50th percentile is also called the median.

The same method of calculation can be used for all percentiles, not just the 50th. As there is no universally correct consensus on how to measure percentiles and quantiles, some numbers may slightly vary by mathematician.

To calculate the interquartile range (the middle 50%), subtract the number found at the 25th percentile from the 75th percentile.

24. B. The confidence interval becomes smaller

As the number of subjects increases, the confidence interval becomes smaller. This is because the confidence interval is calculated by the formula $\overline{x} \pm Z\text{-}score\left(\frac{S.D.}{\sqrt{n}}\right)$, where n represent the number of trials. $\frac{S.D.}{\sqrt{n}}$ represents the standard error of the mean (S.E.M), which describes how far a population mean varies from the true mean. Since n is in the denominator, the S.E.M. will decrease as n increases. As the S.E.M. decreases, the confidence interval decreases.

The NNT is a function of the attributable risk and is unrelated to the number of overall subjects.

As the number of subjects increases, the power increases.

Clinical relevance of a study is unrelated to the number of subjects.

As the prevalence of a disease increases, the PPV increases. Thus, the PPV will only increase if the number of subjects with the disease of interest increases.

25. A. If the drug is tested on 1000 similar sample populations, 950 would lose a mean between 1 and 39 lbs.

The true value of this weight loss medication may never be known. However, confidence intervals provide a range of values that are believed to include this value. Confidence intervals are often preferred to *p*-values because they convey more information. Using the information presented in the vignette, the reader is not only able to tell that the average weight loss is 20 lbs, the reader is also able to know that if this subject is repeated in a similar group 95% of the time, the average weight loss will fall between 1 and 39 lbs.

Confidence intervals are used to describe parameters of the study population, not the individual subjects. If the confidence interval characterizing a risk difference does not include 0, the findings are said to be statistically significant. This is because if 0 is not included in the confidence interval, less than 5% of similar studies would have a mean of 0. A risk difference of 0 means that there is no change. Similarly, if a confidence interval describing an odds ratio (*OR*) or risk ratio (*RR*) includes 1, the results are not statistically significant. This vignette describes the risk difference, not the *OR* or *RR*.

Clinical significance depends on the situation and the person interpreting the data. A 1 lb weight loss may not be considered clinically significant in a 250 lbs person, but a 39 lbs weight loss in a 250 lbs person is significant.

26. C. Larger samples produce smaller confidence intervals

The true value of an intervention within a population may never be known. Confidence intervals provide a range of values that are believed to include this value. They are used to describe parameters of the study population, not the individual subjects. A confidence interval gives the likelihood of future studies to yield a range of results. A 95% confidence interval means that 95 out of 100 studies in similar groups will yield a population statistic that falls within the confidence interval.

Unlike *p*-values, confidence intervals are expressed in units. Confidence intervals are often preferred to *p*-values because they convey more information. Both confidence intervals and *p*-values express a degree of certainty, which may be converted from one approach to the other. For example, a 95% confidence interval is the same as a *p*-value of 0.05.

If the confidence interval characterizing a risk difference does not include 0, the findings are said to be statistically significant. This is because if 0 is not included in the confidence interval, less than 5% of similar studies would have a mean of 0. A risk difference of 0 means that there is no change. Similarly, if a confidence interval describing an odds ratio (*OR*) or risk ratio (*RR*) includes 1, the results are not statistically significant.

Confidence intervals measure precision around a point estimate. Larger studies are more precise and yield more narrow intervals. In homogenous confidence intervals, all values carry equal importance. However, in heterogeneous confidence intervals, certain values hold more significance than others.

27. B. 98%

97.5% of the medical students passed the exam. To better understand this question, one must understand standard deviation within the normal distribution. When the count is normally distributed, 68% will fall within one standard deviation. Half of these (34%) will be greater than the mean and the other half (34%) wall fall below the mean. Roughly 96% of the count will fall between two standard deviations. Half of the 96% (48%) will fall above the mean and half below. The remaining 4% falls outside of the two standard deviations, with 2% above and 2% below the cutoff.

The following illustration demonstrates the distribution of the test scores. All of the students that received a score greater than two standard deviations below the mean have been shaded in gray. Adding together all of the gray segments, one can find that 98% of the medical students passed their art history course. They are now able to continue their medical education and take a course in geography.

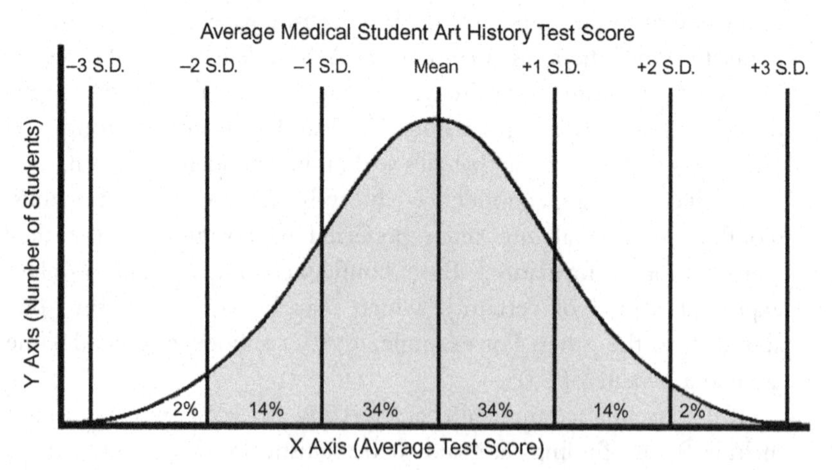

28. C. 0.8

This question requires the understanding of probabilities. For any event, the highest possible probability is 1.0, while the lowest possible probability is 0.

If the number of alcohol (0.6) and opiate (0.5) abusers are added together, the number would be 1.10, which is not a possible percentage. This addition does not account for overlap in the group that abuse both alcohol and opiates. If alcohol and opiate abuse were mutually exclusive, adding these two numbers would yield the correct answer.

To find the probability that the next patient abuses either alcohol or opiates, the Rule of Addition must be used. This requires the addition of the events of interest (alcohol and opiate abuse), minus the common overlap between events. Therefore the answer to this question would be calculated by:

$0.6 + 0.5 - 0.3 = 0.8$

There is a 0.8 probability that the next patient will abuse either alcohol or opiates.

29. E. 1.4

$$\textit{Standardized Mortality Ratio} = \frac{\textit{Observed Number of Deaths}}{\textit{Expected Number of Deaths}} \times 100$$

Calculating the standardized mortality ratio (SMR) is a method dividing the total number of observed deaths in a population to the total number of expected deaths in a population. The number is usually multiplied by 100, leaving the standard population with a value of 100. If the observed number of deaths is greater than the expected number of deaths, the SMR will be >1.

30. E. Indirect adjustment

Adjustment produces fictional numbers that can be used to compare populations with different variables. There are two forms of adjustment: Direct adjustment and indirect adjustment. Direct adjustment requires a second population from the original, which is used to extrapolate rates that create a less biased comparison. Meanwhile, indirect rates are performed when there is no comparison population, so a standard population must be used to accomplish the same goal.

The standardized mortality ratio (SMR) is a form of indirect adjustment used to evaluate the actual versus the expected ratio of deaths and compare this metric between populations.

$$\textit{Standardized Mortality Ratio} = \frac{\textit{Observed Number of Deaths}}{\textit{Expected Number of Deaths}} \times 100$$

An SMR value of 1 indicates that the number of observed deaths is what is expected. Meanwhile, an SMR greater than 1 indicates there have been more observed deaths than expected. Finally, an SMR less than 1 indicates that there have been less observed deaths than expected.

The SMR of each population can be compared to others, while holding the variable of concern constant, in order to determine if the outcome of interest (death) is different between populations.

None of the other options are directly related to the SMR.

31. B. Reading the book is not associated with his test score

Developing a hypothesis is the first step to answering a statistical question. For any given investigation, there is either an association between the variables, or there is not. A null hypothesis (H_0) assumes that there is no difference between the variables being tested. To the contrary, an alternative hypothesis (H_A) assumes that there is a difference between the variables. The alternative hypothesis may be considered the opposite of the null hypothesis.

The null hypothesis is assumed to be true unless stated otherwise. The purpose of a hypothesis test is to determine if the sample results of a study provide enough evidence against the null that it is likely the null would be false in the target population. Once the null is rejected, the alternate hypothesis is accepted as true. If the null cannot be rejected, it does not mean that the null is accepted as true. This is because data that is insufficient to show that a difference between variables is zero does not prove that the difference is zero.

In this case, the null hypothesis would assume that there is no association between reading the book and test score. Meanwhile, alternative hypothesis states that there is an association between reading the book and test score.

32. C. Accept the alternative hypothesis when there is a low *p*-value

The answer builds upon the explanation of hypothesis testing from Answer #31 (directly above).

The strength of evidence to accept or reject the null hypothesis is calculated as the *p*-value. It estimates the probability of finding an association in the target population as large as the association found in the sample, while assuming that the null hypothesis is true. A small *p*-value means that the association found in the sample is unlikely to be due to chance. Furthermore, the null and alternative hypothesis are differentiated by an artificial cut point, known as the significance level. If the *p*-value is less than the significance level, the null hypothesis is rejected and the alternative hypothesis is accepted. Because it is considered to be a measure of strength of evidence against the null, the *p*-value should not be used to infer whether or not the null is true or false.

33. E. More than one of the above

A type I error occurs when a null hypothesis is rejected when it is actually true. It is frequently called a false-positive. If the true null is rejected, that means the alternative hypothesis may falsely accepted when the association may be due to chance. The probability of making a type 1 error is represented by α.

A type II error occurs when a false null hypothesis is not rejected, while the alternative hypothesis is true. This is known as a false-negative. The probability of making this type of error is represented by β. The power $(1 - \beta)$ of a study is the likelihood of committing a type II error.

34. D. This is a type II error because the result is likely negative when the disease is present

This question requires understanding the concept of anergy. Anergy refers to an inadequate immune response and may be a result of several variables including: The age of the patient, the overall health of the patient, immunosuppression, and many other factors. In this problem, the poorly controlled HIV status implies that the patient is immunosuppressed and incapable of demonstrating a response to the PPD. Therefore, the PPD will be negative, despite the patient actually having tuberculosis.

This question is an example of a Type II error. Also known as false-negative error and beta error, type II error occurs when something is declared as false, when it is actually true. This type of error may occur in cases of anergy and testing within the window period, amongst many other examples.

A type I error occurs when something is declared as true, when it is actually false. This type of error is also known as false-positive error and alpha error. An example of a type I error would be a positive PPD due to nontuberculosis mycobacterium exposure. In this instance, a positive PPD would indicate that a patient has been exposed to tuberculosis, when they have not. Another example of a type I error would be a positive RPR while testing for syphilis in a patient that has Lyme disease, lupus, malaria, or is pregnant.

35. D. 78, 82

The confidence interval is derived by using the formula:

$$\textit{Confidence Interval } (Z\text{-}score) = \overline{x} + Z\text{-}score\left(\frac{SD}{\sqrt{n}}\right)$$

Where:

- $\overline{x}$ is the mean
- $\frac{SD}{\sqrt{n}}$ represents the standard error of the mean (SEM), the variation within a sample
- 95% of the scores fall within 1.96 standard deviations of the mean.

The equation is solved by plugging in the numbers as follows:

$$95\% = 80 \pm 1.96\left(\frac{15}{\sqrt{225}}\right) = 80 \pm 1.96(1) = 78.04, 81.96$$

36. A. Increases

As the Prevalence of the disease in the population increases, the PPV increases. Alternatively, when the prevalence increases, the NPV decreases.

37. D. Increasing the null from 65th to 75th percentile

The power of a test is the probability of correctly rejecting the null hypothesis. A study with insufficient power may not detect and accurately identify an important causative effect. Increasing the number of study participants always increases the power. Power can be represented by the equation: Power = 1 − beta, where beta represents the probability of rejecting the null when the null is actually true. Therefore, as beta increases, power decreases (answer C). Increasing the threshold of the null hypothesis (answer D) means that the null is more likely to be rejected, thus increasing power. As the difference between the alternative and mean hypothesis increases, the power will also increase.

A higher alpha level will result in rejecting the null more often, thus increasing power (answer B).

38. E. All of the above are considerations when calculating sample size

Determining minimum sample size for a study is an important stage in the planning process. Having a large enough sample size is important for obtaining power and detecting a clinically meaningful differences with statistical assurance. On the other end, having a large sample size is expensive and exhausts resources. A common goal of clinical researchers is to find the minimum number of study participants necessary to yield meaningful results that are valid, accurate, reliable, and have integrity.

Each study has special considerations when determining the sample size. The type of study and the hypothesis being tested are primary considerations. Other important variables include the degree of precision desired, expected attrition (dropout) rates, size of population under investigation, and the method of sampling adopted.

The larger the sample size, the greater the accuracy and precision. Specific types of studies may anticipate larger attrition. There should be enough study participants to counteract the expected attrition. Larger populations call for larger sample sizes to represent enough of the population to produce accurate study conclusions. Finally, sample size depends on the type of sampling being used. For example, randomly drawn studies will require a larger sample size than a stratified sampling plan.

Sample size calculation should be deliberated with consideration to precision analysis, power analysis, and probability assessment. A large part of calculations are based upon criteria for controlling type I and type II errors. Sample size calculation is typically performed in stages: Size estimation/determination, sample size justification, sample size adjustment, and sample size re-estimation. Each stage carried specific considerations. For example, the adjustment stage must consider factors such as expected attrition rate and covariates.

39. B. Dependent variable

This equation is the key to performing multivariable analysis, which is used to understand the relationship that different independent variables interact upon a dependent variable. This type of analysis is useful for showing the change in a dependent variable when one or more independent variable changes.

In $y = a + b_1x_1 + b_2x_2 + e$,

y = dependent variable

a = regression constant, the starting point where independent variables begin to act on the dependent variable

b = adjustment coefficient that weighs different independent variables according to importance

x = independent variable

e = error

An example of multivariate analysis would be the understanding of how grade point average (GPA), physics test scores, biology test scores, and chemistry test scores correlate with medical school entrance exam scores. If the entrance exam emphasizes biology most of all, the biology test scores will hold a higher adjustment coefficient.

$$
\begin{aligned}
\textit{Entrance Exam Score}\,(y) = {} & \textit{Beginning Knowledge}\,(a) \\
& + \textit{Adjustment Coefficient}(b_1) \times \textit{PhysicsScores}\,(x_1) \\
& + \textit{Adjustment Coefficient}(b_2) \times \textit{Biology Scores}\,(x_2) \\
& + \textit{Adjustment Coefficient}(b_3) \times \textit{ChemistryScores}\,(x_3) \\
& + \textit{Error}(e)
\end{aligned}
$$

40. **C. There is a positive correlation between time of sunset and traffic**

The variable (r) describes the linear correlation between two quantitative variables. Potential values of r span from -1 to $+1$. If $r < 0$, the correlation is negative. If $r = 0$, there is no correlation. If $r > 0$, the correlation is positive. Answer "D" is not correct because correlation does not equal causation. In this example, it is unlikely that the sun setting causes more traffic, or that an increase in traffic causes the sun to set. It is more likely that rush hour occurs during the time the sun is setting.

41. **E. Positive linear**

A line of best fit can be used to help identify a linear relationship between variables. When inserted into the scatterplot from the vignette, it reveals a positive linear relationship.

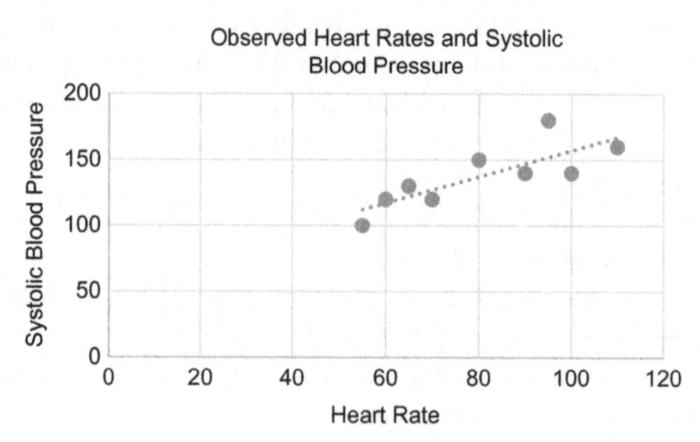

This question is an example of correlation analysis. When the investigator has control over the independent variable, it is known as regression analysis.

42. **D. Mildly positive**

Pearson correlation coefficient is a tool used to estimate the strength of a linear relationship between two normally distributed variables. It is represented by the r value, which varies from -1 to $+1$. Values closer to -1 have a negative association. That is, when one value decreases, the other increases. In contrast, values closer to $+1$ have a positive association, meaning both variables increase together. When r is 0, there is no association between the variables.

43. C. Multiple regression

Multiple regression is a method for examining how a normally distributed dependent variable is influenced by two or more continuous independent variables. If performed correctly, it allows researchers to assess the impact of one variable while controlling others. Multiple regression may be viewed as an extension of simple linear regression.

44. C. Studying survival rates in fixed periods

Survival analysis is used to determine the outcome of dichotomous variables, including live/die and success/failure.

The actuarial method of survival analysis is used to determine the number of survivors in fixed time intervals, such as years and months. A new line of the table is created for every fixed time period. Because of the set time periods, this method as not as good as the Kaplan–Meier method for accounting for censorship and loss to follow-up. This method is useful medical research and the insurance industries. It may be easier to apply this method if the sample size is large.

The Kaplan–Meier method of survival analysis does not have fixed time intervals. A new line of the life table is calculated for every new death. During death free intervals, study participants may be removed from the denominator if these participants are censored or lost to follow-up. This allows for more accurate computation of survival rates. It is easier to apply this method if the sample size is small.

45. D. Meta-analysis

A funnel plot is a tool used to evaluate for publication bias. Publication bias occurs when the results of published studies are different than the results of unpublished studies. There is a tendency of studies demonstrating certain opinions to be published, while studies demonstrating the contrary opinion go unpublished. Consider an example of a study that demonstrates colon cancer screening actually increases the incidence of colon cancer. The researcher may more apprehensive to publish such a study that bucks conventional understanding and may provide a danger to the public. Funnel plots populate a "funnel" of expectations around the mean. Gaps in the funnel may suggest publication bias.

Meta-analysis studies are composed of numerous studies and have an inherent risk of publication bias.

46. D. 20

This question is asking about the NNT, the number of people who would need to be treated to benefit one person. It is calculated by the following equation:

$$\textit{Number Needed to Treat}(NNT) = \frac{1}{\textit{Absolute Risk Reduction}}$$

Where:

$$\textit{Absolute Risk Reduction}(ARR) = \textit{Risk}(\textit{Exposed}) - \textit{Risk}(\textit{Unexposed})$$

Ideally, the length of time that treatment is required to obtain a unit of benefit should be included in the calculation.

In this question, an ARR of 5% (0.05) is given. The NNT is calculated as follows:

$$\mathrm{NNT} = 20 = \frac{1}{0.05}$$

Meanwhile, the number needed to harm (NNH) is calculated as follows:

$$\textit{Number Needed to Harm}\ (NNH) = \frac{1}{\textit{Absolute Risk Increase}}$$

47. E. All of the above

The following nine considerations are widely utilized to distinguish causal from noncausal associations. Although none of the considerations can bring indisputable evidence for causality, together the considerations can have a strong predictive value. Besides temporality, there are no necessary or sufficient criterion to establish a relationship as causal.

1. Consistency of association
2. Strength of association
3. Specificity (only one factor is consistently implicated)
4. Temporal factors
5. Coherence of explanation—does it make logical sense?
6. Biological plausibility
7. Experimental evidence from a controlled trial supports causality
8. Dose-response relationship
9. Analogy (can be applied to similar associations)

48. B. Standard deviation, number of samples

The standard error (SE) is calculated as:

$$SE = \frac{SD}{\sqrt{N}}$$

Where SD = Standard deviation

N = Number of samples

While the *SD* shows the variability of individual observations, the *SE* shows the variability of means of samples.

49. E. None of the above are true

This question is intended to test understanding of statistical inference and validity. Statistical inference is the practice of making general characterizations after analyzing a sample. Part of exercising statistical inference is the assessment of validity. Internal validity describes how well a study represents true associations present within the study. It is dependent on how well the design, data collection, and analysis is performed. Bias and random variation can reduce internal validity. There is nothing to suggest that the study in the vignette lacks internal validity. External validity describes how well results of one study are generalizable to a different population. Due to the small number of subjects, this study lacks external validity and should not be used to infer that patients worldwide will experience similar results.

A study may be statistically significant with or without biological or scientific significance. If this study is found to be internally valid, the weight loss medication not promoting weight loss would be considered clinically significant.

50. B. 2

Degrees of freedom (*df*) is calculated by the following formula:

$$Degrees\ of\ Freedom = (Rows - 1)(Columns - 1)$$

In this problem, there are two rows and three columns.

$$Degrees\ of\ Freedom = (2 - 1)(3 - 1) = 2$$

51. D. 57

Chi-squared is the most commonly used nonparametric test and is best for hypothesis testing between categorical variables. When used appropriately, it tells an investigator whether observations are correlated, or if they are due to chance.

After rounding, the test statistic is 57. Because of rounding, other people calculating this question may reach a slightly different test statistic.

Chi-squared (χ^2) is calculated by using the following formula:

$$\sum \frac{(Observed\ Data - Expected\ Data)^2}{Expected\ Data}$$

The observed data is the number in the original table that was plotted by the health department employee.

The expected data is calculated though the following formula for each box:

$$Expected = \frac{(Rows)(Columns)}{(Total)}$$

This formula is plugged into the original table to get the following:

Expected use of outdoor exercise equipment by gender

	Aerobic	Anaerobic	
Men	$Expected = \frac{(R)(C) = (125)(100)}{(Total) = 275} = 45$	$Expected = \frac{(R)(C) = (125)(175)}{(Total) = 275} = 80$	125
Women	$Expected = \frac{(R)(C) = (150)(100)}{(Total) = 275} = 55$	$Expected = \frac{(R)(C) = (150)(175)}{(Total) = 275} = 95$	150
	100	175	275

The observed and expected values may now be clearly expressed in the following table:

Observed and expected use of outdoor exercise equipment by gender

	Aerobic	Anaerobic
Men	*Observed* 75 *Expected* 45	*Observed* 50 *Expected* 80
Women	*Observed* 25 *Expected* 55	*Observed* 125 *Expected* 95
	100	175

Finally, the chi-squared equation $\left(\chi^2 = \sum \frac{(Observed\ Data - Expected\ Data)^2}{Expected\ Data}\right)$ may be used for each box and yield the following

Use of outdoor exercise equipment by gender

	Aerobic	Anaerobic
Men	$\frac{(75-45)^2}{45} = 20$	$\frac{(50-80)^2}{80} = 11.3$
Women	$\frac{(25-55)^2}{55} = 16.3$	$\frac{(125-95)^2}{95} = 9.5$
	100	175

When these four numbers are added together (20 + 16.3 + 11.3 + 9.5), the test statistic is 57.1. The test statistic is then compared with the χ^2 critical value using the χ^2 table. To use the table, you must choose a significance level and select the degree of freedom (*df*):

$$Degrees\ of\ Freedom = (Rows - 1)(Columns - 1)$$

In this example, *df* is 1. Following the table, using *df* 1 and α 0.05, the critical value (aka significance value) is found to be 3.84. Because the test statistic is higher than the critical value, the null hypothesis is rejected and the alternative hypothesis is accepted. This means that the worker's observations show that men and women have different preferences between aerobic and anaerobic exercise machines.

52. B. 40%

The Kappa ratio is a measurement of the agreement between two parties when accounting for random chance. It is necessary to account for chance because random agreements will happen by chance. Kappa ratios are valued from −1 to +1. The further negative the ratio, the more disagreement, while the further positive, the further the agreement. A Kappa ratio of 0 indicates that the agreement is occurring due to chance alone.

The true or false answers represent a dichotomous variable that can be placed on a 2 × 2 table. The table is set up by the number of true and number of false responses by each resident. Boxes a and d represent agreement, while boxes b and c represent disagreement.

The table below has been filled out with the information extracted from the above vignette.

		Resident B		
		TRUE	FALSE	
Resident A	TRUE	40 (a)	20 (b)	60 (a + b)
	FALSE	10 (c)	30 (d)	40 (c + d)
		50 (a + c)	50 (b + d)	100

From this, one can calculate the Kappa statistic.

$$Kappa = \frac{Observed\ Agreement - Agreement\ Due\ to\ Chance}{Total\ Number - Agreement\ Due\ to\ Chance}$$

Where,

$$Observed\ Agreement = Cell\ a\ Observed + Cell\ d\ Observed$$

$$Agreement\ Due\ to\ Chance = Cell\ a\ Agreement\ Due\ to\ Chance + Cell\ d\ Agreement\ Due\ to\ Chance$$

Individual cell agreement due to chance is found by

$$\frac{(Row\ Total)(Column\ Total)}{Total\ Number}$$

$$\textit{Cell a Agreement Due to Chance} = \frac{(\mathrm{A}+\mathrm{B})(A+C)}{\textit{Total Number}} = \frac{(60)(50)}{100} = 30$$

$$\textit{Cell d Agreement Due to Chance} = \frac{(\mathrm{D}+\mathrm{C})(D+B)}{\textit{Total Number}} = \frac{(40)(50)}{100} = 20$$

$$\text{To solve the equation: } Kappa = \frac{(40+30)-(30+20)}{100-(30+20)} = \frac{(70)-(50)}{100-(50)} = 0.4 = 40\%$$

53. B. Paired t-test

No matter how well a study is matched, it is not possible to find better comparisons than the subjects to match themselves. A paired t-test allows researchers to compare the significance of an intervention on a normally distributed group before and after they experience the intervention. The null hypothesis is that the intervention produces little to no difference from before to after. Meanwhile, the alternative hypothesis is true if the before and after measurements are further apart.

54. C. f-test

Analysis of variance (ANOVA) uses the f-test. This test compares the dispersion within individual variable groups to dispersion between variable groups. An f-test is used when one is comparing three or more variables. It may indicate that one group is statistically different from the others, but it does not exhibit which group is different.

55. A. ANOVA

While t-tests are used to directly compare two groups to each other, analysis of variance (ANOVA) is used to compare multiple groups simultaneously. The null hypothesis when using ANOVA is that there is no difference between groups. To the contrary, the alternative is that the groups are different. ANOVA does not tell how they are different, only that they are different. After a significant effect has been found through ANOVA, post-hoc analysis is used to tell how the variables differ. In this analysis, post-hoc analysis would determine which coffee is best (and worst).

While t-tests could be used to compare the data directly, it is not preferred. This is because it would require each combination of variables to be compared directly. In this case, each coffee would have to be compared against one another. If each t-test has 0.05 risk of error, the overall error rate compounds with each t-test.

ANOVA uses the f-test. This test compares the dispersion within individual variable groups to dispersion between variable groups.

56. D. 64%

The coefficient of determination describes the proportion of variation of a dependent variable that can be explained by an independent variable. It can be calculated by squaring the Pearson correlation coefficient. If the Pearson correlation coefficient (r) is 0.8, the coefficient of determination (r^2) is 0.64, or 64%. This means that hours of sleep (independent variable) is responsible for 64% of point production (dependent variable). Because only 64% of point production is attributed to sleep, 36% (100%−64%) is caused by other factors.

Multiple R squared (R^2) is analogous to r_2, but is used in multiple regression.

57. B. ANCOVA

The ideal test to calculate this problem should be able to compare the means of categorical independent variables (type of insecticide) with a continuous dependent variable (incidence of mosquito-borne illness). <u>An</u>alysis <u>o</u>f <u>va</u>riance (ANOVA), <u>an</u>alysis of <u>cova</u>riance (ANCOVA), and t-test are all capable of performing this calculation. Because there are more than 2 independent variables (3 insecticides), it is less preferable to use the t-test. This is because multiple t-tests would have to be performed, a test comparing each independent variable to one other. Not only does this route take longer, but it also leaves more room for error. ANOVA compares the means of ≥ 2 independent variables on one dependent variable to investigate if there is a significant difference between the independent variables. ANOVA can only tell that a difference exists, but it cannot tell where the difference is. To find where the difference exists, post-hoc tests should be performed. ANCOVA is similar to ANOVA, except that group means are adjusted by a covariate to adjust for confounding in ANCOVA. Confounding occurs when there is an association between the exposure and outcome that is distorted by another variable, such as age.

58. B. Each is a type of time-series analysis

Cohort studies, epidemic curves, and longitudinal data collection are epidemiologic tools used in time-series analysis. A time-series is a sequence of measurements and observations made at successive points in time. Time-series analysis interprets this data by recognizing time as the independent variable. The effect is measured at various times, including before and after suspected cause, but is not necessarily used to demonstrate causation of correlation. Another example of time-series analysis is the multiple time-series study, in which a suspected risk factor is introduced to several groups at different times.

59. C. Interval

Degrees on the Fahrenheit or Celsius scales are examples of interval data. When an interval scale is used, the exact difference between each number is known. However, because there is no true zero, one number that is twice the other number does not have twice the difference. For example, 60°F is not twice as hot as 30°F.

To the contrary ratio numbers have a true zero and exact difference between numbers. For example, 60 meters is exactly twice as long as 30 meters.

A nominal scale does not rank the variables, it merely categorizes them, such as the hair color of people a room.

An ordinal scale categorizes variables by the order they are placed in, even if there is no constant value between the variables, such as the satisfaction scale of patients in a hospital.

60. C. Ordinal

The possible grouping of variables in this question is: Underweight (BMI $<$ 18.5%), ideal weight (BMI between 18.5 and 25), overweight (BMI $>$ 25%), and obese (BMI $>$ 30%). This grouping of variables are considered ordinal. If the question asked for the type of variable according to each individual weight, the correct answer would be ratio.

Ordinal variables have an order, but not necessarily equal values between the variables. Consider the pain scale in a hospital, where the difference between one and two may not be the same as six and seven. Similarly, with regards to BMI classification: Obesity $>$ overweight $>$ ideal weight $>$ underweight.

Interval variables are ordered according to value. The difference between ordinal variables and interval variables is that interval variables have set values between the variables. The difference between 67° and 68° is the same as between 97° and 98°.

Nominal variables do not have an order and are grouped in name only. An example of nominal variables are colors: Red, blue, green, etc.

Ratio variables have a true zero, where zero value actually means there is zero of the variable. In ratio variables, two times of a specific value is actually twice as high. For example, 20K is twice as warm as 10K. To the contrary, 20°C is not twice as warm as 10°C

There is no such thing as numerical variables.

61. D. Ratio

The Kelvin temperature scale is an example of a ratio variable. Other examples include measurements in height (meters, feet) and weight (pounds).

For further explanation of nominal, ordinal, interval, and ratio variables, refer to Answers #59 and #60 (directly above).

62. B. Lower p (success) rate skews the distribution to the right

The binomial distribution curve graphs probabilities of dichotomous, binary variables (not continuous). There is a different binomial distribution for every combination of numbers (n) and probability of success (p). The larger the number of observations (n) and the closer the probability of success (p) is to 0.5, the closer the binomial curve appears to the normal curve. If $n \geq 30$, many statisticians feel comfortable using the normal distribution in place of the binomial distribution.

When the smaller the p, the further the distribution is skewed to the right. Likewise, the larger the p, the further the distribution is skewed to the left. Even with extreme p, a larger n will approximate the normal distribution.

When the binomial distribution approaches the normal distribution, the following tendency measures apply:

$$Mean = (n)(p)$$

$$Variance = (n)(p)(q)$$

$$Standard Deviation = \sqrt{(n)(p)(q)}$$

Where:

n = number of observations

p = probability of success

q = probability of failure

63. B. Chi-square

Chi-square uses a hypergeometric probability distribution, where larger numbers accurately follow the distribution. For this reason, chi-square only provides approximate p-values. If a sample size is sufficiently small, the sample size will not follow a chi-square distribution.

Fisher's exact test shows exact *p*-values. When larger chi-square numbers are used, the two tests approximate each other. If less than 20% of the cells in a chi-square table have an expected count of <5, or any one cell has an expected count of <1, it is recommended to use Fisher's exact test.

64. B. Dichotomous

McNemar's test may be viewed as a special type of chi-squared, in which the variables are not completely independent (variables in chi-square are independent). McNemar's test is used to analyze matched pairs or calculate before and after changes in the same variable. While the t-test analyzes continuous variables, McNemar's test checks for an association between binary/dichotomous variables

65. C. Mann–Whitney U test

The Mann–Whitney U test is a nonparametric test comparable to the two sample t-test. It is used to test the median between two groups. The null hypothesis is that both groups are similar. The alternative hypothesis is that the two populations are different.

66. C. 7

The interquartile range is the difference between the 25th and 75th percentiles of the observations.

To calculate the interquartile range, the observations should be placed in ascending order.

$_{1}59,\ _{2}62,\ _{3}62,\ _{4}62,\ _{5}64,\ _{6}65,\ _{7}65,\ _{8}66,\ _{9}66,\ _{10}67,\ _{11}68,\ _{12}69,\ _{13}70,\ _{14}70,\ _{15}71,\ _{16}71,\ _{17}72,\ _{18}73,\ _{19}74$

The 25th percentile is calculated by using the formula:

$$25th\ Percentile = \frac{(Number\ of\ Observations + 1)}{4}$$

$$= \frac{20}{4} = 5th\ Number = 64$$

The 75th percentile is calculated by using the formula:

$$25th\ Percentile = \frac{3(Number\ of\ Observations + 1)}{4}$$

$$= \frac{60}{4} = 15th\ Number = 71$$

The 25th percentile is then subtracted from the 75th percentile to get the interquartile range: 71 − 64 = 7.

If the interquartile range were to fall between two numbers, an average of these numbers may be chosen to represent a quartile.

As you can see, the interquartile range provides insight to the spread of data, but it ignores a large amount of data.

67. D. Sign test

Due to the small sample population available, normal distribution may not be assumed and a nonparametric test must be used. This immediately eliminates the paired t-test and student's t-test, which are parametric studies. By comparing two otherwise identical populations, a t-test would have been appropriate if there were the assumption of normalcy. Another important observation of the available data is that there are no specific numbers available. The data only denotes that there is a difference between the two groups, but does not describe how much the difference is. Both Wilcoxon signed rank test and chi-squared test (as do both t-tests) require specific number to perform. The sign test is a nonparametric test that compares dichotomous differences (better/worse or +/−) in data from matched otherwise identical pairs and ignores the magnitude of difference. Related to the Wilcoxon signed rank test, the sign test is an analog to the paired t-test. The null hypothesis in a sign test is that the difference between two groups is zero. In this question, the null hypothesis is that each group tested better than the other group five times.

68. B. Paired t-test

Please refer to the following table to depict the relationship between parametric and nonparametric tests

Parametric test	Nonparametric test (alternative to parametric test)
Student's t-test	Mann–Whitney U Test (also known as the Wilcoxon Rank-Sum Test)
Paired t-test	Wilcoxon signed rank test, Sign test
ANOVA	Kruskal–Wallis test
Pearson Correlation	Spearman correlation

Chi-square (χ^2) is a nonparametric test.

69. E. Spearman rank correlation

Spearman rank correlation coefficient (r_s) is the nonparametric alternative test to the Pearson correlation coefficient (used to measure linear strength of association between two variables). It works by ranking the *X* and *Y* variables according to value and inserting these rankings into the formula used for the Pearson correlation coefficient. In addition to being used for nonnormal continuous data, it can also be used for ordinal data.

Analysis of variance (ANOVA) is used to compare multiple groups simultaneously. The null hypothesis when using ANOVA is that there is no difference between groups. To the contrary, the alternative is that the groups are different. ANOVA does not tell how they are different, only that they are different. After a significant effect has been found through ANOVA, post-hoc analysis is used to tell how the variables differ. In this analysis, post-hoc analysis would determine which coffee is best (and worst).

Chi-squared is the most commonly used nonparametric test and is best for hypothesis testing between categorical variables. When used appropriately, it tells an investigator whether observations are correlated, or if they are due to chance.

The Mann–Whitney U test is a nonparametric test comparable to the two sample t-test. It is used to test the median between two groups. The null hypothesis is that both groups are similar. The alternative hypothesis is that the two populations are different.

The sign test is a nonparametric test that compares dichotomous differences (better/worse or +/−) in data from matched otherwise identical pairs and ignores the magnitude of difference. Related to the Wilcoxon signed rank test, the sign test is an analog to the paired t-test. The null hypothesis in a sign test is that the difference between two groups is zero.

Refer to the table in Question #68 (directly above) to see nonparametric alternatives to parametric tests.

70. A. Kruskal–Wallis

Refer to the table in Question #68 (two questions above) to see nonparametric alternatives to parametric tests.

71. D. 20%

The survival table for this question is shown in the following graph:

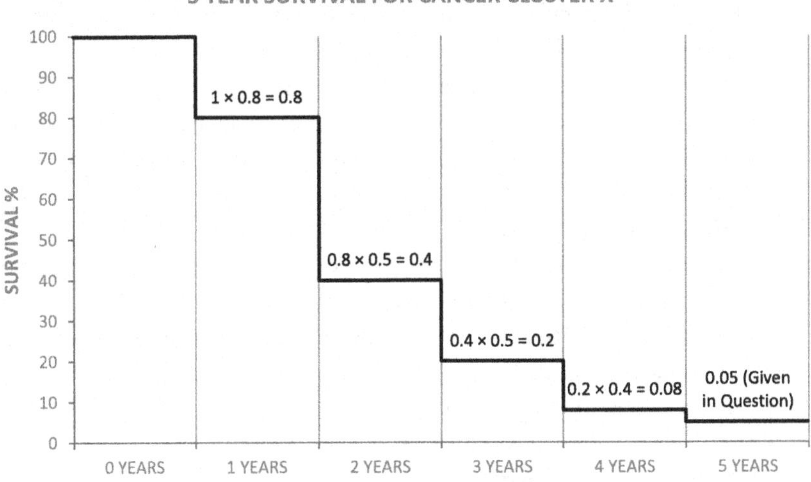

To solve this problem, the survival rate from the end of the first three time periods (years) of interest should be multiplied by each other.

72. C. 2

Logistic regression is used to find the likelihood of an outcome when the outcome is dichotomous. Dichotomous variables are commonly represented in the form of success/failure, improved/unimproved, or alive/dead.

73. D. Multivariate analysis of variance (MANOVA)

This question asks to find the overall effect that one independent variable (new medical office) has on two dependent variables (physician satisfaction and patient satisfaction). When there is more than one dependent variable, the situation is said to be multivariate. Furthermore, multivariate tests involve more than one dependent variable. Multivariate analysis of variance (MANOVA) is a tool used to evaluate multivariate tests and determine significance between groups. MANOVA is an extension of analysis of variance (ANOVA).

When there are multiple dependent variables, they are often times related to one another. In this case, the contentment of each party in the physician–patient relationship likely depends on the contentment of the other party. If one were to perform separate t-tests or ANOVA tests to solve for each dependent variable, this relationship is not properly addressed.

MANOVA considers the correlation between dependent variables, reducing distortion from relationships amongst them.

74. E. Publication bias

Publication bias refers to the tendency of investigators to publish results that have the desired outcome. Studies that achieve desired outcome are more likely to be published. If an investigator performing a meta-analysis only includes studies that have achieved the investigators' preconceived notions, the meta-analysis will have a differential error.

75. C. 20–28 weeks

Fetal death prior to 20 weeks gestation is defined as an early fetal death, commonly called a miscarriage.

Fetal death between 20 and 28 weeks gestation is defined as intermediate fetal death.

Fetal death after 28 weeks gestation is defined as a late fetal death, commonly referred to as stillbirth.

76. B. Number of live births

$$\textit{Maternal Mortality Rate} = \frac{\textit{Number of Pregnancy Related Deaths}}{\textit{Number of Live Births}} \times 100{,}000$$

Although the denominator of this equation should technically be the number of pregnancies, that statistic is not as readily available. For ease of calculation, the number of live births is used. This figure includes pregnancies with more than one child.

77. A. Difference of birth and death rates

The demographic gap is the difference between birth and death rates.

78. C. White Hispanic women

Life expectancy may be calculated at birth or any age afterwards. The overall life expectancy in the United States has been steadily increasing due to public health and medical advancements. Differences of life expectancy between sexes, ethnic groups, and races have been narrowing. Women are expected to live roughly 81.2 years, while men succumb roughly 5 years earlier, at age 76.4. Of course these numbers fluctuate with a plethora of variables accrued over a lifespan.

In the United States overall, the non-Hispanic White population has a higher life expectancy than the non-Hispanic Black population. However, the Hispanic White population has a higher life expectancy than the non-Hispanic White population and likewise for the Hispanic Black population. The difference with Hispanic ethnicity adds on over two years of extended life expectancy to both White and Black populations.

The exact reason for this Hispanic epidemiological paradox is under debate. Although extended life expectancy is typically tied to wealth and education, Hispanics buck this trend. The healthy migrant effect reasons that Hispanic immigrants are generally healthier than those that do not immigrate. Other arguments state that unhealthy Hispanic immigrants to the United States may return to their country of origin prior to death. Other theories suggest that cultural effects may confer a protective risk factor.

The Asian-American population enjoys the longest life-expectancy in the United States.

79. E. A health department epidemiologist calling the local hospital to ask about confirmed HIV cases

Active surveillance occurs when the health department takes action to seek out cases of illness. A health department calling healthcare providers to inquire about cases of illness is an example of active surveillance. Passive surveillance is where a health facility or laboratory notifies the health department of a reportable disease.

80. B. Centers for Disease Control and Prevention (CDC)

The NNDSS is a collaboration between local, state, and federal public health agencies to combat notifiable diseases through surveillance, data collection, data analysis, and sharing of public health data. It does this through numerous media, including the maintenance of the National Electronic Disease Surveillance Syndrome (NEDSS).

NNDSS is supported through the Center of Disease Control and Prevention's Division of Health Informatics and Surveillance (DHIS).

81. D. Council of State and Territorial Epidemiologists

The CSTE is an organization composed of epidemiologists, representing epidemiologists from all of the states and territories of the United States. Together, these epidemiologists collaborate and provide assistance to each other and other public health agencies, such as the CDC.

CTSE maintains a list of notifiable diseases that the states modify and adopt into law. States submit this data to the CDC to help track local trends in infectious diseases.

82. A. When the condition is designated as "notifiable" by the state

The CSTE is an organization that represents public health epidemiologists from states and territories. It maintains a list of notifiable diseases that the states modify and adopt into law. In addition to practitioners and laboratories being mandated to report notifiable diseases to the state in a set time period, the state is also asked to submit their notifications to the CDC in a set time period of either 4 hours, 24 hours, or 7 days depending on the type of disease.

83. B. Sensitivity

The appropriate screening test to identify those at risk of the fatal adverse event would have a high sensitivity [true-positive/(true-positive + false-negative)]. Sensitivity shows the proportion of those that have a disease that are accurately identified as those really having it. In a 2 × 2 tablet, sensitivity is calculated by $\frac{A}{A+C}$. A highly sensitive test helps to rule out a disease because it indicates that a negative test is likely not to have the disease. This can be remembered by the acronym *snout* (*sen*sitivity + rule *out*).

To the contrary, specificity is defined by true-negative/(true-negative + false-positive) and calculated on the 2 × 2 table from $\frac{B}{B+D}$. It represents the proportion of those without a disease that are accurately identified as not having it. A highly specific test helps rule in a disease because it indicates that a positive test is less likely to be a false-positive. Specificity can be remembered by the acronym *spin* (*sp*ecificity + rule *in*).

The PPV identifies the probability of those who test positive for the disease to those that actually have the disease.

The NPV identifies the probability of those that test negative for the disease to those that do not have the disease.

Prevalence is the proportion of those that have the disease in the population.

84. E. All of the above are considerations to make when creating screening recommendations

Implementation of a properly conducted screening test is a complex effort. Prior to initiating a screening test, the test itself must be fully scrutinized. Factors that should be evaluated include ethical consequences, psychological consequences, stigmatization, predictive value of results, test validity (and reliability once validity is established), treatment options, economics, and the risk of false-positives or false-negatives. It is also important to figure out how to properly notify the public about the availability of the screening test so that only those in the target population come forward for screening. After results are available, it must be determined how to appropriately disseminate the results.

Ethical and psychological considerations when considering a screening test come in a variety of forms, depending on the test being deliberated. Consider for example, a test that has no available treatment. What type of mental anguish would it cause for someone to know that they have an ailment that cannot be reversed? On the other hand, what if one sibling tests positive for a genetic trait that leads to cancer in everyone that has that trait. Is it ethical to perform a genetic test on one sibling, while the other would rather not have the test performed? Often times there is no correct answer in bioethics. However, it is important to consider the main bioethics principles of autonomy, beneficence, nonmaleficence, and justice.

There are always benefits and drawbacks in medicine. One common drawback is stress and angst related to undergoing the screening process. Those found to be at risk of disease suffer from psychological stress and identify themselves as weak and vulnerable. Meanwhile, self-perceived health is a predictor of future health status.

Another common drawback is false-positive results. The rate of false-positives depends on the test being conducted. A test yielding a large amount of false-positives may be considered acceptable if the benefit of discovering a serious disease is present. This is currently up for debate with prostate cancer, where it is estimated that for every man whose life is saved from prostate cancer by PSA screening methods, 47 men are over-diagnosed and treated, even though they do not need the treatment.

85. E. Cases of Norovirus in Tulsa by Data of Onset, Tulsa, Oklahoma, July, 2014

When creating any type of graph in epidemiology, it is important to be as descriptive as possible. It is ideal if a reader can understand the context of the graph by looking at the graph alone. The title should include the type of illness, place of outbreak, and when the outbreak occurred. In addition, the table should be labeled with the dates of incidence on the x-axis and the number of cases on the y-axis.

86. B. The middle

When charting the annual incidence of a disease, the incidence pattern is the primary focus. This is best displayed by the epidemiologic year, which spans from the month of lowest incidence from one year to another. By placing the lowest incidence at both sides of the graph, it allows the person viewing the graph to appreciate the time leading up to and after the highest incidence. If the highest incidence is plotted in the beginning or end of the curve, the peak incidence is likely to be broken up between the beginning and end of the curve.

87. A. Number of cases and time

An epidemic curve is an investigative tool that describes the patterns of an outbreak. These patterns help to identify the source of the outbreak and potentially how to address it. To compose an epidemic curve, the number of new cases should be plotted against a unit of time, most often days.

88. A. Common source

Epidemic curves were defined in Answer #87 (directly above).
There are typical outbreak patterns found on epidemic curves:

Common source—A group of people become ill after being exposed to a point source contaminant. All affected persons become ill within one incubation period. There are no secondary waves, in which people fall ill outside of the first incubation period. Example: Radiation toxicity after a nuclear power plant radiation leak.

Continuous common source—A common source continuously affects those with contact. Example: Soft serve ice cream machine contaminated with Listeria.

Propagated—Infection is transmitted from one person to another. May be direct or indirect contact. Often include waves of secondary or tertiary spread outside of the first incubation period. Rate of propagation depends on herd immunity, opportunities for exposure, and secondary attack rate. Example: Influenza outbreak.

Mixed—Occurs when a common source outbreak is complicated by person-to-person spread. Example: A bacterial conjunctivitis outbreak from a telescope that spreads amongst children at daycare.

89. D. Propagated

Different patterns of disease outbreaks were explained in Answers #87 and #88 (directly above).

90. B. Burger

To solve this question, it is necessary to build upon the numbers presented in the table. Variables of interest include the attack rate % (number of ill/total number), the attack rate difference (attack rate in exposed − attack rate in unexposed), and the attack rate ratio (attack rate in exposed/attack rate in unexposed).

Burgers have the highest attack rate %, attack rate difference and attack rate ratio, making them the most likely source of the diarrheal illness.

Food	Ate			Did not eat			Comparison	
	Ill	Well	Attack rate %	Ill	Well	Attack rate %	Attack rate difference	Attack rate ratio
A. Chicken	20	18	52.6	22	10	68.7	−16.1	0.77
B. Burger	32	8	80.0	6	22	21.4	58.6	3.74
C. Hotdog	12	15	44.4	15	28	34.9	9.5	1.27
D. Oranges	25	30	45.5	6	9	40.0	5.5	1.14
E. Egg roll	1	5	16.7	20	44	31.3	−14.6	0.53

$$Attack\ Rate = \frac{Ill}{Ill + Well} \times 100$$

$$Attack\ Rate\ Difference = Attack\ Rate\ of\ Food\ Eaten - Attack\ Rate\ of\ Food\ Not\ Eaten$$

$$Attack\ Rate\ Ratio = \frac{Attack\ Rate\ of\ Food\ Eaten}{Attack\ Rate\ of\ Food\ Not\ Eaten}$$

91. C. Quarantine

Smallpox is a viral infection that can be transmitted by respiratory droplets and fomites, such as blankets. Although there is no antiviral agent that has been proven to be effective against smallpox, there is a very effective vaccination. This vaccination has been used to eradicate the smallpox virus worldwide. The last naturally occurring case of smallpox was in 1977. Because of this, worldwide immunization ceased in 1980. The vaccination is not used to prevent acute infection.

The key to answering this question correctly is understanding the difference between isolation and quarantine. Isolation insulates people with an infectious illness from those without the illness. Meanwhile, quarantine insulates people who have been exposed to a contagious illness from those without the illness. There are several different types of personal and property quarantine measures. In the United States, government entities have the ability to enforce quarantine and isolation measures.

The incubation period for smallpox is typically 10–14 days. Quarantine should last this long at a minimum. If possible, use of a negative pressure room would be advised.

92. C. United States

Although individuals have rights, their liberties may be trumped by the rights of society as a whole to be protected from health threats. For this reason, governments may impose isolation for people showing signs of contagious illness and quarantine for those exposed (and asymptomatic) to the illness. These public health practices protect the public by reducing exposure to infectious disease.

In the United States, legal authority to isolate and quarantine is divided between the states and federal government. If a communicable disease is suspected or present in someone entering the United States, the CDC may issue a federal order to isolate or quarantine. Furthermore, the CDC may issue orders to isolate or quarantine in order to limit spread of disease from one state to another. Each state has their own isolation and quarantine statutes. States may isolate, quarantine, and trace persons with infections disease within their borders. This is commonly performed for tuberculosis.

Public health officials have the legal authority to react swiftly to infectious disease threats. An order to quarantine or isolate does not need advance approval from courts and violation of these orders may result in arrest. Detainees may legally challenge public health orders, but these orders take time and judges have limited jurisdiction and typically defer to medical experts.

For the sake of the rights of society to be protected from health threats, public health officers have the authority to reveal a patient's condition to those exposed. Similarly, hospitals are obligated to inform health departments of names and contacts of those with specific contagious disease.

The WHO governs disease globally and maintains the International Health Regulations (IHR). These voluntary regulations attempt to limit spread of contagious disease by addressing by influencing political, diplomatic, and trade relationships amongst all WHO member states. IHR is not directly enforceable, but insubordinate nations may face economic and social disruptions from other participating nations.

93. C. Geographic information systems (GIS)

Nearly every event on Earth can be spatially referenced into geographic data. The GIS are interactive computer-based applications that map geographic data. It describes the way we study the environment and produces spatial data that is used in a variety of industries, including healthcare and public health. Many aspects of health and well-being have spatial dimensions, including health disorders, disease risk factors, health interventions, and health outcomes.

GIS is an important tool in epidemiology, health administration, and health marketing. It may be used to map out any combination of factors to reveal potential correlations between health events. For example, one may use it to look at a cluster of increased disease incidence and compare it to individual demographic characteristics. Equally interesting, it could identify the exposed and unexposed groups to known risk factors for disease. With this information, GIS may help find appropriate places to institute a health intervention.

Epidemic curves are charts that plot the number of people with an infection versus the time at which they get the infection. It is used to identify the origin of an infection and the speed at which it travels through the population.

Gantt charts are an administrative tool to help create a project schedule. On a Gantt chart, each member is assigned a task and a time period to complete that task.

Ishikawa diagrams are also commonly known as cause-and-effect diagrams and fishbone diagrams. An Ishikawa diagram reads from right to left. At the far right of the diagram is the problem to be addressed. Moving to the left, the diagram identifies root causes of the problem(s). These root causes are further broken into sub-causes. Once the diagram has been drawn out, it takes the shape of fish bones.

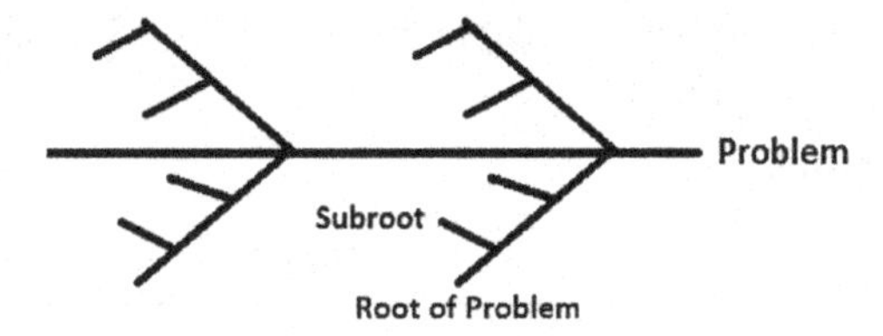

94. C. 0.65

To calculate the answer, it is easiest to use a 2 × 2 table to determine the unknown values from the known values:

Test status	Tobacco use: Positive	Tobacco use: Negative	
Positive	a	b	(a + b)
Negative	c	d	(c + d)
	(a + c)	(b + d and total-prevalence)	(a + b + c + d)

Prevalence is the percentage of a population that has a condition. It is calculated by a + c.

Sensitivity represents the proportion of those that have a disease that are accurately identified as those really having it. In a 2 × 2 tablet, sensitivity is calculated by $\frac{A}{A+C}$.

Specificity represents the proportion of those without a disease that are accurately identified as not having it. In a 2 × 2 tablet, specificity is calculated by $\frac{D}{B+D}$

Test status	Tobacco use: Positive	Tobacco use: Negative	
Positive	153 (0.9 x 170)	83 (830 – 747)	136 (153 + 83)
Negative	17 (170 – 153)	747 (0.9 x 830)	764 (17 + 747)
	170	830	1000

The PPV represents the proportion of those who test positive for the disease to those that actually have the disease. In a 2 × 2 tablet, sensitivity is calculated by $\frac{A}{A+B}$.

Using the numbers available in the 2 × 2 table, the PPV is $\frac{A}{A+B} = \frac{153}{153+83} = 0.648$

95. C. It decreases

Exposure	Disease: Yes	Disease: No
Yes	A	B
No	C	D

In a 2 × 2 table, when prevalence $\frac{A+C}{A+B+C+D}$ increases, the PPV $\frac{A}{(A+B)}$ is increased and the NPV $\frac{C}{C+D}$ is decreased.

The sensitivity $\frac{A}{(A+C)}$ and specificity $\frac{D}{(B+D)}$ are not affected by changes in prevalence.

96. A. It stays the same

		Disease: Yes	Disease: No
Exposure	Yes	A	B
	No	C	D

In a 2 × 2 table, when prevalence $\frac{A+C}{A+B+C+D}$ increases, the sensitivity $\frac{A}{(A+C)}$ and specificity $\frac{D}{(B+D)}$ remain unchanged.

However, when prevalence increases, the PPV $\frac{A}{(A+B)}$ increases and the NPV $\frac{C}{C+D}$ decreases.

97. D. False-positive

Specificity is represented by the formula $\frac{D}{(B+D)}$. Meanwhile, false-positive is represented as $\frac{B}{(B+D)}$. When added together, the specificity and false-positive error rate is equal to 1 or 100%. The false-positive rate is often calculated as (1 − specificity).

98. A. False-negative

Sensitivity is represented by the formula $\frac{A}{(A+C)}$. Meanwhile, false-negative error rate is represented as $\frac{C}{(A+C)}$.

When added together, the sensitivity and the false-negative error rate is 1.

99. D. Sensitivity

Sensitivity ($\frac{A}{(A+C)}$) is the proportion of those with a disease that test positive. As stated in the vignette, nine out of every ten people with the disease test positive. Meanwhile, specificity ($\frac{D}{(B+D)}$) is the proportion of those without a disease that test negative for it. These two metrics are not to be confused with the PPV and NPV. The PPV ($\frac{A}{(A+B)}$) is the proportion of those with a positive test result that actually have the disease. Finally, the NPV ($\frac{C}{C+D}$) is the proportion of those with a negative result that do not have the disease.

100. D. False-positive error rate

The receiver operating characteristic (ROC) curve can be considered a graph of positive likelihood ratios. The y-axis is the sensitivity, while the x-axis is 1 − specificity, which is the proportion of false-positive results.

The closer the cutoff point is to the upper left corner of an ROC Curve, the higher is the sensitivity and the lower is the false-positive error rate.

When measuring a continuous variable, such as the amount of potassium in a serum metabolic panel, setting the cutoff point between abnormal and normal limits can be a challenge. If the cutoff is too high, there are a lot of false-negative results. If the cutoff is too low, there will be more false-positive results. The ROC curve is a tool used to determine the best cutoff point for a continuous variable. It is a graph composed of the sensitivity along the y-axis and the false-positive error rate along the x-axis.

The false-positive error rate is equal to (1 − specificity). The sensitivity divided by the false-positive error rate is the likelihood ratio + (LR+). Therefore, the ROC curve graphs the LR+.

101. D. National Vital Statistics System

Recording vital statistics (birth, marriage, divorce, and death) is a responsibility of the states, two select cities (Washington D.C. and New York City) and United States territories. These entities share their vital statistics with the National Vital Statistics System (within the NCHS).

102. D. Myalgias

The US Outpatient Influenza-like Illness Surveillance Network (ILINet) receives information regarding the number of patients seen overall and the number of Influenza-Like Illness (ILI) seen from roughly 2000 outpatient healthcare providers weekly. This information is stratified by age group and evaluated for changes in trends.

ILINet case definition includes the following: Fever > 100°F (37.8°C), cough, and/or sore throat. If a patient has these signs/symptoms and is found to have a noninfluenza illness, it is not reported to be ILI. If a flu test is positive, it will be reported as influenza-like illness.

103. D. Options A and C

The NHANES is a program designed to assess the health and nutritional status of residents of the United States. It is a survey conducted by the NCHS, which is a part of the CDC. The data gathered from NHANES helps determine risk factors and prevalence of disease seen in the United States. This input has been used to influence health policy affecting individuals and the general public. Examples of changes influenced by NHANES include removing lead from gasoline, establishing baseline estimates for serum cholesterol and use of height/weight percentiles.

Information gathered for NHANES is exclusively through home interviews and standardized physical exams in mobile exam centers. Physical exams vary depending on age, gender, and medical history.

NHANES participants are chosen at random, based on their community, which is further divided into neighborhoods. After being screened for eligibility, there are nearly 5000 people surveyed each year. This means each participant represents approximately 50,000 US residents.

The BRFSS is performed exclusively through a telephone survey.

104. B. Access the Pregnancy Medication Exposure Registry

The best option listed for this student is to access a pregnancy medication exposure registry. With restricted resources, the student will be limited in conducting his own study. Furthermore, conducting an RCT to find medication adverse events is controversial in pregnant women. Pregnant women are usually excluded from pharmaceutical clinical trials, so information on medication adverse events is limited. Therefore, the Food and Drug Administration supports pregnancy exposure registries for reporting of adverse events discovered after taking specific drugs.

Departments of vital statistics are often times mandated to only cover a combination of statistics related to birth, death, fetal deaths, marriages, and separations. This data would not be adequate to evaluate the risk of adverse events related to pharmaceuticals taken during pregnancy.

105. E. World Health Organization (WHO)

The International Classification of Diseases (ICD) is published by the WHO. This classification outlines an international catalog of diseases, disorders, and injuries into a universal common language. It is used by epidemiologist, health administrators, and clinicians in over 100 countries around the world to study population-wide disease patterns and healthcare outcomes. This data is used to adjust the way healthcare is provisioned and practiced. Health-related variables recorded include vital records, reason for physician encounters, morbidity, and mortality. ICD is also widely used as a basis for resource allocation, including reimbursements in the United States.

106. C. Black

In the United States, Black individuals experience the highest IMR, while the lowest IMR is experienced by Asians and Pacific Islanders.

107. C. 28 days

Obstetrical-related data are important metrics used to compare health status among different populations. The most widely compared obstetric metric is the IMR, which widely serves as a surrogate marker for the overall status of a health system.

The neonatal mortality rate takes into consideration deaths that occur within the first 28 days of life. This is an important metric, as the majority of infant deaths typically occur shortly after birth.

A list of obstetric oriented rates has been listed below:

- *Fetal Death Rate*

$$= \frac{\textit{Total Number of Fetal Deaths (Intermediate \& Late) in a given time period}}{\textit{Total Number of Live Births During the Same Period of Time}} \times 1000$$

- *Infant Mortality Rate*

$$= \frac{\textit{Total Number of Deaths of Infants (< 1-year Old) in a Given Time Period}}{\textit{Total Number of Live Births During the Same Period of Time}} \times 1000$$

- *Maternal Mortality Rate*

$$= \frac{\textit{Deaths Due to Pregnancy} - \textit{Related Illness in a Given Time Period}}{\textit{Total Number of Live Births During the Same Period of Time}} \times 100{,}000$$

- *Neonatal Mortality Rate*

$$= \frac{\textit{Total Number of Deaths of Neonates (< 28 Days Old) in a Given Time Period}}{\textit{Total Number of Live Births During the Same Period of Time}} \times 1000$$

- *Perinatal Mortality Rate*

$$= \frac{\textit{Neonatal Deaths} + \textit{Fetal Deaths(Intermediate \& Late) in a Given Time Period}}{\textit{Total Number of Live Births and Fetal Deaths During the Same Period of Time}} \times 1000$$

108. E. 5–7 per 1000

The IMR is an important health statistic internationally. Although it includes many variables, it is used as a key measure to compare health systems between countries, regions, and even cities. IMR is recorded as the number of deaths occurring per thousand live births that occur before the first birthday. The IMR in the United States typically hovers around 6.0. This number has been declining over the decades.

The WHO estimates that the global IMR is roughly 32 deaths per 1000 live births. This number is a vast improvement from 30 years ago, when the IMR was double today's numbers.

Worldwide, nearly 75% of all deaths under five years of age occur within the first year of life.

109. E. Every 10 years

The US Census Bureau is a government organization mandated by the constitution with the mission to serve as the leading source of quality data about the country and economy. The Census Bureau conducts a population and housing census once every ten years. It also conducts an economic and government census every five years. In addition, it conducts a community survey annually. The information is used to distribute resources adequately to fund programs for public health, education, neighborhood improvements, etc.

110. C. Students in school

The Youth Risk Behavior Surveillance System (YRBSS) is a nationwide high and middle school-based survey conducted by the CDC in conjunction with all other levels of government. It monitors behaviors that contribute to morbidity and mortality in America's youth. Classifications of behavior surveyed include actions leading to violence and injury, sexual decision making, alcohol use, tobacco use, drug use, diet, physical activity, and prevalence of asthma.

111. C. Telephone interview

The Behavior Risk Factor Surveillance System (BRFSS) is the largest telephone survey in the world, with over 400,000 surveys annually. It conducts telephone surveys monthly in all 50 states, the District of Columbia, the US Virgin Islands, Guam, American Samoa, and Puerto Rico. Amongst many individual state and federal uses for monitoring trends, BRFSS data monitors progress towards Healthy People objectives. Individual surveys consist of core questions, rotating core questions (asking every other year), and additional standardized modules on specific topics determined by the state.

112. D. Administering a new type of drug to compare it to the safety of the old drug

Each of these answers describes a different type of study. Only answer D is an example of an experimental study. An experimental study is one in which the investigator has control over the exposure of the intervention being studied. The other four options are all examples of observational studies, in which the investigator observes, without intervening.

113. C. The exposure

The difference between experimental studies and nonexperimental studies is that investigators have control over the exposure in experimental studies. For this reason, experimental studies are considered to be the gold standard, secondary to meta-analysis studies, which analyze data from numerous peer-reviewed studies.

It is not always possible to control the exposure. Quasi-experimental studies are utilized when an investigator only has partial control over the study. For example, an investigator may wish to investigate a hypothesis by creating an exposure to a cohort that experienced of a nonreproducible natural disaster and compare it to a control that was not in the disaster.

114. C. Drug A has a higher rate of death

To answer this question, it is important to understand the definitions of both risk and rate. Risk is defined as the number of subject achieving the qualifying event (death in this example) during the defined time period. Assuming there is no attrition, the denominator (subjects at risk of qualifying event) does not change. To the contrary, the rate is the number of qualifying events that occurring during the defined time period divided by the average number of subjects at risk. The average subjects at risk used in this equation is typically the number of subjects at risk during the halfway mark of the time period.

In this question, the rats exposed to Drug A died early in the study. Therefore, there were fewer rats susceptible to death at the halfway point (because they had already died.) When more deaths take place in the early period, the number of subjects at risk during the halfway point is lower. When the denominator is lower, the rate will increase, and vice versa.

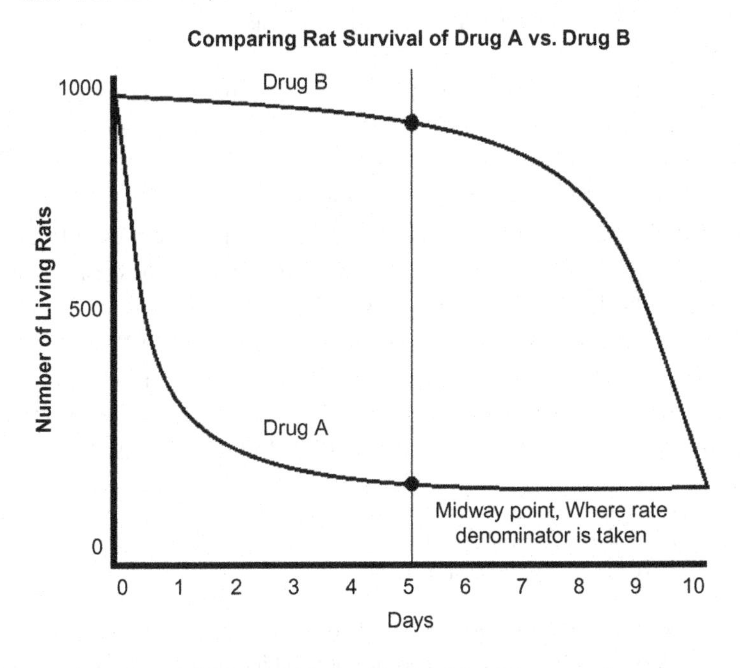

115. B. Number of bicycle injuries in Florida per the state population in a year

A rate must have a numerator and a denominator. The numerator typically displays the event of interest, while the denominator will establish the population at risk. Because there is no denominator in answer A, this statistic is not considered to be a rate.

Rates are either crude or specific. Crude rates use the entire number of events without breaking it down into subgroups. Meanwhile, specific rates are created by using subgroups such as hair color and age.

116. B. Life expectancy is higher at age 60

Life expectancy can be described as the average number of years of life one can expect to live based on mortality rates. It can be calculated from any stage of life. The life expectancy at birth is considerably lower than life expectancy of the same person at a later age. The reason for this is that those still living at the later age will have avoided mortality at a younger age, while others in the same cohort will have died. The life expectancy at birth takes account of all of those that will suffer premature death.

117. C. Unintentional injuries

Between the ages 1 and 44, unintentional injuries are the leading cause of death. Within this group, motor vehicle accidents are the most common specific cause. Heart disease is the leading cause of death in the United States, but typically affects the elderly.

118. A. Cancer

Years of potential life lost (YPLL) is a measure of premature mortality. It statistic emphasizes deaths that occur at an earlier age and deemphasizes deaths that occur later in life.

The YPLL is calculated as follows:

When there are less than 20 deaths, it is easiest to subtract the age of death from the endpoint. In the United States, the endpoint is usually determined to be 75 years of age. The difference between age of death and the endpoint represents the YPLL from one person. YPLLs from all individuals in consideration are then added together to get the aggregate YPLL. If a 17-year-old and 20-year-old die, the total YPLL would be 113. This is calculated as follows:

$75 - 17 = 58$ YPLL(1)
$75 - 20 = 55$ YPLL(2)
YPLL(1) + YPLL(2) = 113

When there are more than 20 people in consideration, it is easier to calculate YPLL through frequency tables. It is calculated by dividing ages into the following groups: Under 1 year, 1–14 years, 15–24 years, 25–34 years, 45–54 years, 55–64 years, and 65–74 years. Each age group is then identified by the midpoint. That midpoint is then subtracted from the endpoint (usually 75). The difference between the midpoint and endpoint is then multiplied by each person that falls within the group. For example, if a 17-year-old and 20-year-old were to die in a motor vehicle accident, this would tally into the 15–24 group. The midpoint of this group is 19.5, which is subtracted from 75 to get 55.5. Finally, this number is multiplied by two, because there are two people in this age group being calculated. Therefore, the YPLL from this group could be calculated as 110.

The leading causes of YPLLs in descending order are cancer, unintentional injuries, heart disease, and respiratory infections. Note that although heart disease is the leading overall cause of death, most deaths caused by heart disease are older and do not contribute on an individual basis as greatly to YPLL.

119. B. Drowning of a 10-year-old

Years of potential life lost (YPLL) is explained in Question #118 (directly above).

YPLLs for the five options are listed below:

Option A: Because all 10 patients are older than 75, there are no YPLLs.
Option B: 75 − 10 = 65 YPLL (Total)
Option C: 75 − 50 = 25 YPLL (1)
75 − 48 = 27 YPLL (2)
YPLL (1) + YPLL (2) = 52 YPLL (Total)
Option D: 75 − 35 = 40 YPLL (1)
75 − 55 = 20 YPLL (2)
YPLL (1) + YPLL (2) = 60 YPLL (Total)
Option E: 75 − 68 = 7 YPLL (1)
75 − 68 = 7 YPLL (2)
75 − 68 = 7 YPLL (3)
75 − 68 = 7 YPLL (4)
YPLL (1) + YPLL (2) + YPLL (3) + YPLL (4) = 28 YPLL (Total)

120. C. 1625

When calculating years of potential life lost (YPLL) for more than 20 people in consideration, it is easier to calculate YPLL through frequency tables. It is calculated by dividing ages into the following groups: Under 1 year, 1−14 years, 15−24 years, 25−34 years, 45−54 years, 55−64 years, and 65−74 years. Each age group is then identified by the midpoint. That midpoint is then subtracted from the endpoint (usually 75). The difference between the midpoint and endpoint is then multiplied by each person that falls within the group.

The following table calculates the YPLLs of the 50 gas leak victims.

People killed in gas leak in underground subway

(1) Age group	<1	1−14	15−24	25−34	35−44	45−54	55−64	65−74	>75	Sum
(2) Midpoint of group	0.5	7.5	19.5	29.5	39.5	49.5	59.5	69.5	[a]	
(3) Difference of midpoint to endpoint	74.5	67.5	55.5	45.5	35.5	25.5	15.5	5.5	[a]	
(4) Number of deaths	5	5	5	5	5	5	5	5	10	50
(5) Difference x deaths (lines 3 × 4)	372.5	337.5	277.5	227.5	177.5	127.5	77.5	27.5	[a]	1625

[a]Because these victims were older than 75 years, for calculation purposes, they are considered not to have lost any potential years of life.

121. C. Declining

The population growth of the population pyramid pictured in the question represents a declining population. As is often seen in developed nations, there is a larger population of older individuals than younger ones. A smaller group of younger individuals signifies that the birth rate has slowed down. When the birthrate equals the death rate, the population pyramid exhibits parallel vertical lines on both sides. When a population is growing, as seen in developing nations, the pyramid appears as a triangle with the base at the bottom wider than the peak at the top.

122. A. (−0.1)

Sick building syndrome (SBS) is a constellation of symptoms that occurs in workers continuously exposed to indoor environments. Symptoms include dry skin/mucous membranes, pruritus, mental fatigue, headache, and airway infections that are more pronounced during exposure to the indoor workplace. These symptoms lead to a large economic impact due to absenteeism, presenteeism, litigation, and workers' compensation claims. There are numerous risk factors contributing to SBS, including overly populated buildings, presence of carpet within the building, presence of mold/dust, and psychiatric stress. Due to this varied etiology and lack of specific biological markers, the prevalence of SBS is hard to determine.

This question asks to find the risk difference (RD):

Risk Difference = *Risk in exposed* − *Risk in unexposed*

RD may be performed with point prevalence, cumulative incidence, or incidence rates.

Risk difference is positive when the risk in the exposed population is greater than the risk in the unexposed. It is negative when the risk in the exposed group is less than the risk in the unexposed group. In this problem, the new AC system is considered to be the exposure.

To calculate the RD for this equation, you must first calculate the risk in the exposed and unexposed groups. Knowing that there are 1000 employees, there are 225 employees in the unexposed group. This means that the point prevalence rate of SBS is

$\frac{225}{1000} = 0.225$. After the installation of the new AC unit, the prevalence dropped to $\frac{125}{1000} = 0.125$.

These numbers are the plugged into the risk difference equation:

$$\begin{aligned} \textit{Risk Difference} &= \textit{Risk in Exposed} - \textit{Risk in Unexposed} \\ &= 0.125 - 0.225 = -0.1 \end{aligned}$$

The risk difference is negative when the exposed population is less than those without the exposure.

123. C. 50%

To solve this problem, one must use the following equation:

$$\textit{Vaccine Efficacy}\ \% = \frac{\textit{Attack Rate in Unvaccinated} - \textit{Attack Rate in Vaccinated}}{\textit{Attack Rate in Unvaccinated}} \times 100$$

Where;

$$\text{Attack Rate in Unaccinated} = \frac{\textit{Those with Disease That Were Unvaccinated}}{\text{Number Unvaccinated}} = \frac{80}{100} = 0.8$$

$$\text{Attack Rate in Vaccinated} = \frac{\textit{Those with Disease That Were Vaccinated}}{\text{Number Vaccinated}} = \frac{80}{200} = 0.4$$

Plugging in the available data:

$$\textit{Vaccine Efficacy}\,\% = \frac{\textit{Attack Rate in Unvaccinated} - \textit{Attack Rate in Vaccinated}}{\textit{Attack Rate in Unvaccinated}} = \frac{0.8 - 0.4}{0.8} = 0.5 = 50\%$$

Vaccine efficacy can also be found through the formula:

$\textit{Vaccine Efficacy}\% = 1 - \text{RR} \times 100$

In this case, the data can be summarized into the following table:

	Develop mononucleosis	Do not develop mononucleosis	
Mononucleosis vaccine	80	120	**200**
Placebo	80	20	**100**
	160	**140**	**300**

Using this 2 × 2 table, the risk ratio may be calculated;

$$Risk\ Ratio(RR) = \frac{\frac{a}{a+b}}{\frac{c}{c+d}} = \frac{\frac{80}{200}}{\frac{80}{100}} = 0.5 = 50\%$$

Vaccine efficacy is a useful calculations for ideal conditions, while Vaccine effectiveness better demonstrates how vaccinations work in the real world. Vaccine effectiveness accounts for variables such as vaccine storage, vaccine administration, access to care, and cultural barriers to vaccination.

124. B. Attributable risk percent

The Attributable risk (AR) is also known as the risk difference. It is the risk in the exposed, minus the risk in the unexposed. It does not give the percentage, as asked for in the question. Attributable risk percentage (AR%) is able to give the percentage of risk attributed to a specific risk factor, in those that are exposed to the risk factor, as asked for in the question. Meanwhile, the population attributable risk percentage (PAR%) is able to give the percentage of risk attributed to a specific risk factor in the general population, not just the group exposed to the risk factor. Note that in the second PAR% equation below (fourth equation), the variable "effective proportion" is added to the second AR% equation to specify the population at risk, in this case red meat consumers.

The attributable risk percent is represented by the following two equations;

$$Attributable\ Risk\ Percent(AR\%) = \frac{Incidence\ Exposed - Incidence\ Unexposed}{Incidence\ Exposed} \times 100$$

$$Attributable\ Risk\ Percent\ (AR\%) = \frac{Relative\ Risk - 1}{Relative\ Risk} \times 100$$

The population attributable risk percent is represented by the following two equations;

$$\textit{Population Attributable Risk Percent } (\text{PAR\%}) = \frac{\textit{Risk(Total)} - \textit{Risk(Unexposed)}}{\textit{Incidence Total}} \times 100$$

$$\textit{Population Attributable Risk Percent } (\textit{PAR\%}) = \frac{(\textit{Effective Proportion})(\textit{Relative Risk} - 1)}{1 + (\textit{Effective Proportion})(\textit{Relative Risk} - 1)} \times 100$$

125. C. 1.5

The relative risk (also known as the risk ratio) represents the risk amongst those exposed to a risk factor, compared to those unexposed. The vignette explains that 15% of those exposed had brain cancer, while 10% of the unexposed had brain cancer. The *RR* for this problem could be solved as follows:

$$\textit{Relative Risk(RR)} = \frac{\textit{Risk Among Exposed}}{\textit{Risk Among Unexposed}} = \frac{0.15}{0.1} = 1.5$$

The relative risk may also be completed by competition of a 2 × 2 table:

Relationship of brain cancer to Chemical X

Chemical X exposure	Brain cancer: Yes	Brain cancer: No	
Yes	a	b	a + b
No	c	d	c + d
	a + c	b + d	a + b + c + d

Filling in the information from the vignette:

Relationship of brain cancer to Chemical X

Chemical X exposure	Brain cancer: Yes	Brain cancer: No	
Yes	15 %	b	100
No	10 %	d	c + d
	a + c	b + d	300

Solving for the missing information:

Relationship of brain cancer to Chemical X

		Brain cancer: Yes	Brain cancer: No	
Chemical X exposure	Yes	0.15 x 100 = 15	100 – 15 = 85	100
	No	0.1 x 200 = 20	200 – 20 = 180	300 – 100 = 200
		15 + 20 = 35	85 + 180 = 265	300

$$Relative\ Risk\ (RR) = \frac{Risk\ among\ exposed}{Risk\ among\ unexposed} = \frac{\frac{a}{a+b}}{\frac{c}{c+d}} = \frac{\frac{15}{15+85}}{\frac{20}{20+180}} = \frac{0.15}{0.1} = 1.5$$

The relative risk (risk ratio) shows that those exposed to Chemical X are 1.5 times more likely to be diagnosed with brain cancer than those not exposed to Chemical X. Because the *RR* is >1, there is said to be a positive association between Chemical X and brain cancer. When *RR* is equal to 1, there is no relationship between the two variables. Finally, when $RR < 1$, there is a negative association.

126. C. Odds ratio

When possible, the risk ratio (*RR*) is the preferred method of analysis of risk. Cohort studies are best calculated with *RR*. However, the risk ratio cannot be calculated directly from a case-control study. The odds ratio (*OR*) is the best tool for risk analysis of case-control studies. It is important to remember that as the *OR* is only a reliable estimator of the *RR* if the prevalence of disease is less than 5%.

The *OR* is also used in other methods of analysis, including calculating the logistic regression and Cox regression analysis.

127. C. Pregnancy Risk Assessment Monitoring System (PRAMS)

The PRAMS is operated by the CDC in conjunction with state health departments in order to identify healthcare trends in pregnant women and women that have just given birth. It provides state-specific, population-based data on maternal attitudes and experiences prior-to, during, and after pregnancy. This data is used by health professionals to monitor and track healthcare goals in addition to formulate opportunities for improvements in care.

128. B. If the prevalence is >10%

A	B
C	D

The odds of an event considers the chances that one event will occur compared to another. The odds of square A to square B would be compared as A:B. This is opposed to the risk, which includes the denominator and would be written as $\frac{a}{a+b}$. When the prevalence (A + C) is small, the Odds Ratio $\left(\frac{A \times D}{B \times C}\right)$ is similar to the Risk Ratio $\left(\frac{\frac{a}{a+b}}{\frac{c}{c+d}}\right)$. As prevalence grows, the denominator of the risk ratio also grows (while odds remain the same) and the odds ratio will no longer approximate the risk ratio.

In summary, the odds ratio is a good estimate of the risk ratio when the prevalence of a disease is low. Some investigators are willing to accept a 10% prevalence, while others feel that the odds ratio does not approximate the risk ratio if the prevalence is greater than 5%.

129. B. 52

To answer this question correctly, the table must be completed, as shown below. To better depict all that is truly occurring in a 2 × 2 table, the borders have been added.

Relationship of supplement to constipation over 1-month observation

		Disease		
		Yes	No	
Exposure (risk factor)	Yes	a	b	a + b
	No	c	d	c + d
		a + b	b + d	a + b + c + d

The standard 2 × 2 table is then completed based off of information available in the vignette:

Relationship of supplement to constipation over 1-month observation

Exposure (risk factor)	Disease: Yes	Disease: No	
Yes	7	b (55 – 7)	55
No	13	d (65 – 13)	65
	20	b + d (120 – 20)	120

Once the incomplete squares are calculated, the final table may be completed:

Relationship of supplement to constipation over 1-month observation

Exposure (risk factor)	Disease: Yes	Disease: No	
Yes	7	48	55
No	13	52	65
	20	100	120

In case-control studies, the risk is calculated through the use of the odds ratio

$$\textit{Odds Ratio} = \frac{ad}{bc}$$

When solved for this problem,

$$\textit{Odds Ratio} = \frac{(7)(100)}{(20)(52)} = 0.67$$

The people using the supplement are 0.67 as likely to experience constipation as those not taking the supplement. Because this number is below 1, the supplement is considered to be a protective factor of constipation.

130. C. Not necessary and not sufficient

In a sufficient cause, the disease will always occur if the cause is present. In a necessary Cause, the cause must be present for the disease to occur.

Cirrhosis is not always a result of alcohol and not everyone that consumes alcohol will develop cirrhosis. Therefore, alcohol is not sufficient or necessary for cirrhosis.

131. C. Necessary

Healthcare is necessary, but not sufficient for improved healthcare. This was the conclusion of "Health care: Necessary but not sufficient," a brief from the Center of Society and Health in conjunction with the Robert Wood Johnson Foundation.

Necessary and sufficient are terms used to describe correlation between two variables, once causality has been established. For something to be necessary, it must be present to produce change. The change does not occur unless the necessary factor is present. For something to be sufficient, the presence of that factor will always bring change. Consider Mycobacterium tuberculosis: It is both necessary and sufficient for miliary tuberculosis.

132. A. A case-control study where exposed subjects are misclassified as unexposed and a similar number of unexposed are classified as exposed

Differential and nondifferential error are classified as the two types of misclassification bias.

Nondifferential misclassification occurs when the frequency of errors are the same in both populations being compared. This coincides with option A, where both the exposed and unexposed subjects are misclassified. Nondifferential error typically reduces the effect of the association and brings the measured association back to the null. An example of nondifferential bias would be analysis of incomplete medical records. More specifically, if a dichotomous variable such as cigarette smoking were left blank, the bias would be nondifferential.

Meanwhile, differential misclassification occurs when the misclassification occurs more in one of the groups being compared than the other. Differential misclassification may influence the association either towards or away from the null. Going back to the previous example of incomplete medical records, if each patient's chart was indiscriminately checked off for being a nonsmoker to save time, this would be an example of differential bias.

Option C is an example of recall bias. People that have had a memorable event (such as mothers of children with birth defects) will typically think harder to recall that event than those that did not experience the event. Because the ill and nonill groups may provide different recollection of their experience recall bias is a type of differential misclassification.

133. C. The trial lacks external validity

The sample group in this study represents a small spectrum of the population. In a drug that is potentially marketed to men and women of age that places them at increased risk of cardiovascular disease resulting from dyslipidemia, this study only evaluated men between the ages 57 and 60. External validity (also known as generalizability) occurs when results of an observation hold true in different situation. The sample population in clinical studies should represent the target population that is to be treated. Due to the limited variation in this study sample, the trial lacks external validity.

With the information provided in the vignette, there is every reason to believe that the conclusion of the study accurately represents the population that was studied. When the conclusions accurately represent the population being studied, the study is said to be internally valid. Internal validity depends on the study design, data collection, and analysis of data.

There is great clinical significance in controlling dyslipidemia and reducing cardiovascular events.

Statistical significance is demonstrated via the alpha value.

134. E. All of the above

Vaccine efficacy is obtained through studies, while vaccine effectiveness is how vaccinations perform in the real world. All of the above options are real-life variables that work to decrease the efficacy of vaccines.

135. C. Sensitivity analysis

Sensitivity analysis is the process of examining how expected outcomes change when they are placed under different assumptions.

Attributable risk is used to determine what effect an exposure to a risk factor has on the effect of the population.

Data organization is a general term that is not specific to sensitivity analysis.

Standard deviation is a measure of dispersion, which biostatisticians use to see how far spread out the numbers in a population are.

136. E. Virulence

- Immunogenicity—The ability to produce an immune response and protection from reinfection of a pathogen.
- Infectivity—Ability to cause an infection. Measured by the number of infectious particles required to cause infection.
- Pathogenicity—Ability of microbial agent to induce disease.
- Secondary attack rate—The proportion of susceptible people that contract a disease after exposure from an infected person. It is a measure of the infectivity.
- Virulence—Severity of the infection after the disease occurs. Measured by case fatality or severe morbidity.

137. B. Antigenic shift

Influenza is an RNA virus in the Orthomyxoviridae family. It is composed of hemagglutinin and neuraminidase proteins that are directly involved in the infectivity of the influenza virus. The two component proteins are continuously undergoing changes. When these changes are abrupt the genetic changes generate a new set of amino acids, leading to different proteins and a new influenza strain that may differ greatly from immunity provided from vaccination or previous exposure. Dramatic changes to the influenza virus are known as antigenic shift and are more often responsible for particularly severe influenza seasons. Less dramatic changes are called antigenic drift. In this case, the population typically carries a greater immunity from vaccinations and prior exposure of influenza strains that were similar to the currently circulating one.

138. D. Epizootic

The current outbreak of influenza in chickens can best be described as epizootic. An epizootic is an increase in the usual prevalence of a disease in an animal population, an animal disease outbreak.

An endemic is when the prevalence level of a disease in a human population is regular and constant.

An enzootic is when the prevalence level of a disease in an animal population is regular and constant.

An epidemic is an unusual increase in the occurrence of a disease. Even one case of a disease may be considered epidemic, such as would be the case if there was a confirmed polio diagnosis in the United States.

A pandemic is when a disease affects more people than usual and affects many regions and nations.

Animal and human disorders may interact. Sometimes enzootic trends can trigger epidemics in humans.

139. E. Determine whether or not an epidemic is occurring

The steps in an Epidemiologic Outbreak Investigation are as follows;

1. Confirm that an outbreak is occurring
2. Establish a diagnosis
3. Establish a case definition
4. Investigate the number of cases
5. Analyze the data and characterize the epidemic
6. Develop a hypothesis
7. Test the hypothesis
8. Implement action for control
9. Monitor prevention measures

140. C. The drug has little margin for safety

The ED_{50} represents the smallest dose that is effective in 50% of the test population. Meanwhile, the LD_{50} represents the dose that is lethal in 50% of the test population. If the ED_{50} and the LD_{50} are close together, the same dose (volume of drug) that produces a beneficial effect in half of the population will also cause death in half of the population. Contrarily, if the ED_{50} and LD_{50} are far apart, the drug has more room for error.

The LD_{50} is a poor indicator of health effects because death is one of the most undesirable outcomes of toxicity. Depending on the situation, a more preferable measurement is the TD_{50}, the dose at which the drug is toxic to 50% of the population. As with the LD_{50}, comparing the TD_{50} to the ED_{50} can yield important safety information.

Other important acronyms to know are NOAEL and LOAEL. NOAEL stands for "no observed adverse effect level," while LOAEL stands for "lowest observed adverse effect level."

141. E. Vaccine Adverse Event Reporting System (VAERS)

The National Childhood Vaccine Injury Act requires clinical professionals and vaccine manufacturers to report vaccine associated events to the US DHHS. To accomplish this, the VAERS was developed by the CDC and FDA.

The objectives of VAERS are as follows:

- Detect new vaccine adverse events
- Monitor trends in vaccine adverse events
- Identify risk factors for vaccine adverse events

- Identify vaccinations with high adverse event rate that were manufactured in the same batch
- Assess and monitor the safety of newly licensed vaccines

142. C. Review the ethical considerations of a research study

Organizations that conduct research on human subjects are required to create an IRB to evaluate the potential risks and benefits of human experimental research. In order to approve human research experiments, the IRB must evaluate the full proposed study. The IRB members must have an understanding of the science behind the research and the legislation regarding human research. The IRBs also assure that human subjects receive appropriate informed consent and fully understand their involvement in the research process.

143. A. IRB approval

Informed consent is an educational process by which a person makes an educated decision to participate or not participate in a procedure. To adequately obtain informed consent, several factors must be in place.

The individual must be legally allowed to make a decision. For example, minors are typically unable to make a decisions that lead to informed consent.

Informed consent requires presumption of competence. This presumption implies that a person can comprehend information, understand risks and benefits, exercise judgement, and make a decision based off the information.

Informed consent requires voluntary decision making that is free of coercion. If there are external factors influencing the decision making process, the decision may not be independent and cannot be considered appropriate for informed consent.

Finally, informed consent requires the full disclosure of all relevant information. An informed decision cannot be made if there is missing information.

144. C. It would be unethical to not screen patients for a curable disease

It is considered unethical not to use a screening test that has been shown to save lives. Consider mammography as an analogous example to the screening test in this question. Because it is known to be effective at detecting breast cancer, many groups would consider it unethical to not use screening mammography to screen breast cancer for research purposes. On the other hand, mammography often leads to harms in the form of stress to the patient and unnecessary surgical procedures.

145. B. Prison inmates

The NSDUH is housed within the Substance Abuse and Mental Health Administration (SAMHSA). It is the nation's primary source of information on patterns, prevalence, and consequence of drug use and mental disorders in the noninstitutionalized population, age 12 and older. NSDUH questions include alcohol, marijuana, tobacco, and all other illicit drugs. The study gathers data through face-to-face interviews at the place of residence and does not include incarcerated prisoners, homeless people not living in shelters, or military personnel on active duty.

The NSDUH is a key source of information to provide complimentary information to the BRFSS.

146. E. Substance Abuse and Mental Health Services Administration

SAMHSA is the operating division under the DHHS that aims to reduce the public health impact of mental illness and substance abuse in the United States. The DAWN records hospital emergency room information in order to provide surveillance of trends in drug use. Meanwhile, SAMHSA's National Study on Drug Use and Health (NSDUH) tracks patterns and consequences of alcohol, tobacco, illicit drugs, and mental illness in the United States through random interviews.

The BRFSS is conducted by the CDC to monitor health-related risk behaviors, use of preventive health services and status of chronic health conditions. Like SAMHSA, CDC is one of the operating divisions under the DHHS.

The DOJ contains the Drug Enforcement Agency, which originally founded DAWN. However, DAWN has been fully transferred to SAMHSA. The DOJ used to maintain the National Drug Intelligence Center (NDIC), which predicted future drug use trends via a national drug threat assessment. The NDIC is no longer in operation.

The ONDCP is a component of the Executive Office of the President of the United States. ONDCP advises the president on drug issues and coordinates activities to control illicit drug use. Additionally, the ONDCP composes the annual National Drug Control Strategy, which describes efforts to reduce drug use, drug distribution, drug-related violence, and health problems related to illicit drugs.

147. A. Congenital malformations

The IMR in the United States hovers around six deaths per 1000 live births. Of these deaths, the largest percentage is due to congenital malformations. Low birth weight is the second leading contributor to IMR in the United States. After these top two causes of infant morbidity, there is a significant drop in infant specific causes of death. Other common contributors to IMR include maternal complications, SIDS, unintentional injuries, placental complications, sepsis, and respiratory distress.

BIBLIOGRAPHY

[1] Katz DL, Elmore JG, Wild DMG, Lucan SC. Jekel's epidemiology, biostatistics, preventive medicine, and public health. 4th ed. Philadelphia, PA: Saunders; 2014. p. 62.

[2] Riffenburgh RH. Statistics in medicine. 3rd ed. San Diego, CA: Elsevier/Academic Press; 2014. p. 7–8.

[3a] CDC. Cancer registries' value for you. National Program of Cancer Registries (NPCR), <http://www.cdc.gov/cancer/npcr/value/>; [accessed 09.10.2016].

[3b] CDC. National Vital Statistics System. National Center for Health Statistics, <http://www.cdc.gov/nchs/nvss.htm>; [accessed 09.10.2016].

[4] CDC. NCHS overview. National Center for Health Statistics, <http://www.cdc.gov/nchs/data/factsheets/factsheet_overview.htm>; [accessed 09.10.2016].

[5] Merrill RM. Brigham Young University Utah Ray M Merrill. Fundamentals of epidemiology and biostatistics: combining the basics. Sudbury, MA, United States: Jones & Bartlett Publishers; 2013. p. 264.

[6] Riffenburgh RH. Statistics in medicine. 3rd ed. San Diego, CA: Elsevier/Academic Press; 2014. p. 7–8.

[7] Aschengrau A, Seage GR. Essentials of epidemiology in public health. 3rd ed. Burlington, MA, United States: Jones & Bartlett Publishers; 2014. p. 432–4.

[8] Szklo M, Nieto J. Epidemiology: beyond the basics. 3rd ed. Burlington, MA, United States: Jones & Bartlett Publishers; 2014. p. 121–33.

[9] Katz DL, Elmore JG, Wild DMG, Lucan SC. Jekel's epidemiology, biostatistics, preventive medicine, and public health. 4th ed. Philadelphia, PA: Saunders; 2014. p. 200–1.

[10] Tulchinsky TH, Varavikova EA. 3rd ed. The new public health, 517. San Diego, CA: Elsevier Science; 2009. p. 621.

[11] Friedman LM, Furberg CD, DeMets DL. September 21 Fundamentals of clinical trials. 4th ed. New York: Springer-Verlag; 2010. p. 175

[12–13] Bonita R, Beaglehole R, Kjellström T. Basic epidemiology. 2nd ed. Geneva, Switzerland: World Health Organization; 2006. p. 55–7.

[14] Katz DL, Elmore JG, Wild DMG, Lucan SC. Jekel's epidemiology, biostatistics, preventive medicine, and public health. 4th ed. Philadelphia, PA: Saunders; 2014. p. 93–103.

[15–16] Fletcher R, Fletcher S, Fletcher G. Clinical epidemiology: the essentials. 5th ed. Baltimore, MD: Lippincott Williams & Wilkins; 2014. p. 22–3.

[17] Forthofer RN, Lee ES, Hernandez M. Biostatistics: a guide to design, analysis, and discovery. 2nd ed. San Diego, CA, United States: Elsevier Academic Press; 2007. p. 157.

[18] Haith MM. In: Benson JB, editor. Diseases and disorders in infancy and early childhood. San Diego, CA: Academic Press; 2009. p. 251–2.

[19] Wassertheil-Smoller S. Biostatistics and epidemiology a primer for health professionals. New York, NY: Springer New York; 1990. p. 30–1.

[20] Forthofer RN, Lee ES, Hernandez M. Biostatistics: a guide to design, analysis, and discovery. 2nd ed. San Diego, CA, United States: Elsevier Academic Press; 2007. p. 64.

[21] U.S. Department of Agriculture and U.S. Department of Health and Human Services. Dietary Guidelines for Americans, 2010. 7th ed. Washington, DC: U.S. Government Printing Office, December 2010.

[22] Perrin K. In: Riegelman R, editor. Essentials of planning and evaluation for public health. Burlington, MA: Jones & Bartlett Publishers; 2016. p. 141–3.

[23] Upton G, Cook I. Introducing statistics. 2nd ed. Oxford: Oxford University Press; 2001. p. 49–50.

[24] Peat J, Barton B, Elliott E, Jennifer P. August 29 Statistics workbook for evidence-based healthcare. Hoboken, NJ: Wiley-Blackwell (an imprint of John Wiley & Sons Ltd); 2008. p. 1–5

[25–26] Lang TA, Secic M. How to report statistics in medicine. 2nd ed. New York: American College of Physicians; 2006. p. 37–43.

[27] Norman GR, Streiner DL, editors. Biostatistics: the bare essentials. 3rd ed. Hamilton: B.C. Decker; 2008.

[28] Glover T, Mitchell K. An introduction to biostatistics. 3rd ed. Lang Grove, IL: Waveland Press; 2016. p. 44–9.

[29] Katz DL, Elmore JG, Wild DMG, Lucan SC. Jekel's epidemiology, biostatistics, preventive medicine, and public health. 4th ed. Philadelphia, PA: Saunders; 2014. p. 38.

[30] Macera C, Shaffer R, Shaffer P. In: Stoskoph C, editor. Introduction to epidemiology. Clifton Park, NY: Cengage Learning US; 2012. p. 116–8.

[31–32] Dorey F. In brief: the P value: what is it and what does it tell you? Clin Orthop Relat R 2010;468(8):2297–8. Available from: http://dx.doi.org/10.1007/s11999-010-1402-9.

[33] Yarnell JWG, editor. Epidemiology and prevention: a systems based approach. New York: Oxford University Press; 2007.

[34] Katz DL, Elmore JG, Wild DMG, Lucan SC. Jekel's epidemiology, biostatistics, preventive medicine, and public health. 4th ed. Philadelphia, PA: Saunders; 2014. p. 107–8.

[35] Polit DF, Beck CT. Nursing research: generating and assessing evidence for nursing practice. 8th ed. Philadelphia: Wolters Kluwer Health/ Lippincott Williams & Wilkins; 2008. p. 586–7.

[36–37] Trochim W, Donnelly J, Arora K. Research methods: the essential knowledge base. Cincinnati, OH: Cengage Learning; 2015. p. 286–7.

[38] Shao J, Wang H, Chow S-C. In: Boca Raton FL, editor. Sample size calculations in clinical research.. 2nd ed. Chapman & Hall/CRC; 2008.

[39] Katz DL, Elmore JG, Wild DMG, Lucan SC. Jekel's epidemiology, biostatistics, preventive medicine, and public health. 4th ed. Philadelphia, PA: Saunders; 2014. p. 163–4.

[40] Bush HM. In: Scutchfield D, editor. Biostatistics: an applied introduction for the public health practitioner. 1st ed. United States: Cengage Learning; 2011. p. 93–5.

[41] Glover T, Mitchell K. An introduction to biostatistics. 3rd ed Long Grove, IL: Waveland Press; 2016. p. 315–20.

[42] Oleckno WA. Epidemiology: concepts and methods. Long Grove, IL: Waveland Press; 2008. p. 241–5.

[43] Forthofer RN, Lee ES, Hernandez M. Biostatistics: a guide to design, analysis, and discovery. 2nd ed. San Diego, CA, United States: Elsevier Academic Press; 2007. p. 368–71.

[44] Katz DL, Elmore JG, Wild DMG, Lucan SC. Jekel's epidemiology, biostatistics, preventive medicine, and public health. 4th ed. Philadelphia, PA: Saunders; 2014. p. 147–51.

[45] Rubin A. Statistics for evidence-based practice and evaluation. Belmont, CA: Cengage Learning; 2013. p. 153.

[46] Katz DL, Elmore JG, Wild DMG, Lucan SC. Jekel's epidemiology, biostatistics, preventive medicine, and public health. 4th ed. Philadelphia, PA: Saunders; 2014. p. 78–9.

[47] Rothman KJ, Greenland S, Lash TL. Modern epidemiology. 3rd ed. Philadelphia, PA: Lippincott Williams & Wilkins; 2008. p. 26–31.

[48] Polit DF, Beck CT. Nursing research: generating and assessing evidence for nursing practice. 8th ed. Philadelphia: Wolters Kluwer Health/ Lippincott Williams & Wilkins; 2008. p. 586–7.

[49] Fletcher R, Fletcher S, Fletcher G. Clinical epidemiology: the essentials. 5th ed. Baltimore, MD: Lippincott Williams & Wilkins; 2014. p. 11–2.

[50–51] Boslaugh S, editor. Encyclopedia of epidemiology, vol. 1. Thousand Oaks, CA: SAGE Publications; 2008. p. 140–1.

[52] Belle G, Fisher LD, Heagerty PJ, Lumley T. Biostatistics: a methodology for the health sciences. 2nd ed. Hoboken, NJ: John Wiley; 2004. p. 217–9.

[53] Katz DL. Clinical epidemiology and evidence-based medicine: fundamental principles of clinical reasoning and research.. Thousand Oaks, CA: SAGE Publications; 2001. p. 131–2.

[54–55] Norman GR, Streiner DL, editors. Biostatistics: the bare essentials. 3rd ed. Hamilton: B.C. Decker; 2008.

[56] Oleckno WA. Epidemiology: concepts and methods. Long Grove, IL: Waveland Press; 2008. p. 241–5.

[57] Bush HM. In: Scutchfield D, editor. Biostatistics: an applied introduction for the public health practitioner. 1st ed. United States: Cengage Learning; 2011. p. 140–1.

[58] Merrill RM. Environmental epidemiology: principles and methods. Sudbury, MA: Jones and Bartlett; 2008. p. 202–21.

[59–61] Norman GR, Streiner DL, editors. Biostatistics: the bare essentials. 3rd ed. Hamilton: B.C. Decker; 2008.

[62] Norman GR, Streiner DL, editors. Biostatistics: the bare essentials. 3rd ed. Hamilton: B.C. Decker; 2008.

[63] Weinberg SL, Abramowitz SK. Statistics using SPSS: an integrative approach. 2nd ed. Cambridge: Cambridge University Press; 2008. p. 499.

[64] Rubin A. Statistics for evidence-based practice and evaluation. Belmont, CA: Cengage Learning; 2013. p. 197–8.

[65] Bland M. An introduction to medical statistics. 4th ed. New York, NY: Oxford University Press; 2015. p. 177–82.

[66] Forthofer RN, Lee ES, Hernandez M. Biostatistics: a guide to design, analysis, and discovery. 2nd ed. San Diego, CA, United States: Elsevier Academic Press; 2007. p. 68–9.

[67] Chernick MR. The essentials of biostatistics for physicians, nurses, and clinicians. Hoboken, NJ: Wiley-Blackwell; 2011. p. 149–50.

[68] Chattopadhyay A. Oral health epidemiology: principles and practice. Sudbury, MA: Jones & Bartlett Publishers; 2011. p. 148–53.

[69–70] Forthofer RN, Lee ES, Hernandez M. Biostatistics: a guide to design, analysis, and discovery. 2nd ed. San Diego, CA, United States: Elsevier Academic Press; 2007. p. 63–4.

[71] Katz DL, Elmore JG, Wild DMG, Lucan SC. Jekel's epidemiology, biostatistics, preventive medicine, and public health. 4th ed. Philadelphia, PA: Saunders; 2014. p. 147–51.

[72] Norman GR, Streiner DL, editors. Biostatistics: the bare essentials. 3rd ed. Hamilton: B.C. Decker; 2008.

[73] Norman GR, Streiner DL, editors. Biostatistics: the bare essentials. 3rd ed. Hamilton: B.C. Decker; 2008.

[74] La Torre G. Applied epidemiology and biostatistics. 1st ed. Torino, It: SEEd; 2010. p. 173.

[75] Katz DL, Elmore JG, Wild DMG, Lucan SC. Jekel's epidemiology, biostatistics, preventive medicine, and public health. 4th ed. Philadelphia, PA: Saunders; 2014. p. 28–9.

[76] Katz DL, Elmore JG, Wild DMG, Lucan SC. Jekel's epidemiology, biostatistics, preventive medicine, and public health. 4th ed. Philadelphia, PA: Saunders; 2014. p. 35.

[77] Sundar PSS, Richard J, Rao R. An introduction to biostatistics and research methods. 5th ed. New Delhi: Prentice-Hall of India Pvt; 2012. p. 172.

[78a] Xu JQ, Murphy SL, Kochanek KD, Bastian BA. no 2 Deaths: final data for 2013. National vital statistics reports, vol 64. Hyattsville, MD: National Center for Health Statistics; 2016.

[78b] Mapping the Measure of America, <http://www.measureofamerica.org/maps/>; [accessed 10.10.2016].

[79] Review of and recommendations for the National Notifiable Disease Surveillance System: A State and Local Health Department Perspective. Council of State and Territorial Epidemiologists, <http://c.ymcdn.com/sites/www.cste.org/resource/resmgr/PDFs/NNDSS_Report.pdf>; [accessed 10.10.2016].

[80] National Notifiable Diseases Surveillance System (NNDSS). Centers for Disease Control and Prevention, <https://wwwn.cdc.gov/nndss/>; [accessed 1010.2016].

[81–82a] Council od State and Territorial Epidemiologist. About CSTE, <http://www.cste.org/?page = About_CSTE>; [accessed 10.10.2016].

[82b] Data collection and reporting. National Notifiable Diseases Surveillance System (NNDSS), <https://wwwn.cdc.gov/nndss/data-collection.html>; [accessed 10.10.2016].

[83] Goeree R, Hopkins R. Health technology assessment: using biostatistics to break the barriers of adopting new medicines. Boca Raton, FL: CRC Press; 2015. p. 92.

[84] Soini S. Public health—ethical issues. Copenhagen: Nordic Council of Ministers; 2011. p. 35–59.

[85] CDC. Create an Epi curve. Quick Learn Lesson, <http://www.cdc.gov/training/quicklearns/createepi/>; [accessed 30.10.2016].

[86] Katz DL, Elmore JG, Wild DMG, Lucan SC. Jekel's epidemiology, biostatistics, preventive medicine, and public health. 4th ed. Philadelphia, PA: Saunders; 2014. p. 35.

[87–89] Abubakar I, Stagg HR, Cohen T, Rodrigues LC. Infectious disease epidemiology. New York, NY: Oxford University Press; 2016. p. 37–9.

[90] Procedures to investigate foodborne illness. New York, United States: International Association for Food Protection; 2011. p. 52–7.

[91a] Bope ET, Kellerman RD. Conn's current therapy: 2014. Philadelphia, PA: Saunders; 2014. p. 181–4.

[91b] CDC. Quarantine and isolation, <https://www.cdc.gov/quarantine/>; [accessed 30.10.2016].

[92] Price PJ. Ebola and the law in the United States: a short guide to public health authority and practical limits. Emory legal studies research Paper No. 14–299, <https://ssrn.com/abstract = 2538187> or <http://dx.doi.org/10.2139/ssrn.2538187>; [accessed 21.04.2015].

[93] Cromley EK, McLafferty S. GIS and public health. 2nd ed. New York: Guilford Press; 2012. p. 1–15.

[94] Goeree R, Hopkins R. Health technology assessment: using biostatistics to break the barriers of adopting new medicines. Boca Raton, FL: CRC Press; 2015. p. 92.

[95–96] Howlett B, Rogo E, Shelton TG. Evidence based practice for health professionals. Burlington, MA: Jones & Bartlett Publishers; 2013. p. 204.

[97–98] Katz DL. Clinical epidemiology and evidence-based medicine: fundamental principles of clinical reasoning and research.. Thousand Oaks, CA: SAGE Publications; 2001. p. 14.

[99] Kestenbaum B, Shoben A. In: Adeney K, Weiss N, editors. Epidemiology and biostatistics: an introduction to clinical research. New York, NY: Springer Science + Business Media; 2009. p. 125–6.

[100] Lang TA, Secic M. How to report statistics in medicine. 2nd ed. New York: American College of Physicians; 2006. p. 137–8.

[101] CDC. National vital statistics system. National Center for Health Statistics (NCHS), <http://www.cdc.gov/nchs/nvss/>; [accessed 11.10.2016].

[102] CDC. Overview of influenza surveillance in the United States, <http://www.cdc.gov/flu/weekly/overview.htm#Outpatient>; [accessed 11.10.2016].

[103] CDC. National Health and Nutrition Examination Survey. National Center for Health Statistics, <http://www.cdc.gov/nchs/nhanes.htm>; [accessed 11.10.2016].

[104] Dellicour S, ter Kuile FO, Stergachis A. Pregnancy exposure registries for assessing antimalarial drug safety in pregnancy in malaria-endemic countries. PLOS Med 2008;5(9):e187. Available from: http://dx.doi.org/10.1371/journal.pmed.0050187.

[105] International Classification of Diseases (ICD) information sheet. World Health Organization, <http://www.who.int/classifications/icd/factsheet/en/>; [accessed 11.10.2016].

[106–108a] Mathews TJ, MacDorman MF. Infant mortality statistics from the 2010 period linked birth/infant death data set no 8 National vital statistics reports, Vol. 62. Hyattsville, MD: National Center for Health Statistics; 2013.

[108b] WHO. Infant mortality. World Health Organization, <http://www.who.int/gho/child_health/mortality/neonatal_infant_text/en/>.

[109] United States Census Bureau. About the Bureau, <http://www.census.gov/about.html>; [accessed 11.10.2016].

[110] CDC. Youth Risk Behavior Surveillance System (YRBSS), <http://www.cdc.gov/HealthyYouth/yrbs/index.htm>; [accessed 11.10.2016].

[111] CDC. BRFSS, <http://www.cdc.gov/brfss/>; [accessed 11.10.2016].

[112–113] Brownson RC, Baker E, Leet T, Gillespie K, True W. Evidence-based public health. 2nd ed. New York: Oxford University Press; 2011. p. 148.

[114] Katz DL, Elmore JG, Wild DMG, Lucan SC. Jekel's epidemiology, biostatistics, preventive medicine, and public health. 4th ed. Philadelphia, PA: Saunders; 2014. p. 16–31.

[115] Forthofer RN, Lee ES, Hernandez M. Biostatistics: a guide to design, analysis, and discovery. 2nd ed. San Diego, CA, United States: Elsevier Academic Press; 2007. p. 78.

[116] Johnson NB, Hayes LD, Brown K, Hoo EC, Ethier KA, Centers for Disease Control and Prevention (CDC). CDC National Health Report: leading causes of morbidity and mortality and associated behavioral risk and protective factors—United States, 2005–13. MMWR 2014;63(04):3–27.

[117] CDC. Ten leading causes of death and injury. Injury Prevention and Control: Data and Statistics, <http://www.cdc.gov/injury/wisqars/leadingcauses.html>; [accessed 11.10.2016].

[118–120] Dicker R, et al. Principles of epidemiology in Public Health Practice. 3rd ed. Atlanta, GA: CDC; 2012 [chapter 3].

[121] Howden LM, Meyer JA. United States Census Bureau. Age and Sex Composition: 2010, <http://www.census.gov/prod/cen2010/briefs/c2010br-03.pdf>; May 2011 [accessed 11.10.2016].

[122a] Friis RH, Sellers T. Epidemiology for public health practice. 5th ed. Burlington, MA: Jones & Bartlett Publishers; 2014. p. 411.

[122b] Sullivan LM. Essentials of biostatistics in public health. 2nd ed. Sudbury, MA: Jones & Bartlett; 2012. p. 27–8.

[123] Weinberg GA, Szilagyi PG. Vaccine epidemiology: efficacy, effectiveness, and the translational research roadmap. J Infect Dis 2010;201(11):1607–10. Available from: http://dx.doi.org/10.1086/652404.

[124] Katz DL, Elmore JG, Wild DMG, Lucan SC. Jekel's epidemiology, biostatistics, preventive medicine, and public health. 4th ed. Philadelphia, PA: Saunders; 2014. p. 94.

[125] Gordis L. Epidemiology. 4th ed. Philadelphia, PA: Elsevier Health Sciences; 2009. p. 203–5.

[126] Katz DL, Elmore JG, Wild DMG, Lucan SC. Jekel's epidemiology, biostatistics, preventive medicine, and public health. 4th ed. Philadelphia, PA: Saunders; 2014. p. 73.

[127] CDC. What is PRAMS? <http://www.cdc.gov/prams/>; [accessed 11.10.2016].

[128] Katz DL, Elmore JG, Wild DMG, Lucan SC. Jekel's epidemiology, biostatistics, preventive medicine, and public health. 4th ed. Philadelphia, PA: Saunders; 2014. p. 73.

[129] Gordis L. Epidemiology. 4th ed. Philadelphia, PA: Elsevier Health Sciences; 2009. p. 206–7.

[130] Katz DL, Elmore JG, Wild DMG, Lucan SC. Jekel's epidemiology, biostatistics, preventive medicine, and public health. 4th ed. Philadelphia, PA: Saunders; 2014. p. 65.

[131a] Center on Society and Health, RWJF. Health care: necessary but not sufficient, <http://www.rwjf.org/content/dam/farm/reports/issue_briefs/2014/rwjf415715>; [accessed 11.10.2016].
[131b] Katz DL, Elmore JG, Wild DMG, Lucan SC. Jekel's epidemiology, biostatistics, preventive medicine, and public health. 4th ed. Philadelphia, PA: Saunders; 2014. p. 65.
[132] Szklo M, Nieto J. Epidemiology: beyond the basics. 3rd ed. Burlington, MA, United States: Jones & Bartlett Publishers; 2014. p. 121–33.
[133] Fletcher R, Fletcher S, Fletcher G. Clinical epidemiology: the essentials. 5th ed. Baltimore, MD: Lippincott Williams & Wilkins; 2014. p. 10–2.
[134] Weinberg GA, Szilagyi PG. Vaccine epidemiology: efficacy, effectiveness, and the Translational research Roadmap. J Infect Dis 2010;201(11):1607–10. Available from: http://dx.doi.org/10.1086/652404.
[135] Brownson RC, Baker E, Leet T, Gillespie K, True W. Evidence-based public health. 2nd ed. New York, NY: Oxford University Press; 2011. p. 281.
[136] Nelson KE, Williams C. Infectious disease epidemiology. 3rd ed. Burlington, MA: Jones & Bartlett Publishers; 2014. p. 27.
[137] Shors T. Understanding viruses. 2nd ed. Burlington, MA: Jones & Bartlett; 2013. p. 327.
[138] Webb P, Bain C. Essential epidemiology: an introduction for students and health professionals. 2nd ed. New York: Cambridge University Press; 2011. p. 278–9.
[139] Brachman PS, Thacker SB. Centers for Disease Control and Prevention (CDC) evolution of epidemic investigations and field epidemiology during the MMWR era at CDC—1961–2011. MMWR 2011;60(04):22–6.
[140] Richards IS, Bourgeois M. Principles and practice of toxicology in public health. 2nd ed. Burlington, MA: Jones & Bartlett Publishers; 2014. p. 372–80.
[141] Vaccine adverse event reporting system, <https://vaers.hhs.gov/index>; [accessed 11.10.2016].
[142] Goodman RA, Hoffman R, Lopez W, Matthews G, Rothstein M, Foster K, editors. Law in public health practice. 2nd ed. New York, NY: Oxford University Press; 2007.
[143] Work PHS, Keefe RH, Jurkowski ET. The American Public Health Association. Handbook of social work and pubic health. New York, NY: Springer Publishing Company; 2013. p. 28–9.
[144] Goodman RA, Hoffman R, Lopez W, Matthews G, Rothstein M, Foster K, editors. Law in public health practice. 2nd ed. New York, NY: Oxford University Press; 2007.
[145] Population data / NSDUH. Substance Abuse and Mental Health Services Administration, <http://www.samhsa.gov/data/population-data-nsduh>; [accessed 11.10.2016].
[146a] Emergency department data. Substance Abuse and Mental Health Services Administration, <http://www.samhsa.gov/data/emergency-department-data-dawn>; [accessed 11.10.2016].
[146b] Office of National Drug Control Strategy. About ONDCP, <https://www.whitehouse.gov/ondcp/about>; [accessed 11.10.2016].
[147] Murphy SL, Kochanek KD, Xu JQ, Arias E. vol. 229 Mortality in the United States, 2014. NCHS data brief. Hyattsville, MD: National Center for Health Statistics; 2015.

CHAPTER FOUR

Environmental Medicine

4.1 ENVIRONMENTAL MEDICINE QUESTIONS

1. Antibiotics target which member of the epidemic triangle?
   A. Agent
   B. Environment
   C. Host
   D. Time
   E. Vector
2. As a whole, which of the following is the greatest source of emitting criteria pollutants?
   A. Automobiles
   B. Coal power plants
   C. Farms
   D. Hydraulic fracturing (fracking)
   E. Industrial plants
3. Which of the following was not permitted by the Clean Air Act (CAA) and subsequent amendments?
   A. Creating air quality standards
   B. Give the Environmental Protection Agency role to monitor air pollution
   C. Mining of natural gas
   D. Phasing out of compounds harmful to individuals and the environment
   E. Requiring permits to vent pollutants into the atmosphere
4. What is the most reliable way to remove *Cryptosporidium* from drinking water?
   A. Chlorine
   B. Flocculation
   C. Filtration
   D. Antibiotics
   E. Coagulation

Board Review in Preventive Medicine and Public Health.
DOI: http://dx.doi.org/10.1016/B978-0-12-813778-9.00004-9

5. Which of the following sources of water is not protected under the Clean Water Act (CWA)?
   A. Lake water
   B. Ocean water
   C. River water
   D. Stormwater discharge
   E. All of the above are protected
6. How much fluoride does the Centers for Disease Control and Prevention (CDC) recommend to be in drinking water?
   A. 0.3 ppm
   B. 0.7 ppm
   C. 1.4 ppm
   D. 2.0 ppm
   E. 2.5 ppm
7. Which of the following compounds is mostly responsible for causing acid rain?
   A. Carbon monoxide
   B. Hydrogen
   C. Lead
   D. Magnesium
   E. Sulfur-oxides
8. Which of the following is addressed by the Delaney Clause?
   A. Emergency preparedness
   B. Food additives
   C. Highway safety
   D. Medical errors
   E. Pharmaceutical distribution
9. As the director of epidemiology of a state health department, you become aware of an outbreak of pneumonia in a long-term care facility. You send your most experienced epidemiologist to the facility, and she discovers that the air conditioning system is not maintained and is in violation of modern building code.

   Which agent grows in air conditioning that grows in air conditioning units is likely causing this outbreak?
   A. *Coccidiodes*
   B. *Histoplasma*
   C. Influenza virus
   D. *Legionella*
   E. Respiratory syncytial virus

10. Which organization oversees the transportation of hazardous wastes?
  A. Centers for Disease Control and Prevention (CDC)
  B. Department of Transportation (DOT)
  C. Environmental Protection Agency (EPA)
  D. Food and Drug Administration (FDA)
  E. Individual states and municipalities

11. Which of the following organizations is tasked with the oversight of nuclear waste from power plants in the United States?
  A. Centers for Disease Control and Prevention (CDC)
  B. Environmental Protection Agency (EPA)
  C. Food and Drug Administration (FDA)
  D. Individual states and municipalities
  E. Nuclear Regulatory Commission (NRC)

12. Which federal organization is responsible for federal oversight of emergency preparedness for nuclear power plants?
  A. Centers for Disease Control and Prevention (CDC)
  B. Federal Bureau of Investigation (FBI)
  C. Federal Emergency Management Agency (FEMA)
  D. Nuclear Regulatory Commission (NRC)
  E. Both C and D

13. Noise may increase risk of which of the following?
  A. Angina
  B. Insomnia
  C. Hypertension
  D. Stroke
  E. All of the above

14. Which of the following approaches is not recognized as one of the four E's of injury prevention?
  A. Economics
  B. Education
  C. Enforcement
  D. Engineering
  E. Establish

15. Approximately what percentage of traffic fatalities in the United States involve a driver with a blood alcohol content (BAC) of 0.08 or higher?
  A. 5%
  B. 10%
  C. 20%

D. 30%
E. 40%

16. An intoxicated driver loses control of his vehicle after and crashes into the center median. Using the Haddon matrix below, which box would correctly categorize the cause of this unintentional incident?

	Host	Agent/vehicle	Environment	Social/behavioral
Pre-event	A		B	
During event				C
Post-event	D		E	

A. A
B. B
C. C
D. D
E. E

17. A driver hits a large pothole in the road, causing her car to lose control and strike another car. Using the Haddon matrix below, which box would correctly categorize the cause of this unintentional incident?

	Host	Agent/vehicle	Environment	Social/behavioral
Pre-event		A		B
During event		C	D	
Post-event			E	

A. A
B. B
C. C
D. D
E. E

18. Which of the following types of seafood is most likely to contain the highest levels of mercury?
A. Crab
B. Oyster
C. Salmon
D. Shark
E. Shrimp

19. One of your patients presents to your clinic after purchasing home. Because the home was constructed in the 1920s, she is concerned about the adverse effects that her newborn child may suffer. In what blood lead level (BLL) range is cognitive impairment first seen in children as a result of lead exposure?

A. 0–10 μg/dL
B. 11–20 μg/dL
C. 21–30 μg/dL
D. 31–40 μg/dL
E. >45 μg/dL

20. At what marker is chelation therapy recommended for elevated BLL in asymptomatic children?

A. 10–14 μg/dL
B. 15–19 μg/dL
C. 20–44 μg/dL
D. 45–69 μg/dL
E. ≥70 μg/dL

21. Which type of mercury, typically, bioaccumulates in marine life?

A. Elemental mercury
B. Inorganic mercury
C. Methyl mercury
D. Compound mercury
E. Liquid mercury

22. Intake of arsenic from the environment often occurs from which source?

A. Inhaling secondhand smoke
B. Inhaling fumes from car exhaust
C. Drinking well water
D. Eating vegetables farmed with pesticides
E. Eating red meat

23. Which of the following is an example of point source pollution?

A. Agricultural fertilizer runoff
B. Bacteria from farm animal waste runoff
C. Coal plant smokestack
D. Motor vehicle emissions
E. Oil runoff from roadways

24. Within which layer of the atmosphere do humans reside?

A. Exosphere
B. Mesosphere
C. Stratosphere
D. Thermosphere (Ionosphere)
E. Troposphere

25. Which of the following statements about firearm violence in the United States is false?
    A. Alcohol abuse is a predictor of firearm violence
    B. Black men have the highest rate of mortality from firearm homicide of any group
    C. Homicide from firearm use is more common than suicide from firearms
    D. Suicide rates from firearms are higher in rural settings than urban settings
    E. There is a strong statistical association between firearm ownership and firearm violence
26. What is the most practical way to eliminate risk of contracting brucellosis from cow milk?
    A. Add antibiotics to livestock food
    B. Clean the cow's utters and machinery prior to obtaining the milk
    C. Educate farmers on dangers of infected animals
    D. Pasteurization
    E. Vaccinate milk consumers
27. Which gas has the highest composition in the atmosphere?
    A. Carbon dioxide
    B. Helium
    C. Hydrogen
    D. Nitrogen
    E. Oxygen
28. Polychlorinated biphenyls (PCBs) have been banned in the United States since 1977. Where are they found in the United States today?
    A. Air
    B. Food
    C. Water
    D. Wildlife
    E. All of the above
29. Which of the following is a criteria air pollutant, as set forth in the Clean Air Act (CAA)?
    A. Ozone
    B. Asbestos
    C. Carbon dioxide
    D. Methane
    E. Toluene

30. In the United States, which federal organization is responsible for ensuring that imported foods meet federal standards and are safe for human consumption?
A. Centers for Disease Control and Prevention (CDC)
B. Food and Drug Administration (FDA)
C. US Customs and Border Protection (CBP)
D. US Department of Agriculture (USDA)
E. Imported foods are not regulated by any federal organization

31. Which federal organization is the primary organization that assures the nation's supply of meat and poultry is safe for human consumption?
A. Centers for Disease Control and Prevention (CDC)
B. Department of Food Safety (DFS)
C. Department of Health and Human Services (DHHS)
D. Food and Drug Administration (FDA)
E. US Department of Agriculture (USDA)

32. What is the FDA's role in performing food inspections at restaurants?
A. FDA and local health departments have equally shared responsibility in the inspection process
B. FDA does not have any input in the inspection process
C. FDA is the agency responsible for conducting inspections at restaurants
D. FDA provides input to the CDC to conduct the inspection process
E. FDA publishes guidelines for local governments to consider in the inspection process

33. Prior to a blizzard, local public health physicians urged government officials to open shelters for the city's homeless population. Due to budget cuts, the government officials were only able to open a single shelter. The day after the blizzard struck, the hospital on the opposite side of town from the shelter reported several cases of hypothermia.

At what body temperature can hypothermia first be diagnosed?
A. $\leq$82°F (28°C)
B. $\leq$85°F (29°C)
C. $\leq$90°F (32°C)
D. $\leq$95°F (35°C)
E. $\leq$97°F (36°C)

34. What is the fluid waste product that forms within a landfill?
A. Compost
B. Effluent

C. Flocculent
D. Leachate
E. Methane

35. Which legislation calls for supervision of hazardous waste during its lifespan (generation, transportation, treatment, storage, and disposal)?
    A. Comprehensive Environmental Response, Compensation, and Liability Act (CERCLA)
    B. National Environmental Policy Act (NEPA)
    C. Resource Conservation and Recovery Act (RCRA)
    D. Superfund Amendments and Reauthorization Act (SARA)
    E. Toxic Substances Control Act (ToSCA)
36. Which agency is primarily tasked with conducting the processes set forth in the CERCLA, RCRA, and SARA?
    A. Agency for Toxic Substances and Disease Registry (ATSDR)
    B. American Conference of Governmental Industrial Hygienists (ACGIH)
    C. Centers for Disease Control and Prevention (CDC)
    D. Food and Drug Administration (FDA)
    E. World Health Organization (WHO)
37. A scientist discovers a new cleaning chemical that he would like to mass produce. Which law requires this chemical to be evaluated for safety of people and the environment?
    A. Comprehensive Environmental Response, Compensation, and Liability Act (CERCLA)
    B. Federal Insecticide, Fungicide, and Rodenticide Act (FIFRA)
    C. Hill–Burton Act
    D. Superfund Amendments and Reauthorization Act (SARA)
    E. Toxic Substances Control Act (ToSCA)
38. Which act lays the foundation for the EPA to monitor and control pesticide use?
    A. Comprehensive Environmental Response, Compensation, and Liability Act
    B. Delaney Clause
    C. Federal Insecticide, Fungicide, and Rodenticide Act
    D. Food, Drug, and Cosmetic Act
    E. Toxic Substances Control Act
39. A real estate developer is considering building homes in a narrow strip of land alongside a major highway. As a local public health officer, you recommend that this developer not proceed with his plan to

build homes in that location. Which of the following is not true of residing near busy roads?

A. Automobile emissions include criteria air pollutants

B. Economically disadvantaged, typically, experience more traffic-related air pollution than the wealthy

C. Minority populations, typically, experience more traffic-related air pollution than nonminorities

D. Rates of asthma are increased in people residing near major roads

E. Rates of cardiovascular disease are decreased in people residing near major roads

40. The US EPA recommends intervention to reduce radon exposure within the home if the radon exposure reaches which level?

A. 2 pCi/L

B. 4 pCi/L

C. 6 pCi/L

D. 8 pCi/L

E. 10 pCi/L

41. What most likely happens if the effluent of sewage treatment has a biological oxygen demand (BOD) that is greater than the dissolved oxygen of the waterway it is released in?

A. Nothing

B. The sewage system will work optimally

C. Fish kills and algal blooms may occur

D. The excess oxygen will promote aquatic life

E. The result depends on the local environment

42. Which of the following countries experiences the highest rate of death due to gun usage?

A. Austria

B. Finland

C. France

D. Switzerland

E. United States

43. Which of the following is an appropriate chlorine level for a recreational swimming pool?

A. 2 ppm

B. 20 ppm

C. 200 ppm

D. 2000 ppm

E. 20,000 ppm

44. Which of the following is the leading cause of diarrheal illness from swimming pools in the United States?

A. *Cryptosporidium*
B. *Escherichia coli*
C. *Giardia*
D. Hepatitis A
E. Norovirus

45. Which of the following environmental factors has been shown to contribute to obesity?

A. High density of fast food restaurants
B. Lack of accessible sidewalks
C. Large portions
D. Unsafe neighborhoods
E. All of the above

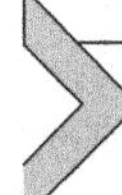

4.2 ENVIRONMENTAL MEDICINE ANSWERS

1. **A. Agent**

   The epidemiologic triangle is a concept that can be used to track the interaction of the variables that result in illness. Traditionally, the three components of the epidemiologic triangle are the host, agent, and environment. These three variables may interact to produce the optimal opportunity for infection. Many epidemiologists also include a disease vector as a fourth variable.

   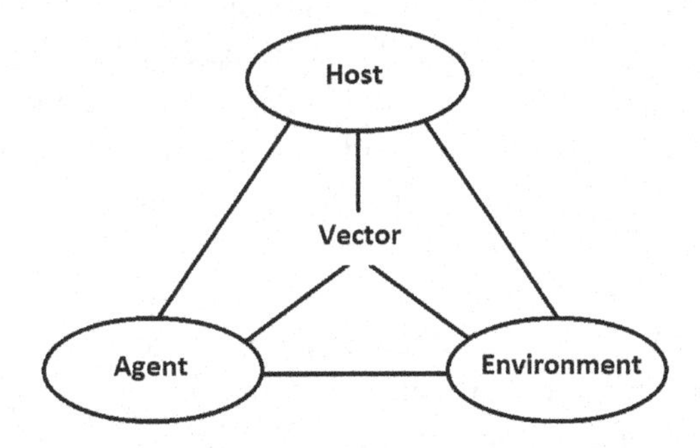

2. **A. Automobiles**

   The six criteria air pollutants, as appointed by National Ambient Air Quality Standards (NAAQS) of the Clean Air Act (CAA) are ozone, carbon monoxide, lead, sulfur oxides, nitrogen oxides, and particulate matter. Automobiles are the greatest overall contributors of the six pollutants. They also produce the largest portion of carbon monoxide in the atmosphere and are one of the top polluters of nitrogen-oxides. Although not a criteria pollutant, automobiles are one of the largest producers of volatile hydrocarbons.

   Automobile pollution is a public health problem that will continue to be addressed. Through tougher government and private industry standards, current-day automobiles emit up to 80% less pollutants than automobiles in 1960. This accomplishment is partly negated by an increase in the number of cars on the road.

3. **C. Mining of natural gas**

   The CAA and subsequent amendments dramatically changed the way air pollutants are monitored and controlled in the United States. It gave federal authority to the US EPA to develop air quality criteria and emissions standards to protect public health. It requires those that produce air pollution to obtain permits in order to vent pollutants into the atmosphere. In some cases, it created a marketplace where polluters could barter their pollution allowances to other polluters. Through the CAA, the EPA gained the ability to administer subpoenas to gather compliance data and handout penalties for air pollution violators. The EPA requires states to develop their own implementation plans that are at least as strict as the EPA guidelines.

The CAA established National Air Quality Standards that targeted six major air pollutants for control. The six pollutants are ozone, carbon monoxide, sulfur-oxides, nitrogen-oxides, lead, and particulate matter. The CAA and subsequent amendments identified hazardous air pollutants (HAP) as substances that are hazardous to the environment or human health, but are not addressed in other parts of the CAA. The EPA identifies HAP sources and requires producers to utilize maximum achievable control technology to limit HAPs released into the air. Some examples of hazardous air pollutants include arsenic, asbestos, benzene, and mercury.

The CAA specifically addresses several major sources of air pollution. It takes measures to reduce acid rain by addressing emissions of sulfur-oxides and nitrogen-oxides at utility plants. Furthermore, it creates emission standards for automobiles and requires phasing out of chemicals that are known to cause harm to individuals and the environment, specifically pertaining to depletion of the Ozone layer.

4. **C. Filtration**

Cryptosporidium live a complex life cycle through a variety of animal hosts, including humans. The only species of *Cryptosporidium* shown to cause infection in humans is *C. parvum*. After ingestion, they are excreted from the GI Tract of an infected host through 4–6 micrometer oocysts.

Cryptosporidium are only reliably removed from drinking water through filtration. Dissolved air floatation (DAF) and coagulation help improve conditions to remove *C. parvum*, but these two methods are not reliable on their own. Chlorination is not effective for removing *Cryptosporidium*. Antibiotics are not used when cleaning drinking water in the United States. Boiling and freezing also reduce infectiousness.

5. **E. All of the above are protected**

The CWA aims to eliminate discharge of toxic pollutants into surface waters. In doing so, the CWA aims to clean up water for protection of sea life and water recreation. The CWA applies only to surface waters and other waters that are hydrostatically connected to surface water. This may include groundwater and stormwater.

The CWA describes three categories of pollutants: Conventional, toxic, and nonconventional. It also established a permitting system to regulate discharge of pollutants into water.

Groundwater enjoys direct and indirect protection from a plethora of sources including the RCRA, the CERCLA, the Safe Drinking Water Act (SDWA), the FIFRA, the ToSCA, and the Pollution Protection Act.

6. **B. 0.7 ppm**

The CDC currently states that fluoride in public drinking water (systems that serve >25 people) should contain 0.7–1.2 ppm of fluoride. However, due to increased exposure to fluoride from other sources (mainly toothpaste and mouthwash), the DHHS has proposed that the CDC change this level to 0.7.

On the other extreme, the EPA has set a secondary standard to keep fluoride below 2.0 ppm due to risk of fluorosis. Secondary standards are not enforceable. The EPA has set the enforceable maximum contaminant level (MCL) at 4.0 ppm. The MCL is enforceable.

Fluoride is temporarily retained in tooth and gum tissues. Amongst other mechanisms, it alters the pH to allow calcium and phosphate ions to reduce tooth demineralization in the enamel and dentin and enhance remineralization.

7. **E. Sulfur-oxides**

Sulfur-oxides react with moisture, oxidizing agents, and other environmental compounds to form sulfates, sulfites, and sulfuric acid. These compounds precipitate into acid deposition: Acid rain, acid sleet, and acid snow. Rain, sleet, and snow are naturally acidic, with a pH near 5.0, due to the presence of carbon dioxide in the atmosphere that yields carbonic acid. Nitrogen-oxides are the second leading cause of acid deposition. Sulfur dioxide and nitrogen dioxide are two of the EPA's six criteria pollutants of NAAQS.

Hydrogen is not responsible for acid deposition.

Magnesium presence in soil may act to buffer acid deposition, but magnesium is not responsible for causing acid deposition.

The remaining four NAAQS pollutants are carbon monoxide, lead, ozone, and particular matter.

8. B. Food additives

The Delaney Clause was originally written within the Food, Drug, and Cosmetic Act. The intent of the act was to prohibit use of carcinogenic food additives. The act stated that "No additive shall be deemed to be safe if it is found to induce cancer when ingested by man or animal, or if it is found, after tests which are appropriate for the evaluation of the safety of food additives, to induce cancer in man or animals."

The bill was the first of its type to recognize the carcinogenic effects of food additives. However, the way in which the law was written created problems. The Delaney Clause originally took an all or nothing approach, where additives would be banned from food regardless of the potency or amount needed to ingest to cause carcinogenesis. The law was amended numerous times before finally being repealed in 1996.

9. D. *Legionella*

Legionella is a bacteria that is found naturally in aquatic environments. When aerosolized and inhaled it may cause Legionnaires disease, a form of pneumonia typically indistinguishable from community acquired pneumonia and characterized by fevers, myalgias, headaches, shortness of breath, and cough. Immunocompromised populations are more susceptible to infection. *Legionella* may grow in wet manmade environments such as air cooling towers, hot water systems, shower heads, spas, and respiratory ventilators. If any one of the nearly 60 *Legionella* species (*L. pneumophilia* is most common) are aerosolized and vented into the air breathed by a group of people, they may cause an outbreak after a 2–10-day incubation period.

Legionellosis (Legionnaires disease without pneumonia) is a notifiable disease, which allows quick identification and correction of outbreaks. After the discovery of Legionnaire's disease federal AC standards were changed to reduce the risk of future outbreaks.

Coccidiodes and *Histoplasma* are responsible for causing coccydiomycosis and histoplasmosis respectively. Both pathogens are fungi that primarily live in soil.

Influenza and RSV are typically spread through respiratory droplets and not air conditioning units.

10. B. Department of Transportation

The Hazardous Materials Transportation Act (HMTA) establishes specific requirements for the packing, handling, and delivery of hazardous materials, including oil, nuclear waste, and chemical substances. Prior to the passage of HMTA, transportation of hazardous wastes was regulated by states and local municipalities. The HMTA created national standardization of safety precautions and guidelines for the transport of hazardous materials. This legislation is mainly overseen by the Department of Transportation, although it is also influenced by the Department of Energy, the NRC, the Occupational Safety and Health Administration (OSHA).

11. E. Nuclear Regulatory Commission (NRC)

The NRC is responsible for the oversight of licensing, safety, and security of radioactive material in the United States. Although the ToSCA allows the EPA to oversee most chemical substances, radioactive material is a specific exception. Other exceptions include biocides (under the FIFRA), tobacco products, foods, and drugs.

Nuclear waste can be divided into three distinct categories: Extraction waste, low-level waste, and high-level waste. Extraction waste is a byproduct of uranium mining. Low-level waste includes items that have become radioactive through proximity to radiation. High-level waste includes radioactive byproducts of nuclear weapons and power plants. The NRC plays a large role in managing high-level wastes. Every active nuclear power plant has inspectors that monitor operations for the NRC. Once the nuclear fuel is spent, the NRC helps find placement for the radioactive remains. These remains may have a half-life of one million years, potentially creating long-term problems in a world with a rapidly growing population.

12. E. Both C and D

Federal oversight of emergency preparedness for licensed nuclear power plants is shared by the NRC and the FEMA.

On the individual level, each plants must maintain their own plans for preparation, evacuation, and sheltering to protect the public in case of an accident.

13. E. All of the above

Noise is an unwanted sound that influences sleep, relaxation, and concentration. Studies have shown direct correlation between noise and cardiovascular disease (angina, hypertension, stroke, and ischemic heart disease), even in people that do not complain of noise annoyance. The impact of noise depends on the duration, frequency, intensity, time of day, and source of noise exposure.

14. E. Establish

Unintentional injuries are the leading cause of death among children between the ages of 1 and 19. They are considered to be predictable and, therefore, preventable. A guideline to preventing unintentional injuries involves utilizing the four E's: Economics, education, enforcement, and engineering. Many scholars interchangeably substitute engineering with environment.

By placing economic barriers, undesirable actions by be averted. By increasing the price of alcohol, less alcohol might be consumed.

Educational barriers include informing people about the risks of a behavior. By educating teenagers on risks of drinking alcohol, they may decide decrease alcohol consumption.

Enforcing laws ensures that the laws will be followed. By enforcing drinking and driving laws, there are less drunk drivers on the road.

Engineering the environment can lead to a reduction of unintentional injuries. By selling liquor with lower ethanol content, people may not consume as much ethanol. Additionally, by retrofitting a person's car to not start if they have alcohol in their breath, it will discourage that person from drinking.

15. D. 30%

Each state has introduced legislation making it illegal to drive with a BAC of 0.08 or higher, with many imposing more stringent laws. Although driving under the influence of alcohol has been on the decline, it remains to be a major public health issue. Fatal auto accidents involving drunk drivers are much more likely to occur both on weekends and during the night, especially in the 21–24 year-old age group.

16. A. A

Accidents are more predictable than previously thought. Therefore, the term unintentional injury is often preferred to the word accident. An injury occurs when the body sustains a transfer of energy that is sufficient to damage tissue. Effective injury prevention requires a multidisciplinary approach. The Haddon matrix is a tool used to identify the anatomy of an incident that causes injury, allowing for intervention. The matrix categorizes the time phase of the injury (pre-event, during event, and post-event) into rows and the factors of the injury (host, agent/vehicle, environment, and social behaviors) into columns.

The Haddon matrix was originally created for highway incidents, such as the one in this question. Because the intoxication was the primary cause of the crash, the main contributor would be the host.

	Host	Agent/ Vehicle	Environment	Social/ Behavioral
Pre-event	• Vision impaired • Driving skill • Drug impairment	• Car maintenance • Abilities of car • Broken headlight	• Road design • Speed limit • Poorly timed traffic light	• Speed limit • Compliance with driving laws • Enforcement of driving laws
During event	• Seatbelt • Car seat	• Airbags • Vehicle size • Vehicle safety rating	• Quality of road • Guard rails • Weather conditions	• EMS quality
Post-event	• Ability to call for help	• Gas tank designed to minimize fires	• EMS access to site • Distance to hospital	• Funding for EMS • State Policy • Rehabilitation program

Furthermore, because the driver became intoxicated before the crash, the incident is categorized as pre-event.

The completed Haddon matrix below further illustrates how to correctly use this tool.

17. D. D

The concept and definition of a Haddon matrix was explained in Answer #16 (directly above). The completed Haddon matrix in Answer #16 can also be used to solve this problem.

Because the pothole was the primary cause of the crash, the main contributor would be the environment. Poor road quality that directly leads to a crash would be categorized as during the event.

18. D. Shark

Methyl mercury is a type of organic mercury present in large concentrations in water bodies. The majority is present as a result of industrial dumping of inorganic mercury, which is converted to methyl mercury by water organisms. Small fish eat these organisms and then larger fish eat these fish. The mercury bioaccumulates as it reaches peak predators. Fish with the highest concentrations of mercury include tilefish, shark, tuna, mackerel, and swordfish. Salmon consumed in the United States is often farm raised and not subject to the same mercury levels as other fish on the list.

Mercury toxicity may lead to stomatitis, intention tremors, irritability, and personality changes. In addition to the central nervous system, mercury can damage the glomerulus and tubules of the kidneys, leading to proteinuria and reduced creatinine clearance.

19. A. 0–10 μg/dL

Although many professional guidelines recommend to limit lead exposure to a BLL of less than 10 μg/dL, lead has been shown to cause cognitive impairment at a BLL of less than 5 μg/dL. Furthermore, lead exposure to a woman during pregnancy may interfere with the cognitive development of her child. Organic lead is lipid soluble and rapidly crosses the blood–brain barrier. It may also contribute to insomnia, mania, and encephalopathy in severe cases. Elsewhere in the body, lead may cause seizures, abdominal pain, hyporeflexia, lethargy, peripheral neuropathy, renal dysfunction, and anemia amongst many other ailments.

Older homes carry a larger risk for lead exposure. This is because lead may be found in old paints and pipes. Children may ingest paint chips, inhale paint dust, and consume water irrigated through lead pipes. After getting into the body, lead is absorbed in the GI tract and in the lungs. Furthermore, it accumulates in bone and may be re-released back into the body gradually.

20. D. 45–69 μg/dL

The Advisory Committee on Childhood Lead Poisoning Prevention (ACCLPP) recommends that chelation therapy be initiated in all children with a BLL $\geq$ 45 μg/dL.

The CDC now recommends categorizing children with BLL $\geq$ 5 μg/dL as those with increased exposure to lead. This was reduced from 10 μg/dL after reviewing National Health and Nutrition Examination Survey (NHANES) data.

21. C. Methyl mercury

Inorganic mercury is methylated to methyl (organic) mercury by microorganisms that are present in the environment. After one animal consumes another animal that has been exposed to methyl mercury, the levels of mercury increases (bioaccumulates) in the tissues. After ingestion, methyl mercury crosses the blood–brain barrier and effects brain function.

Humans are most frequently exposed to elemental mercury through inhalation where it finds its way through the blood–brain barrier. If ingested, it is typically excreted through the GI tract.

Elemental mercury used to be found in batteries, dental fillings, light bulbs, and thermometers.

Inorganic mercury is a gas at room temperature and is, thus, more likely to be inhaled. It can also be absorbed through the skin during prolonged contact.

22. C. Drinking well water

Arsenic may be found in the environment in both natural occurring (organic) and manmade (inorganic) forms. Arsenic blocks the Krebs cycle, an important cellular process in energy production. Toxicity from organic arsenic is typically less severe than the inorganic form. It can be absorbed from food, air, or water. Naturally occurring forms are found mainly in drinking water (notoriously from wells) and seafood. Inorganic arsenic is found in the metal smelting, pyrotechnics, pigmentation, pharmaceuticals, and electronics manufacturing industries. Symptoms of chronic arsenic toxicity include skin hyperpigmentation, skin keratosis, anemia, ischemic heart disease, diabetes mellitus, neuropathy, hypertension, and skin/liver/lung cancer. Meanwhile, acute arsenic toxicity typically occurs from inorganic arsenic and may cause the above symptoms in addition to dysphagia, abdominal pain, and diarrhea. Arsenic is typically monitored through urine levels, which do not differentiate between organic or inorganic origin.

23. C. Coal plant smokestack

Point source pollution occurs when pollutants are released into the environment at a distinct discharge point. Examples include smokestacks and discharge from wastewater treatment plants. Meanwhile, nonpoint source pollution occurs when pollutants are released into the environment from numerous widespread locations. Examples include agricultural chemical, biological, and fertilizer water runoff; motor vehicle emissions; and individual littering. Logistically, point source pollution is easier to control than nonpoint source pollution because a single pollutant is easier to address than multiple ones.

24. E. Troposphere

The troposphere is the atmospheric region that is closest to the surface of the Earth. It is roughly 12 km thick at the equator and 8 km thick at the poles. Within this region, the air is composed of 78% nitrogen, 21% oxygen, and the remaining 1% is composed by numerous other gasses. Life exists solely within the troposphere. Nearly all weather patterns occur in the troposphere. Conventional aircrafts operate within the troposphere.

Extending outwards from the troposphere, the next regions are called the stratosphere, mesosphere, thermosphere, and exosphere. At the outer border of each layer is a "-pause", the tropopause, stratopause, mesopause, and thermopause. The Ozone layer is in the Stratosphere.

25. C. Homicide from firearm use is more common than suicide from firearms

Firearm violence is a multifactorial public health challenge. The United States experiences more firearm-related mortality than other industrialized nations.

All of the answers to this question are true, with the exception that firearm suicide deaths far outnumber firearm homicide deaths. Homicide deaths typically occur in younger people, while there is a higher rate of suicides in the elderly. The risk of firearm suicide greatly increases as men age (especially White men), but the same is not true for women. To the contrary, the largest number of firearm homicide deaths occur in young Black men. In fact, firearm homicide is the leading cause of death for Black men aged 15–34. Young Black women are also more likely to die of firearm homicide than other groups of young women. Firearm suicide rates are higher in rural settings, while firearm homicide rates are higher in urban settings.

Although more research is needed, there are clear risk factors associated with firearm violence. Ownership of a gun is positively correlated with firearm harm to oneself and others. Additionally, a history of violence is predictive of future violence. Mental disorders are not as strong of a predictor of firearm violence as is widely believed, but certain subsets of mental disorders, such as substance abuse are recognized risk factors.

Interventions to reduce firearm injuries include background checks and restriction of sales. Primary preventions to reduce firearm violence include firearm injury surveillance, identification of gun violence risk factors, developing interventions to curb gun violence, and putting these interventions into place.

26. D. Pasteurization

Milk is nutrition rich and serves as a great medium for microorganisms to proliferate. Pasteurization is the most practical way to remove pathogens from milk. It is important to pasteurize milk, as it is produced by mammals and, hence, carries bacteria that are more likely to harm other mammals (including humans).

The pasteurization process can be described as applying heat to a product in order to kill or injure microorganisms. Outside of this, pasteurization extends the shelf life of a product and typically does not harm taste or quality. In the case of milk, pasteurization denatures enzymes that eventually causes a foul taste. Different types of pasteurization protocol are used to achieve different goals. Only ultra-high temperature (UHT) pasteurization can yield sterilization.

Pasteurization may be used on a wide variety of agricultural products outside of milk. Some of the products include juice, eggs, beer, honey, and canned foods.

After being pasteurized, food products should be continuously monitored for bacterial presence. Each food product is capable of carrying a wide array of microorganisms. Milk commonly carries the following: *Achromobacter, Alcaligenes, Bacillus, Brucella, Coxiella, Flavobacterium, Lactobacillus, Mycobacterium, Pseudomonas*, and yeast.

Outside of pasteurization, further approaches should be taken. It is preferable that only the highest quality of milk is used. Sick cows should be taken off the production line until recovered. Many cows are fed antibiotics to assure increased health and production. Antibiotic byproducts are eventually consumed by humans.

Prior to standardized pasteurization, it was common for people to boil milk at home. The development of pasteurization techniques was welcomed, as the boiling process alters the taste of the milk.

27. D. Nitrogen

Although altitude and temperature changes may slightly affect the concentrations of atmospheric gasses, nitrogen is by far the most abundant gas in the atmosphere. The combination of nitrogen and oxygen accounts for nearly 99% of the atmosphere. The remaining 1% is composed of other gasses including argon, carbon dioxide, neon, helium, methane, carbon monoxide, and nitrogenous combinations.

In descending order, the most composition of elements in the atmosphere are shown in the table below:

Element	Percent in atmosphere
Nitrogen	78%
Oxygen	21%
Argon	0.9%
Carbon dioxide	0.03%

28. E. All of the above

From 1929–77, PCBs were produced in the United States as solvents, sealants, paints, and protective coating for electrical wiring. They were banned in the United States due to widespread bioaccumulation and health concerns. Despite being banned, they are still ubiquitous in the environment. Current sources of PCB exposure include air (attached to dust particles), food from wildlife (>90% of fish collected in the United States contain PCB residue), water, and old electrical equipment. Numerous countries around the globe continue to produce PCBs.

29. A. Ozone

The EPA has set NAAQS for six "criteria" pollutants. These pollutants are carbon monoxide, lead, nitrogen dioxide, ozone, particulate matter, and sulfur dioxide.

30. B. Food and Drug Administration (FDA)

Food safety is one of the nation's top public health advancements. Foodborne illness is often preventable. Three main government entities are responsible for maintaining food quality. These organizations are the US FDA, USDA, and CDC.

The FDA is responsible for maintaining the majority of food consumed within the nation. This includes domestically obtained and imported foods. The FDA has been empowered by the Food Safety Modernization Act (FSMA) to regulate the nation's food through many enforcement authorities. The FSMA grants the FDA with the authorities inscribed in the following table:

The FDA powers granted by the FSMA

Category	Specific authority
Prevention	Food facilities must create a written plan for preventive controls
	Food facilities must abide by safety standards
	FDA has authority to prevent contamination
Inspection and compliance	Mandatory inspections (frequency based on risk)
	FDA has access to facility records
	Testing must be performed by accredited laboratories
Response	FDA may issue mandatory recall when company refuses a voluntary recall
	FDA may restrict shipment of substandard foods
	FDA may suspend registration of a facility

(*continued*)

(Continued)

The FDA powers granted by the FSMA

Category	Specific authority
	Create a system to track domestic and imported foods
Imports	Importers are responsible to make sure their suppliers have prevention control
	Third parties certify that foreign food facilities comply with safety standards
	FDA can refuse food entry into the United States
Partnerships	FDA must develop strategies to utilize food safety capacities of state and local agencies

31. E. US Department of Agriculture (USDA)

The Food Safety and Inspection Service (FSIS) of the USDA has been given the responsibility of assuring that the nation's supply meat, poultry, and egg products are safe for consumption and appropriately packaged and labeled. The FSIS maintains the National Advisory Committee on Meat and Poultry Inspection (NACMPI) and the National Advisory Committee on Microbiological Criteria for Foods (NACMCF).

The FDA is responsible for maintaining the safety of the majority of food consumed within the nation.

Meanwhile, the CDC conducts foodborne illness surveillance systems that provide important information for the USDA and FDA.

32. E. FDA publishes guidelines for local governments to consider in the inspection process

While food production is regulated by the FDA and USDA, regulation of food storage and preparation falls into the jurisdiction of states, counties, and local municipalities. The structure of the food inspection process varies by state.

The FDA periodically publishes the Food Code, a document that provides recommended guidelines for state, city, county, and tribal agencies that regulate food service operations. All 50 states have retail codes pattered from these recommended guidelines. The Food Code contains practical, science-based guidance for best practices to reduce foodborne illness that serves as a model for all levels of government.

33. D. $\leq$95°F (35°C)

Extreme weather events act as physical stressors that fall into the domain of public health. Cold environments can lead to significant morbidity and mortality. Hypothermia results when the core body temperature drops below 95°F (35°C). Mild hypothermia is defined as a temperature between 90°F and 95°F. Moderate hypothermia is a core body temperature between 82.4°F and 90°F, and severe hypothermia is a fall in core body temperature below 82.4°F. Symptoms of hypothermia increase with decreasing core body temperature. Hypothermia causes cardiac arrhythmias, decreased respiratory drive, and neurological changes (sensory changes, ataxia, decreased reflexes, altered mental status, and altered consciousness).

34. D. Leachate

Leachate is the liquid that forms from the physical, chemical, and biological processes within a landfill. It is a thick brown fluid largely hurried along by surface and ground water that penetrates into the landfill. Because the leachate is toxic, safeguards (collection systems and bottom liners) are taken to prevent the contamination of groundwater. These safeguards are ultimately expected to fail due to sediment clogging pipes, physical weight of the landfill altering the structure, and chemical degradation. This is one of the many reasons that landfills are falling out of favor.

35. C. Resource Conservation and Recovery Act (RCRA)

The RCRA gives the EPA power to oversee hazardous wastes from "cradle to grave," including generation, transportation, treatment, storage, and disposal. An easy trick to remember this is by associating the acronym RCRA with the initial three letters from "cradle to grave."

The NEPA was the first of the above legislation to be passed. The purpose of NEPA was to declare a national policy that encourages harmony between man and the environment and promote efforts to prevent or eliminate damage to the environment.

The CERCLA, also known as Superfund, was passed in 1980 to provide a framework to address poorly controlled and abandoned hazardous waste sites. It made companies more responsible for funding cleanup for their wastes and created a tax on specific industries to fund a trust fund to clean up orphan waste sites. The cleanup priority list can be found on the EPA's publically posted National Priority List.

Superfund (CERCLA) was amended by the SARA. Several key components of SARA include: New Superfund enforcement authorities and enforcement tools, increased state involvement in Superfund, increased focus on health implications from hazardous waste sites and increase the size of the Superfund trust fund. Importantly, SARA also made changes to the hazard ranking system to better categorize sites on the National Priority List.

The ToSCA gives the EPA authority to better monitor and restrict use of certain chemical substances. Some examples of substances covered by the ToSCA include PCBs, asbestos, radon, and lead-based paint. Notable exceptions from the ToSCA include food, drugs, cosmetics, and pesticides.

36. A. Agency for Toxic Substances and Disease Registry (ATSDR)

The ATSDR is a division within the DHHS that is concerned with the public health aspects of hazardous wastes. It was created by the CERCLA (aka Superfund) to expand knowledge about hazardous chemicals, assess the characteristics of chemical hazards at Superfund sites, and to help prevent sickness from hazardous chemical exposure at Superfund sites. RCRA further authorized ATSDR to evaluate hazardous waste sites and assist the EPA in determining with waste products should be regulated. Furthermore, SARA expanded ATSDR's responsibilities to maintain toxicological databases and disseminate information on hazardous waste products.

The ATSDR is not a regulatory agency and does not have the ability to enforce laws. However, the ATSDR may make recommendations to the EPA, which is a regulatory agency.

37. E. Toxic Substances Control Act (ToSCA)

The ToSCA gives the EPA authority to screen and monitor chemicals for personal and environmental safety. The EPA is then required to act when chemicals post hazards. Appropriate actions include changing of warning labels, regulating disposal, and restricting the manufacturing of specific chemicals, such as asbestos. Under the ToSCA, the EPA holds an inventory list composed of a database of all registered chemicals.

Exceptions from the ToSCA include food, drugs, cosmetics, and pesticides. These compounds are monitored by different agencies.

38. C. Federal Insecticide, Fungicide, and Rodenticide Act (FIFRA)

The FIFRA is administered by the EPA and delivers the framework for pesticide control in the United States. FIFRA ensures that pesticides are handled, applied, and stored properly. Proper storage includes placement in a location that is known by workers, emergency services, and medical personnel.

The CERCLA (aka Superfund) is legislation to clean up hazardous waste sites.

The Delaney Clause assures that food is free of dangerous additives.

The Food, Drug, and Cosmetic Act allows the FDA to monitor the safety of food, drugs, and cosmetics.

The ToSCA allows the EPA to restrict the use of chemical substances. Notable exceptions include the pesticides, food, drugs, and cosmetics.

39. E. Rates of cardiovascular disease are decreased in people residing near major roads

Traffic-related air and noise pollution is a major contributor adverse health effects. Automobiles vent an overabundance pollutants into the environment including criteria air pollutants. Each of these pollutants has been shown to have an adverse effect on human health. This air contamination contributes to development and exacerbation of impaired lung function, cardiovascular mortality and all-cause mortality in people residing near busy roads. Minority and economically disadvantaged persons experience a disproportionate amount of risk of exposure to traffic-related air and noise pollution.

40. B. 4 pCi/L

Radon is a radioactive gas that comes from decay of uranium in soil. The EPA recommends that homes with ≥4 pCi/L undergo mitigation to reduce radon exposure.

For homeowners, radon can be measured with an inexpensive home kit. These kits come in two different forms: Short-term and long-term. The short-term kits may yield results as soon as two days, whereas long-term kits typically take closer to 90 days. Because environmental factors such as barometric pressure and snowfall may effect radon levels, long-term tests are more accurate.

The National Radon Action Plan is a collaborative public health measurement aimed to eliminate radon-induced lung cancer in the United States. Participating agencies include the American Lung Association, the EPA, the DHHS, US Department of Housing and Urban Development, and many others. Goals include increased radon testing, radon mitigation, and radon-resistant construction.

41. C. Fish kills and algal blooms may occur

After processing municipal sewage, treated effluent water may be released into waterways. If this effluent fluid is high in organic material, microorganisms will increase in number during the decomposition process. If the waterway has sufficient dissolved oxygen to meet the BOD of the microorganisms, the waterway can support life. If the dissolved oxygen is not enough to meet BOD, fish will die and an algal bloom may occur. A BOD test can be utilized to test the BOD of the effluent product.

42. E. The United States

The United States has roughly nine gun-related deaths per 100,000 of the population. This includes both homicides and suicides. By a large margin, the United States carries the highest burden of mortality rates in industrialized nations due to gun violence.

The other countries listed are the following four with the highest rates of gun-related mortality in industrialized nations. All four have a higher percentage of the population reporting assault than the United States. This may be attributed to the fact that most firearm-related deaths in the United States are due to suicide and not homicide.

43. A. 1ppm

If public aquatic facilities are not appropriately maintained, recreational water-related illness (infections of GI tract, skin, ear, eye, and respiratory tract) may occur. Public swimming facilities need a multifaceted approach to reduce the risk of transmitting disease. This approach includes filtration, chemical disinfectant, control of pH, and appropriate hygiene of those using the facility.

Chlorine is the most common chemical disinfectant used in recreational swimming pools in the United States. Free chlorine is able to bind with (combine) and disinfect various biological agents, rendering them less harmful. Combined chlorine does not have the same disinfection properties. Free chlorine should be present in a minimum concentration of 1 part per million (ppm) in swimming pools and typically no more than 3 ppm. Levels two or three times higher than these recommended guidelines may cause adverse events in humans.

State and local municipalities are the governing bodies in charge of developing and maintaining public pool and aquatic facility standards, including chlorine levels. The CDC maintains the Model Aquatic Health Code (MAHC) handbook to act as guidelines for the states. States may choose to follow the guidelines presented in MAHC.

44. A. *Cryptosporidium*

All of the options listed have been known to cause pool-associated diarrheal illness, with *Cryptosporidium* being responsible for nearly two-thirds of cases. This is largely due to *Cryptosporidium's* chlorine-resistant properties. Chlorine is the most common chemical disinfectant used in recreational swimming pools in the United States. It is easily capable of killing *E. coli* and moderately efficient at killing *Giardia* and Hepatitis A. *Cryptosporidium* oocytes can survive for days at CDC recommended chlorine levels. A chlorine concentration capable of effectively killing *Cryptosporidium* would be high enough to force closure of a pool due to adverse effects on humans.

Both *Cryptosporidium* and *Giardia* have low infectious doses and are excreted in high concentrations in stool. Cells of these pathogens may be spread for weeks after resolution of diarrheal illness. When present in a pool, it only takes a mouthful or two of contaminated water to yield disease transmission.

45. E. All of the above

The external environment can have both positive and negative effects on calories consumed and expended. Numerous studies have shown that the neighborhoods with a higher density of fast food restaurants is associated with a lower socioeconomic status (SES) and increased obesity. Likewise, neighborhoods that are not conducive to exercise are also associated with higher body mass index (BMI). This includes neighborhoods with lack of access to sidewalks and neighborhoods where residents feel unsafe to go outside and exercise. Additional environmental factors linked to obesity include family culture, large serving portions, larger plates/bowls, and longer time in front of television/computer screens.

BIBLIOGRAPHY

[1] Nelson KE, Williams C. Infectious disease epidemiology. 3rd ed. Burlington, MA: Jones & Bartlett; 2014. p. 46.

[2] Moore GS. Living with the earth: concepts in environmental health science. 2nd ed. Boca Raton, FL: CRC Press; 2002. p. 410−2.

[3] Moore GS. Living with the earth: concepts in environmental health science. 2nd ed. Boca Raton, FL: CRC Press; 2002. p. 535−9.

[4] Lingireddy S, editor. Control of microorganisms in drinking water. The United States: ASCE Publications; 2002.

[5] Moore GS. Living with the earth: concepts in environmental health science. 2nd ed. Boca Raton, FL: CRC Press; 2002. p. 539−40.

[6a] Centers for Disease Control and Prevention. Water fluoridation guidelines & recommendations. Community Water Fluoridation, <http://www.cdc.gov/fluoridation/guidelines/index.htm>; [accessed 15.10.2016].

[6b] Lynch RJM, Navada R, Walia R. Low-levels of fluoride in plaque and saliva and their effects on the demineralisation and remineralisation of enamel; role of fluoride toothpastes. Int Dent J 2004;54(S5):304−9. Available from: http://dx.doi.org/10.1111/j.1875-595x.2004.tb00003.x.

[7] Moore GS. Living with the earth: concepts in environmental health science. 2nd ed. Boca Raton, FL: CRC Press; 2002. p. 429–35.

[8] Hammit J, Rogers M, Sand P. The reality of precaution: comparing risk regulation in the United States and Europe. New York, NY: Routledge; 2011. p. 422.

[9] White F, Stallones L, Last JM. Global public health: ecological foundations. New York, NY: Oxford University Press; 2013. p. 47–8.

[10–11] Zimring CA, Rathje WL, editors. Encyclopedia of consumption and waste: the social science of garbage. Thousand Oaks, CA: SAGE Publications; 2012.

[12] Emergency Preparedness at Nuclear Power Plants. Nuclear Regulatory Commission-Backgrounder, <http://www.nrc.gov/reading-rm/doc-collections/fact-sheets/emerg-plan-prep-nuc-power.pdf>; 2014 [accessed 15.10.2016].

[13] Valuing quiet: an economic assessment of U.S. Environmental noise as a cardiovascular health hazard. Am J Prev Med 2015;49(3):A4. doi:10.1016/s0749-3797(15)00367-0.

[14] Centers for Disease Control and Prevention. National Center for Injury Prevention and Control. National Action Plan for Child Injury Prevention. Atlanta (GA): CDC, NCIPC; 2012.

[15] Alcohol-Impaired Driving. US Department of Transportation. National Highway Traffic Safety Administration. Traffic Safety Facts, 2012 Data, <https://crashstats.nhtsa.dot.gov/Api/Public/ViewPublication/811870>; December, 2013 [accessed 15.10.2016].

[16–17] Friis RH. Occupational health and safety for the 21st century. Burlington, MA: Jones & Bartlett Learning; 2016. p. 327–8.

[18–19] Levy BS, Wegman DH, Baron SL, Sokas RK, editors. Occupational and environmental health: recognizing and preventing disease. 6th ed. New York, NY: Oxford University Press; 2011.

[20] Centers for Disease Control and Prevention. What do parents need to know to protect their children? Lead, <http://www.cdc.gov/nceh/lead/acclpp/blood_lead_levels.htm>; [accessed 16.10.2016].

[21] Moore GS. Living with the earth: concepts in environmental health science. 2nd ed. Boca Raton, FL: CRC Press; 2002. p. 213–4.

[22] Levy BS, Wegman DH, Baron SL, Sokas RK, editors. Occupational and environmental health: recognizing and preventing disease. 6th ed. New York, NY: Oxford University Press; 2011.

[23] Cromley EK, McLafferty S. GIS and public health. 2nd ed. New York, NY: Guilford Press; 2012. p. 188.

[24] Moore GS. Living with the earth: concepts in environmental health science. 2nd ed. Boca Raton, FL: CRC Press; 2002. p. 202–3.

[25] Wintemute GJ. The Epidemiology of firearm violence in the twenty-first century United States. Annu Rev Public Health 2015;36(1):5–19. Available from: http://dx.doi.org/10.1146/annurev-publhealth-031914-122535.

[26] Burlage RS. Principles of public health microbiology. Sudbury, MA: Jones & Barlett Learning; 2012. p. 209–20.

[27] Hilgenkamp K. Environmental health: ecological perspectives. Sudbury, MA: Jones & Bartlett Learning; 2006. p. 135.

[28] Moore GS. Living with the earth: concepts in environmental health science. 2nd ed. Boca Raton, FL: CRC Press; 2002. p. 202–3.

[29] Moore GS. Living with the earth: concepts in environmental health science. 2nd ed. Boca Raton, FL: CRC Press; 2002. p. 535.

[30] U.S. Food & Drug Administration. Background on the FDA food safety modernization act (FSMA), <http://www.fda.gov/Food/GuidanceRegulation/FSMA/ucm239907.htm>; [accessed 20.10.2016].

[31] U.S. Food & Drug Administration. FSIS—About us, <http://www.fsis.usda.gov/wps/portal/informational/aboutfsis>; [accessed 20.10.2016].

[32] College Park, MD: Public Health Service & Food and Drug Administration of the U.S. Department of Health and Human Services. Food Code, <http://www.fda.gov/downloads/Food/GuidanceRegulation/RetailFoodProtection/FoodCode/UCM374510.pdf>; 2013 [accessed 20.10.2016].

[33] Buttaro TM, Trybulski J, Baily PP, Sandberg-Cook J. Primary care: a collaborative practice. 4th ed. St. Louis, MO: Elsevier Health Sciences; 2013. p. 225–6.

[34] Moore GS. Living with the earth: concepts in environmental health science. 2nd ed. Boca Raton, FL: : CRC Press; 2002. p. 470.

[35] Collin RW. The environmental protection agency: cleaning up America's act. Westport, CT: Greenwood Publishing Group; 2006. p. 41.

[36] Agency for Toxic Substances and Disease Registry. About ATSDR, <http://www.atsdr.cdc.gov/about/index.html>; [accessed 20.10.2016].

[37–38] Moore GS. Living with the earth: concepts in environmental health science. 2nd ed. Boca Raton, FL: CRC Press; 2002. p. 532–5.

[39] Centers for Disease Control and Prevention. Residential proximity to major highways—United States, 2010. In: CDC health disparities and inequalities report—United States, 2013. MMWR 2013;62(No. Suppl 3).

[40] US Environmental Protection Agency. Radon, <https://www.epa.gov/radon>; [Accessed 20.10.2016].

[41] Moore GS. Living with the earth: concepts in environmental health science. 2nd ed. Boca Raton, FL: CRC Press; 2002. p. 383.

[42] Wintemute GJ. The Epidemiology of firearm violence in the twenty-first century United States. Annu Rev Public Health 2015;36(1):5–19. Available from: http://dx.doi.org/10.1146/annurev-publhealth-031914-122535.

[43] Centers for Disease Control and Prevention. 2016 Model Aquatic Health Code. 2nd Ed, <http://www.cdc.gov/mahc/pdf/2016-mahc-code-final.pdf>; July 15, 2016 [accessed 28.11.2016].

[44a] Centers for Disease Control and Prevention. Disinfection and testing, <http://www.cdc.gov/healthywater/swimming/residential/disinfection-testing.html>; [accessed 28.11.2016].

[44b] Castor ML, Beach MJ. Reducing illness transmission from disinfected recreational water venues. Pediatr Infect Dis J 2004;23(9):866–70. Available from: http://dx.doi.org/10.1097/01.inf.0000138081.84891.30.

[45] U.S. Department of Agriculture and U.S. Department of Health and Human Services. Dietary guidelines for Americans, 2010. 7th ed. Washington, DC: U.S. Government Printing Office; December, 2010.

CHAPTER FIVE

Occupational and Aerospace Medicine

5.1 OCCUPATIONAL AND AEROSPACE MEDICINE QUESTIONS

1. Which of the following closest fits the definition of a fume?
   A. A substance that expands to the size of the enclosure and can be cooled into liquids or solids
   B. Particles with length multiple times longer than their diameter
   C. A liquid suspended in the atmosphere
   D. Solid particles thrown into the air by processes such as crushing and grinding rocks
   E. Solid particles formed from welding and other high-temperature processes
2. Which types of radiation does the WHO consider carcinogenic to humans?
   A. Alpha
   B. Beta
   C. Gamma
   D. Neutron
   E. All of the above
3. Which of the following measures absorbed radiation dose in humans?
   A. Becquerel (Bq)
   B. Curie (Cu)
   C. Gray (Gy)
   D. Roentgen (R)
   E. None of the above
4. A young thrill seeker ignores warnings to evacuate a region during nuclear bomb testing. After the blast, he begins to feel ill and dials 911. If the thrill seeker is exposed to 50 rad of radiation from the bomb, which is he likely to experience first?

Board Review in Preventive Medicine and Public Health.
DOI: http://dx.doi.org/10.1016/B978-0-12-813778-9.00005-0

A. Elevated intracranial pressure
B. Petechiae
C. Sprue
D. Thyroid cancer
E. Vomiting

5. What is the primary purpose of the wet bulb globe temperature (WBGT)?
A. Act as a gauge for metal volatility when welding
B. Describe effects of pollutants in bodies of water
C. Identify temperature where aircrafts have different flight tendencies
D. Predict air concentration of particulate matter
E. Use as a guideline to prevent heat-related illness

6. While attending a high school football game, you see a young player collapse. You run to his aid and find him to be unconscious. When you pull off his helmet, you notice that his skin is dry despite the sweltering heat. His pulse is rapid. His coach is standing nearby and mentions that he was complaining of a headache and weakness, but refused to be taken out of the game.
Which diagnosis is the most likely?
A. Heat stroke
B. Hypoglycemia
C. Hypertrophic cardiomyopathy
D. Pulmonary embolism
E. Vasovagal syncope

7. At which sound frequency does noise-induced hearing loss typically first occur?
A. 4 Hz
B. 40 Hz
C. 400 Hz
D. 4000 Hz
E. 40,000 Hz

8. An airplane mechanic presents to your office with pain and decreased hearing in his right ear after his coworker unexpectantly turned on a jet engine the day prior. The patient was not prepared for the engine to be turned on and was not wearing personal protective equipment.
On physical examination, the patient has marked inability to hear through the right ear during finger rub and whisper tests. During Rinne test, he can hear bone conduction longer than air conduction.

On otoscopic examination, the tympanic membrane is ruptured and there is dried blood lining the external auditory canal.

What type of hearing loss is your patient experiencing?

A. Central hearing loss
B. Conductive hearing loss
C. Mixed hearing loss
D. Sensorineural hearing loss
E. None of the above

9. Mitchell is a 17-year-old male several months away from graduating high school. He plans on joining the workforce immediately after graduation and is looking at potential career paths. His mother is concerned about the safety of the industries he is looking at. Which of the following industries have the lowest rate of occupational fatalities?

A. Agriculture, forestry, fishing, and hunting
B. Construction
C. Manufacturing
D. Mining
E. Transportation

10. Which group publishes recommended exposure limits (RELs) of hazardous chemicals?

A. Environmental Protection Agency (EPA)
B. Department of Justice (DOJ)
C. Occupational Safety and Health Administration (OSHA)
D. National Institute for Occupational Safety and Health (NIOSH)
E. American Conference of Governmental Industrial Hygienists (ACGIH)

11. Which organization is empowered to legally enforce permissible exposure limits (PELs)?

A. National Institute for Occupational Safety and Health (NIOSH)
B. Occupational Safety and Health Administration (OSHA)
C. Environmental Protection Agency (EPA)
D. Federal Bureau of Investigation (FBI)
E. Department of Justice (DOJ)

12. Which of the following is stated by the OSHA's General Duty Clause?

A. Employers have a duty to provide workers' compensation
B. Employers must make sure their employees are safe
C. Employers must offer health insurance

D. Workers must perform activities within their general duties to be covered by workers' compensation
E. Workers must be fit for general duty

13. A 72-year-old migrant worker presents to your office to complain of severe jaw pain that has been worsening over the past year. She is new to the region, as she continuously moves to new farms where she can find work distributing fertilizers, feeding animals, and performing other agricultural duties. Past medical history is significant for osteoporosis, for which she takes an unknown medication. On physical examination, you note that her jaw has become disfigured and has developed an abscess. Her hands and forearms have scarring the look like healed burns.

What is most likely the cause of this patient's symptoms?
A. Arsenic
B. Chromium
C. Nitrates
D. Phosphorus
E. Psoriasis

14. A newly opened munitions plant calls your occupational medicine clinic to ask about a specific illness that has been affecting employees since the plant opened. Every Monday morning the employees in the department, responsible for making trinitrotoluene (TNT), complain of severe headaches and a flushing sensation that resolves throughout the day. The ill employees vary widely in demographics and past medical history. None of the employees complain of headaches on their weekends off and management is beginning to feel that the workers are malingering.

What is the most likely cause of the workers' complaints?
A. Beryllium exposure at work
B. Carbon monoxide toxicity
C. Excessive drinking over the weekends
D. Malingering
E. Nitrate exposure at work

15. An elderly woman presents to your clinic. She works at the local textile mill where she tans and dyes clothing. She would like you to evaluate a rash on her hands that has been there for a couple months. When you look at the rash, you see numerous well-demarcated circular lesions that look like punched out holes. There is minimal erythema and no appearance of secondary infection. There are no lesions elsewhere on her body. The patient states that these lesions are slightly painful and heal without any intervention, but new ones continue to show up. Upon further questioning, the patient states that she has chronic rhinitis and conjunctivitis that does not respond to antihistamines. When looking within the nares, you notice septal perforation. Her eyes are red and dry.

What is the most likely cause of the workers' complaints?

A. Arsenic
B. Chromium
C. Magnesium
D. Scabies
E. Secondary syphilis

16. Which of the following causes a worker to have a green tongue?

A. Cadmium
B. Melamine
C. Nickel
D. Vanadium
E. Zinc

17. Which disorder does beryllium toxicity mimic?

A. Mesothelioma
B. Sarcoidosis
C. Organophosphate toxicity
D. Influenza
E. Chikungunya

18. Which material causes stannosis when inhaled?

A. Aluminum
B. Barium
C. Iron
D. Silica & quartz
E. Tin

19. A 72-year-old man presents to your clinic complaining of chronic shortness of breath and cough. Several years ago, his old physician attempted to treat his ailment with albuterol, but the patient only experienced minimal relief. During the history, taking part of the examination, this man mentions that he worked in a sugar mill for over 30 years. Which of the following disorders might this patient be suffering from?

A. Bagassosis
B. Byssinosis
C. Farmer's lung

D. Suberosis
E. Woodworker's lung

20. Which of the following current day occupational settings pose the highest risk for lead toxicity?
A. Farm
B. Gas station
C. Gun range
D. Hospital
E. Office building built after 1992

21. What is the most common adverse event from contact with nickel?
A. Arthritis
B. Asthma
C. Cancer
D. Contact dermatitis
E. Hypertension

22. Benzene exposure is most commonly associated with which type of cancer?
A. Brain
B. Hematopoietic
C. Pancreatic
D. Skin
E. Testicular

23. A 45-year-old welder presents to your clinic complaining of tremors resembling Parkinson's disease that have worsened over the past 3 years. He has a noncontributory past medical history and family history. He denies any history of head trauma or contact sports. The patient had quit smoking cigarettes 10 years ago and denies ever using recreational drugs.

Which hazardous occupational exposure is causing his symptoms?
A. Beryllium
B. Cadmium
C. Manganese
D. Mercury
E. Nickel

24. Which of the following is not true of ergonomics?
A. Chronic conditions resulting from poor ergonomics are typically reversible through appropriate ergonomics
B. Improper ergonomics can lead to acute and chronic conditions
C. It is widely understood that proper ergonomics decreases productivity

D. Proper ergonomics should always optimize an employee's work environment

E. Stress-related illness contributes a significant burden to ergonomic-related illness

25. A worker presents to your occupational clinic complaining of numbness and white tips of his upper extremities. Which of the following tools is contributing most to this finding?

A. Hammer

B. Screwdriver

C. Jackhammer

D. Handsaw

E. Shovel

26. Which type of radiation is associated with increased incidence of lung cancer in miners?

A. Alpha

B. Beta

C. Gamma

D. Epsilon

E. Neutron

27. How much louder is 100 decibels than 80 decibels?

A. 5 ×

B. 10 ×

C. 20 ×

D. 100 ×

E. 1000 ×

28. As the director of a new emergency department, you have been tasked with creating the schedule for the clinicians. Their contracts dictate that they must work four, 10 hours shifts each week. After listening to each practitioner's proclaim of why they want to work the daytime shift, you decide that to make things fair; each clinician must complete shifts beginning and ending at all hours of the day to provide adequate 24 hours coverage in the emergency department.

Which of the following schedules would be best for the overall sleep schedules and well-being of the clinicians? The time listed represents the beginning of the schedule.

A. Shift 1—8:00 a.m.–6:00 p.m.
Shift 2—2:00 p.m.–12:00 a.m.
Shift 3—8:00 p.m.–6:00 a.m.
Shift 4—2:00 a.m.–10:00 a.m.

B. Shift 1—8:00 a.m.–6:00 p.m.
Shift 2—2:00 a.m.–12:00 p.m.
Shift 3—8:00 a.m.–6:00 p.m.
Shift 4—8:00 a.m.–6:00 p.m.

C. Shift 1—8:00 a.m.–6:00 p.m.
Shift 2—8:00 p.m.–6:00 a.m.
Shift 3—2:00 p.m.–10:00 p.m.
Shift 4—2:00 a.m.–10:00 a.m.

D. Shift 1—8:00 a.m.–6:00 p.m.
Shift 2—2:00 a.m.–10:00 a.m.
Shift 3—2:00 a.m.–10:00 a.m.
Shift 4—2:00 p.m.–10:00 p.m.

E. There is no difference between these shifts

29. Which of the following flights will likely cause the most symptoms related to jet lag?

A. Traveling east 3 time zones
B. Traveling east 6 time zones
C. Traveling west 3 time zones
D. Traveling west 6 time zones
E. Traveling south 8 hours in the same time zone

30. An industrial painting company contacts your occupational medicine clinic to provide preemployment physical exams to ensure that the workers are fit for duty. What is the most common agent used for bronchoprovocation testing?

A. Albuterol
B. Benzonatate
C. Methacholine
D. Nitrous oxide
E. Smoke

31. What is the name of the resource that describes hazardous chemicals in the workplace? Chemical manufacturers and distributors must provide it to employers to make available for employees.

A. 300 log
B. Emergency Exposure Protocol
C. General Chemical Database
D. Hazardous Chemical Pamphlet
E. Safety Data Sheet

32. The manager of a sugar mill recently became aware that the new machinery in his factory exceeds company issued noise limits.

Assuming that each option adequately reduces noise to safer levels and maintains sugar mill efficiency, which of the following approaches to reduce noise exposure to employees is least preferred?

A. Construct a barrier around the machine that reduces sound
B. Have machine operate at night when employees are not around
C. Designate a radius around the machine where the noise dissipates to appropriate levels. Do not allow workers within this radius
D. Redesign the workplace where workers are isolated in rooms away noise
E. Require all employees to wear earplugs

33. Which of the following professionals is the preferred person for reviewing results from employee drug screens?

A. Board certified pathologist
B. Clinical Laboratory Improvement Amendments (CLIA) supervisor
C. Employer HR personnel
D. Medical review officer (MRO)
E. Public health officer

34. A chef at a restaurant accidently places her left hand on a stovetop, causing second degree burns. The workers' compensation physician determines that she is able to go back to work immediately, but she is not to use her left hand until further specified in a follow-up visit. The employing restaurant asked her to fill the role of hostess (at a lesser pay) until the physician allows her to use her hand. During the third visit with the workers' compensation physician, the limitation is lifted and the employee may go back to full duty. Through her workers' compensation program, she is able to receive some financial compensation to make up for the difference between her usual and temporary pay.

Which type of classification best describes the worker's injury?

A. Temporary total disability
B. Temporary partial disability
C. Permanent total disability
D. Permanent partial disability
E. Permanent total disability

35. A miner presents to your clinic for disability evaluation. He was injured 1 year prior by a work-related underground explosion that left him with bilateral conductive hearing loss. The audiologist has assured him that this hearing loss is not reversible. What type of disability would you classify this patient as having?

A. Temporary total disability
B. Temporary partial disability
C. Permanent total disability
D. Permanent partial disability
E. No disability

36. Which of the following body parts is most sensitive to damage from laser exposure?
A. Brain
B. Eyes
C. Lungs
D. Ovary/testicle
E. Skin

37. An employee within an aerospace manufacturing facility has developed a granulomatous lung disease. Your physician colleague is concerned that it may be sarcoidosis. However, you feel that it may be due to an occupational exposure. Which laboratory test is most useful in determining whether this lung condition is due to beryllium?
A. Lymphocyte proliferation test
B. Eosinophils
C. Neutrophils
D. Serum beryllium levels
E. Breath samples for beryllium

38. A patient presents to your occupational medicine facility complaining of dyspnea. A pulmonary function test (PFT) demonstrates a restrictive pattern. Eggshell calcifications are seen in the chest X-ray. Which agent is responsible for this patient's dyspnea?
A. Asbestos
B. Coal dust
C. Cotton
D. Silica
E. Tin

39. Which of the following lung diseases is most likely to be found in someone that works in a cotton textile mill?
A. Asbestosis
B. Berylliosis
C. Byssinosis
D. Coal worker's pneumoconiosis
E. Sarcoidosis

40. A patient presents to your clinic complaining of myalgias, fatigue, headache, and nausea. The patient works as a welder and is subsequently diagnosed with metal fume fever. Which of the following is the most appropriate treatment for an uncomplicated case of metal fume fever?

A. Antibiotics
B. Chelation therapy
C. Diuretics
D. Steroids
E. Supportive treatment

41. Which of the following is not a type of hypoxia?

A. Inhalation hypoxia
B. Histotoxic hypoxia
C. Hypemic hypoxia
D. Stagnant hypoxia
E. Hypoxic hypoxia

42. Which of the following occupational settings is associated with an increased risk of bladder cancer?

A. Computer programming
B. Construction
C. Dye manufacturing
D. Fire rescue
E. Military

43. Carpal tunnel syndrome occurs from compression of which nerve?

A. Brachial
B. Median
C. Musculocutaneous
D. Radial
E. Ulnar

44. You are the lone public health officer in a rural farming county. Several clinicians at the local federally qualified health centers (FQHCs) report a sharp increase in the number of farm workers that are experiencing a group of symptoms including changes in salivation, lacrimation, defecation, and urination. You suspect that these workers may be suffering from organophosphate toxicity. Which of the following tests would confirm your theory?

A. Pulse oximeter
B. RBC cholinesterase

C. Serum organophosphate
D. Serum potassium
E. None of the above

45. Which of the following statements about working during pregnancy is false?
A. Occupational exposure may produce different effects at different gestational ages
B. Every woman that works during pregnancy should be screened for occupational risks
C. Physiologic interaction with toxins during pregnancy may be different than outside of pregnancy
D. Working during pregnancy increases the overall risk of adverse pregnancy outcomes
E. Work-related psychological stress during pregnancy may yield low birthweight and prematurity

46. Koebner phenomenon is most likely to lead to which of the following skin lesions at the site of physical contact?
A. Allergic dermatitis
B. Hyperpigmentation
C. Onychomycosis
D. Psoriatic lesion
E. Ulcer formation

47. A young factory worker presents to your clinic complaining of a flushing sensation associated with facial redness. It has happened three times in the last month. All three times occurred at the restaurant across the street from his work after finishing his shift. His coworker that works in the same department also experiences this sensation, but the other coworkers do not experience these symptoms. You diagnose the patient with degreaser's flush. Which of the following is triggering your patient's symptoms?
A. Alcohol
B. Hand soap in the sports bar restroom
C. Hamburger cooked in vegetable oil
D. Peanuts
E. Water with lemon

48. Which of the following is true regarding the Americans with Disability Act (ADA)?
A. Alcoholism is covered by the ADA
B. Illicit drug use is covered by the ADA

C. Employers are responsible for providing all accommodations that a disabled employee needs to perform the job
D. The ADA is under the Department of Labor
E. An employer may inquire about an applicant's workers' compensation history prior to making an offer for employment.

49. Which of the following employees is not covered under the Americans with Disabilities Act (ADA)?
A. 40-year-old alcoholic
B. 22-year-old heroin addict
C. 33-year-old pregnant woman with preeclampsia
D. 60-year-old with epilepsy
E. 50-year-old with HIV infection

50. A public health officer notices that his assistant develops a pattern of showing up late for work. As the tardiness progresses, the public health officer occasionally notices the smell of alcohol on his assistant's breath. After months of not addressing these observations, the secretary comes clean about his pending divorce. The public health officer responds by telling his assistant that he is fired unless he attends an employee assistance program. Which of the following is true of employee assistance programs?
A. Employee assistance programs are hosted by state organizations
B. Employee assistance program outcomes may be shared with supervisors
C. Employee assistance programs are an employee benefit that do not benefit the employer
D. Employee assistance programs are primarily for problems that arise outside of work
E. Supervisors may require an employee to use an employee assistance program

51. If there is a chemical explosion in a factory and two workers end up hospitalized, how soon must it be reported to the OSHA?
A. Within 1 hour
B. Within 24 hours
C. Within 72 hours
D. Within 1 week
E. It is not required to report

52. Which form is required by the Occupational Safety and Health Association to be completed by the employer when a worker suffers injury or illness as a result of employment?

A. 300 Form
B. DWC-25
C. Injured Worker Report
D. Safety Data Sheet
E. Workplace Injury Report

53. A concrete worker accidentally crushed his finger beneath a large pipe and received subsequent workers' compensation. After the initial visit, the employer asked for the worker's paperwork. Which of the following pieces of information should not be shared with the employer and the workers' compensation insurer?
A. Medications prescribed at the visit
B. History of hypertension
C. X-ray results
D. All of the above should be shared
E. None of the above should be shared

54. Which of the following employment settings are not covered under the federal OSHA?
A. Private for profit organization
B. Private not for profit organization
C. Public for profit
D. Public nonprofit
E. State government

55. If a private organization has ___ or more employees, they must provide leave benefits for their employees under the Family Medical Leave Act (FMLA).
A. 10
B. 20
C. 30
D. 40
E. 50

56. If a worker at a company of 2000 employees request Family and Medical Leave Act (FMLA) time for a sick niece, how many weeks of leave is the worker entitled to?
A. 0 weeks
B. 86 weeks
C. 12 weeks
D. 18 weeks
E. 24 weeks

57. Under the OSHA, when is a hearing conservation program required?
A. 8 hour time-weighted average (TWA) of >80 decibels
B. 8 hour time-weighted average (TWA) of >85 decibels
C. 8 hour time-weighted average (TWA) of >90 decibels
D. 8 hour time-weighted average (TWA) of >95 decibels
E. 8 hour time-weighted average (TWA) of >100 decibels

58. After the on-site death of an employee, how long does a company have to report the incident to the OSHA?
A. 8 hours
B. 24 hours
C. 1 week
D. 1 month
E. It is not necessary to report an on-site death of an employee

59. Under the Occupational Safety and Health Act, how long must employee medical records be retained?
A. Length of employment
B. Length of employment plus 1 year
C. Length of employment plus 5 years
D. Length of employment plus 10 years
E. Length of employment plus 30 years

60. How many employees must a private organization have to be covered under the Americans with Disability Act (ADA)?
A. 0
B. 15
C. 30
D. 45
E. 60

61. An employee at a retail store tripped in the store's parking lot and suffers a broken bone in her ankle. The incident happened during the employee's lunch break while she was off the clock. Should this incident be considered workers' compensation?
A. No: Incidents outside of the workplace are not considered workers' compensation
B. No: The employee was not on the clock
C. Yes: But it is not recordable
D. Yes: It should be recorded
E. None of the above

62. What is the workers' compensation term that describes when an injured worker recovers to the full potential they are capable of recovering?
    A. Complete Medical Recovery (CMR)
    B. Full Medical Improvement (FMI)
    C. Full Medical Recovery (FMR)
    D. Maximum Medical Improvement (MMI)
    E. Maximal Medical Recovery (MMR)
63. Which of the following national organizations mandates that semi-truck drivers in the United States undergo certification to prove they are healthy enough to drive?
    A. American Conference of Governmental Industrial Hygienists (ACGIH)
    B. Department of Labor (DOL)
    C. Federal Motor Carrier Safety Administration (FMCSA)
    D. Occupational Safety and Health Administration (OSHA)
    E. Truck drivers are certified by the states and not a unified national organization
64. What is the optimal volume of air a jet fighter pilot should have in the breathing apparatus?
    A. As large as possible
    B. As small as possible
    C. Depends on the pilot
    D. Depends on the flight situation
    E. Volume does not matter
65. Which of the following is not a way of increasing tolerance to G-forces (G tolerance)?
    A. Wearing a G-suit
    B. Weight lifting program
    C. Adjust the seating to sit straight up
    D. Using 100% oxygen
    E. Valsalva maneuver
66. A patient presents to your clinic after returning home from a vacation. This was the first flight that she had ever taken. She feels fine now, but complains that she experienced abdominal bloating and ear discomfort on both the departing and arriving flights. Which of the following is mostly responsible for your patient's experience?
    A. Boyle's law
    B. Charles' law
    C. Dalton's law

D. Henry's law
E. Law of partial pressures

67. A scuba diver is hospitalized with decompression sickness after suffering an embolic event during a flight home from a diving expedition. Which of the following gas laws is mostly responsible for this patient's experience?
A. Boyle's law
B. Charles' law
C. Dalton's law
D. Henry's law
E. Law of partial pressures

68. Which sensory function is most important for spatial orientation?
A. Proprioceptive
B. Tactile
C. Vestibular
D. Vision
E. All are equally important

69. A 28-year-old woman presents to a rural emergency room with chest pain immediately after landing at the airport from a scuba diving vacation. She is also experiencing shortness of breath, heart palpitations, visual changes, and joint pain in her elbows, knees, and shoulders. Besides an appendectomy 5 years prior, her past medical history is noncontributory. She has no history of cardiopulmonary problems or arthritis. On physical examination, breath sounds are bilaterally clear to auscultation. There are no heart murmurs and the heart has a regular rate and rhythm.

Which of the following is the most appropriate immediate treatment for this patient?
A. Morphine
B. Antibiotics
C. Administration of 100% oxygen
D. Needle decompression
E. Cardiac catheterization

70. If a diver is 99 ft (30 m) deep in freshwater, how many extra atmospheres of pressure is the diver exposed to?
A. 1
B. 2
C. 3
D. 4
E. 5

71. During the first several hours after leaving orbit, four of the five astronauts aboard a new commercial space flight begin to develop dizziness, nausea, headache, pallor, and dyspepsia. The fifth astronaut, in addition to one of the other four also develops a headache. What is the most appropriate action to take?

A. Immediate reentry to Earth
B. Intramuscular injection of promethazine
C. No treatment necessary
D. Oral scopolamine
E. Place crew members on 100% oxygen

72. An automobile painter of 2 years presents to your occupational medicine clinic to complain of worsening cough that began one year ago. It typically occurs two hours into his shift and lasts for an hour after leaving work. Associated symptoms include wheezing and shortness of breath.

Which occupational exposure is the most likely cause of your patient's symptoms?

A. Carbon monoxide
B. Isocyanate
C. Manganese
D. Methacholine
E. Nickel

73. Which agency is the lead organization in the investigation of civil aviation crashes in the United States?

A. Department of Transportation (DOT)
B. Federal Aviation Administration (FAA)
C. International Civil Aviation Organization (ICAO)
D. National Aeronautics and Space Administration (NASA)
E. National Transportation Safety Board (NTSB)

74. Which agency is responsible for assuring the qualifications of airline pilots and issuing certification in the United States?

A. Department of Transportation (DOT)
B. Federal Aviation Administration (FAA)
C. International Civil Aviation Organization (ICAO)
D. National Aeronautics and Space Administration (NASA)
E. National Transportation Safety Board (NTSB)

75. Which of the following patients has an absolute contraindication to flight?

A. 3-week-old healthy infant
B. 7-year-old child with pressure equalizing (myringotomy) tubes in the ears

C. 16-year-old with latent tuberculosis infection
D. 25-year-old male with pneumothorax
E. 35-year-old woman that is 28 weeks pregnant

76. An airline passenger should receive supplemental oxygen during flight if their oxygen saturation on room air at sea level is at or below which number?
A. 94%
B. 92%
C. 90%
D. 88%
E. 86%

77. While sitting on a commercial airplane, you become concerned for your health because there is a nearby passenger continuously coughing. You are seated in the 10th row aisle seat. Which seat has the highest likelihood of sharing expired air with you?
A. Eighth row aisle seat
B. Eighth row window seat
C. 10th row window seat
D. 12th row aisle seat
E. 12th row window seat

78. Your patient with sickle cell anemia is planning a flight to see her cousin. She is concerned that the lack of oxygen at the altitudes of the flight will decrease her ability to breath. Most commercial airliners in the United States pressurize the cabin to the pressure found at which maximum altitude?
A. 0 ft (sea level)
B. 5000 ft
C. 8000 ft
D. 11,000 ft
E. 15,000 ft

79. Which organization publishes Permissible Exposure Limits (PELs)
A. American College of Occupational and Environmental Medicine
B. American Conference of Governmental Industrial Hygienists
C. Environmental Protection Agency
D. National Institute for Occupational Safety and Health
E. Occupational Safety and Health Administration

80. An agent such as asbestos, which is known to cause cancer in humans, receives which classification by the International Agency for Research on Cancer (IARC)?

A. 1
B. 2
C. 3
D. 4a
E. 4b

81. You are seeing a 34-year-old male patient in an occupational medicine clinic. He complains of back pain following a 15-foot fall yesterday. During the exam, the patient has a decreased right Achilles deep tendon reflex (DTR). In addition, the lateral side of his foot has decreased sensory function.

Where is the level of disc herniation?

A. L2–L3
B. L3–L4
C. L4–S5
D. L5–S1
E. S1–S2

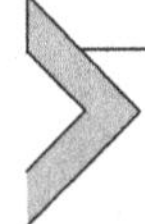

5.2 OCCUPATIONAL AND AEROSPACE MEDICINE ANSWERS

1. **E. Solid particles formed from welding and other high-temperature processes**

   Fumes are just one of many types of air contaminants. A fume is a solid particle suspended in the air. The particles may combine with the air to form an oxide. It may be easy to remember this by thinking of the diagnosis, metal fume fever, where welders experience flu-like symptoms mainly from aerosolized zinc during welding procedures.

   Choice A describes a gas, a substance that may be in atomic or molecular form.

   Choice B describes suspended fibers such as asbestos.

   Choice C describes a mist. This is created when liquids condense into a vapor back to a liquid or by the splashing of liquid. Aerosols are a form of a mist that are highly respirable.

   Choice D describes dust. This is created during the crushing, grinding, and heating of inorganic materials.

2. **E. All of the above**

   The WHO, International Agency for Research on Cancer (IARC) considers all types of ionizing radiation (alpha, beta, gamma, neutron, X-ray) to be carcinogenic. The IARC also considers both solar radiation and ultraviolet (UV) radiation in the 100–400 nm spectrum (including UVA, UVB, and UVC) to be carcinogenic.

3. **C. Gray**

   The amount of radiation absorbed by an organism is represented by the rad (an acronym for radiation absorbed dose). The Gray (Gy) is the international system equivalent of the rad. One Gy equals 100 rad.

   Curies (Ci) quantify the amount of radioactive material, not the amount of radiation emitted. The Bq is the international system united equivalent of the Ci.

   The Roentgen describes the quantity of radiation exposure. The Coulomb is the international system counterpart of the Roentgen.

   Rem is an acronym for rad-equivalent man. It expresses the quantity of radiation received by a person (rad) multiplied by an adjusting factor that represents the severity of the type of radiation.

For example, alpha-particle radiation from radon has a quality factor of 20. Therefore, an exposure of 1 rad of alpha-particle radiation is equivalent to 20 rem. Meanwhile, medical X-rays have a quality factor of one, so rem and rad are equivalent when medical X-rays are used. The Sievert (Sv) is the international system counterpart of the rem. One Sv equals 100 rem.

4. **E. Vomiting**

For acute radiation syndrome to occur, the radiation must come from rapid delivery of tissue-penetrating radiation from an external source that exposes a majority of the body. After exposure to sufficient radiation, the body experiences a prodromal stage with symptoms of anorexia, nausea, vomiting, and diarrhea. This stage may last from minutes to days. After completion of the prodromal stage begins the latent stage, where symptoms may be absent up to a couple of weeks. When symptoms start to develop, the illness manifestation stage begins. If the victim survives the illness manifestation stage, they enter the recovery stage.

Classic syndromes of the illness manifestation stage include bone marrow (hematopoietic) syndrome, gastrointestinal syndrome, and cardiovascular/central nervous system syndrome. Death may occur due to hemorrhage or infection. The bone marrow syndrome typically occurs after exposure to 70–1000 rads. Gastrointestinal syndrome typically occurs after exposure to >1000 rads where the GI tract is irreparably damaged. Dehydration and electrolyte imbalances typically occur, in addition to bone marrow syndrome, leading to certain death. After exposure to >5000 rads, cardiovascular/central nervous system syndrome occurs. The circulatory system collapses and intracranial pressure increases to levels incompatible with life.

Cancer takes years to develop after exposure to radiation.

5. **E. Use as a guideline to prevent heat-related illness**

The WBGT is an equation that produces an index used to reduce the likelihood of heat/cold-related illness. It provides guidance for the duration and intensity of physical activity performed by workers and athletes when exposed to weather elements. The WBGT formula includes the air temperature, humidity, temperature of surfaces (i.e., sun creating a hot surface), and air motion (wind) to create a single variable used to predict illness from environmental conditions. Once calculated, the WGBT may also be adjusted for different types of clothing worn.

WBGT is the most widely used index for heat stress and is commonly used by organizations such as the ACGIH and OSHA.

The WBGT provides only a general guide to the adverse effects of environmental conditions and is not intended to be the resource. It is important to take into account the individual variables when making decisions about worker/athlete exertion.

6. **A. Heat stroke**

This football player is suffering from heat-related illness most likely in the form of a heat stroke. Heat-related illness runs the spectrum of severity all the way from minor annoyance to fatal consequences. It affects people of all ages, sexes, and physical conditions although it is more common in those not acclimatized to the heat. From least severe to most severe, the types of heat-related illness are as follows:

Heat cramps result from an electrolyte imbalance from dehydration and electrolyte loss. Furthermore, the balance may be thrown off further when water is consumed without electrolytes. Cramping typically occurs in the muscles of the legs and abdomen. Treatment is discontinuation of physical activity and hydration with electrolytes.

Heat edema results from vasodilation that occurs due to heat. Fluids accumulate in areas further away from the heart, particularly the legs.

Heat syncope occurs when there is loss of consciousness due to heat exposure. Vasodilation leads to orthostatic hypotension and decreased delivery of oxygen to the brain.

Heat exhaustion arises from altered plasma electrolytes compounded with orthostatic hypotension from heat. Core body temperature may fall within normal range or may be slightly elevated. Symptoms include malaise, weakness, headache, vomiting, tachycardia, and hypotension. Treatment consists of moving to a cooler environment and administration of fluids supplemented with electrolytes.

Heat stroke is the most severe of the heat-related illnesses. It is sometimes preceded by heat exhaustion. Symptoms include dizziness, headache, hallucinations, and weakness. It may also cause permanent neurological damage and multiorgan failure. Body temperature may exceed 104°F and the skin is typically dry to the touch as sweating is absent. In severe cases, heat stroke can even lead to death. Treatment includes emergency measures to get the body hydrated and bring the core temperature down.

7. D. 4000 Hz

Noise-induced hearing loss is a type of bilateral sensorineural hearing loss that is caused by damage to the hair cells in the middle ear. It first occurs in the 3000–6000 Hz frequency. Serial audiograms over time show a characteristic dip at the 4000 Hz mark as noise exposure continues. Lower frequencies may be affected in extreme cases but it is uncommon.

In causing noise-induced hearing loss, continuous noise exposure over many years is worse than intermittent noise exposure with breaks. Once noise exposure is removed, there is no progression of hearing loss. Meanwhile, age-related hearing loss continues over time.

Conductive and mixed hearing loss occurs across all sound frequencies.

8. B. Conductive hearing loss

A perforated tympanic membrane causes conductive hearing loss. This type of hearing loss results from interference of the conduction of sound from the external ear to the structures of the middle ear. In addition to tympanic membrane perforation, other examples of conductive hearing loss include cerumen impaction, external ear infection, and an effusion from otitis media.

Central hearing loss occurs when pathology in the brain or brainstem prevents conduction of hearing signals.

Mixed hearing loss occurs when there are characteristics of both conductive and sensorineural hearing loss. The most common form of mixed hearing loss is otosclerosis.

Sensorineural hearing loss results from damage to the cochlea or cranial nerve VIII. It is the most common form of hearing loss. Examples include presbycusis, sequela from mumps, and acoustic neuroma.

9. C. Manufacturing

Outside of manufacturing, the other four listed industries have the dubious distinction of having the highest rates of fatalities. Construction has the most occupational-related fatalities overall due to the large number of construction workers, but it has a significantly lower rate of fatalities compared to the other three industries.

Older workers typically experience higher rates of occupational fatalities, while younger workers experience higher rates of nonfatal occupational injury.

Occupational injuries and fatalities are recorded by the Census of Fatal Occupational Injuries (CFOI) and the National Traumatic

Occupational Fatalities surveillance system. CFOI uses numerous data sources to verify the work-relatedness of deaths while the National Traumatic Occupational Fatalities system is based on death certificates.

Workplace safety was identified by the CDC as one of the 10 great public health accomplishments of the 20th century.

10. D. National Institute for Occupational Safety and Health (NIOSH)

The NIOSH publishes RELs of chemicals that may be hazardous to the environment, which the OSHA takes into consideration when establishing PELs. NIOSH does not have the power to enforce RELs, but OSHA has the power to enforce PELs.

Meanwhile, the ACGIH produces threshold limit values (TLVs). EPA publishes maximum contaminant level (MCL).

11. B. Occupational Safety and Health Administration (OSHA)

Please see the explanation for Question #10 (directly above).

12. B. Employers must make sure their employees are safe

The General Duty Clause [Section 5(a) (1) of the Occupational Safety and Health Act (OSHA)] requires employers to create a work environment that "is free from recognizable hazards that are causing or likely to cause death or serious harm to employees." In other words, OSHA's General Duty Clause states that employers must make sure their employees are safe.

13. D. Phosphorus

This patient is suffering from phosphorus toxicity. In appropriate quantities, phosphorus is an element that is essential to natural biological processes in the body. There are three forms of phosphorus: White/yellow (most toxic), red, and black. It can be inhaled, absorbed through the skin, or ingested. Exposure of phosphorus to the skin can cause deep burns. Continued phosphorus exposure leads phossy jaw, a condition characterized by necrosis of the maxilla and mandible, which may be associated with abscess. Phosphorus can also lead to hepatic necrosis.

Prior to being recognized as a hazard, phosphorous was found in many industries including agriculture, animal foods, pharmaceuticals, metal cleaning, engraving, electroplating, and semiconductors. Thanks to decreased occupational exposure nowadays, it is rare to see symptoms of phosphorus toxicity in the United States. The patient in this vignette not only suffered occupational exposure to phosphorus, but also used bisphosphonates for her osteoporosis.

14. E. Nitrate exposure at work

The workers at this munitions plant are suffering from Monday morning headaches due to exposure to nitrates. Nitrates are a common component in munitions. Nitrates trigger GMP to cause vasodilation that leads to venous pooling, reduced venous return, and lower end diastolic pressure. Because of these properties, it is used medically to improve the supply–demand balance of the heart in congestive heart failure, angina, and acute coronary syndrome. Common side effects of nitrates include headaches, flushing sensation, orthostatic hypotension, and syncope.

Tolerance to nitrates builds with continuous exposure. Once the exposure decreases, the tolerance disbands and both the effects and side effects return. Monday morning headaches due to nitrates occur when workers are exposed to nitrates after losing their tolerance over the weekend.

Contraindications to nitrates include hypotension, elevated intracranial pressure, and PDE5 inhibitors for erectile dysfunction. Special attention should be given to workers with these contraindications.

Although carbon monoxide toxicity may lead to headaches and a flushing sensation, it is unlikely that workers continuously exposed to carbon monoxide will only experience symptoms 1 day per week. Tolerance does not develop to carbon monoxide as it does to nitrates.

15. B. Chromium

Chromium is an essential element that is vital to metabolism. When in contact with skin (especially open lesions), it is capable of producing contact dermatitis and chrome holes. The patient in this vignette is suffering from chrome holes. They are best described as circular well-demarcated lesions that look like punched out holes. They are more common on the hands where the majority of chromium contact occurs and there is frequently small lesions in the skin. Additionally, chromium may contribute to occupational asthma, septal perforation, chronic rhinitis, chronic bronchitis, conjunctivitis, keratitis, and lung cancer.

Chromium is found in the textile and electroplating industries.

16. D. Vanadium

Vanadium is a compound used in manufacturing steel alloys, glass, dyes, and pesticides. At subtoxic levels, it may cause a metallic taste and/or a green tongue.

17. B. Sarcoidosis

As soon as one year after exposure, beryllium inhalation may lead to chronic t-cell mediated noncaseating granulomatous illness with all of the same symptoms of sarcoidosis. Many patients with beryllium disease are mistakenly diagnosed with sarcoidosis. The disease is confirmed by using a beryllium lymphocyte proliferation test.

18. E. Tin

Pneumoconiosis is a fibrotic lung disease resulting from the inhalation of respirable particles typically occurring over a long period of time.

Inhaled agent	Type of pneumoconiosis
Aluminum	Shaver's Disease
Barium	Baritosis
Iron	Siderosis
Silica	Silicosis
Tin	Stannosis

19. A. Bagassosis

This patient is suffering from bagassosis after years of moldy sugarcane exposure. The fibers cause a hypersensitivity pneumonitis resulting in interstitial inflammation, fibrosis, and granulomas.

There are many types of hypersensitivity pneumonitis. The potential answers from the question have been placed in the table below.

Source of inhaled agent	Type of hypersensitivity pneumonitis
Sugarcane	Bagassosis
Cotton	Byssinosis
Hay and grain	Farmer's lung
Cork	Suberosis
Wood	Woodworker's lung

20. C. Gun Range

Gun ranges are the work setting with the highest risk of lead exposure. Lead is a component of bullets which produces lead dust when fired. This lead dust may be inhaled, absorbed through the skin, or ingested when it transfers from hands to food. Law enforcement is a notable occupation with increased exposure to lead from shooting ranges.

Indoor ranges produce a higher risk of lead exposure than outdoor ranges because the lead dust does not disperse as well indoors.

Both indoor and outdoor ranges are monitored by OSHA to enforce the PEL of lead.

In the United States, lead is no longer used in gasoline or residential/home paint. The Residential Lead-Based Paint Hazard Reduction Act of 1992 allowed for OSHA to increase protection of workers.

21. **D. Contact dermatitis**

Nickel is found in metal alloys, batteries, coins, and jewelry. When it contacts skin, between 10% and 20% of people will experience contact dermatitis. Other than that, nickel may cause adverse effects through inhalation and ingestion. Workers that perform electroplating work are at increased risk of nickel inhalation. Over time, this increases the risk of nasal and lung cancers.

Nickel exposure does not cause arthritis or hypertension.

22. **B. Hematopoietic**

Benzene is an aromatic compound commonly used as a solvent and found abundantly in the petrochemical industry (including gasoline). Although it can be absorbed through the skin, exposure most often occurs through inhalation. Once in the body, benzene suppresses bone marrow leading to pancytopenia and anemia. It rapidly crosses the blood–brain barrier and induces central nervous system changes leading to depression, headache, convulsions, and even death. Longer exposure can lead to hematopoietic cancers such as leukemia. Urinary phenol is the preferred marker for measuring benzene exposure.

23. **C. Manganese**

Manganese is notorious for causing parkinsonian syndromes. It does this by harmfully effecting the basal ganglia after inhalation of manganese dust or fumes. Welders are at increased risk of exposure to metal fumes due to using processes that allow metals to transform into fumes and gases.

24. **C. It is widely understood that proper ergonomics decreases productivity**

Ergonomics aims to create the best fit for a worker in the workplace. While improper ergonomics contributes to physical and mental harm, proper ergonomics reverses these trends and helps to maintain workers' health. When proper ergonomics are in place, workers are able to perform their duties, leading to increases in productivity.

The proper use of ergonomics has been shown to reduce accident-related injuries, musculoskeletal disease (acute and chronic), and stress-related illness amongst other ailments. Common problems contributing to physical stress from a poor ergonomic environment include incorrect posture, repetitive movements, background noise, and extremes in temperature. Meanwhile, common problems contributing to mental stress from a poor ergonomic environment include demanding workload, ambiguous work expectations, and conflict between coworkers. Harm from any of these factors may cumulate over time to produce chronic problems.

25. C. Jackhammer

Hand-arm vibration syndrome (HAVS) is a Raynaud's phenomenon-like trauma induced vasospasm that occurs from repetitive vibration exposure to the upper extremities. Initial symptoms may include finger blanching, pain, and flushing. In severe cases, patients with HAVS may develop arterial thrombosis and digital necrosis. The exact mechanism of injury is unclear.

Outside of repetitive vibration exposure to the hands, risk factors for HAVS include personal or family history of Raynaud's syndrome, smoking, prior arm history, and exposure to cold weather. The severity of HAVS worsens with continued exposure to vibration.

Treatment for HAVS is multifactorial. The preferred approach is to remove exposure from the offending agent. If that is not possible, it is recommended to use tools with vibration dampeners, wear dampening gloves that maintain warmth, quit smoking, and grip tools lightly.

26. A. Alpha

Miners are exposed to naturally occurring alpha radiation emitted from radon deposits below ground. Although alpha radiation is not capable of penetrating clothing or skin, it is able to effect the lungs after inhalation. Alpha particles enter the lungs attached to droplets, dusts, and particles. Although the majority of alpha particles are exhaled without transmission in the body, some alpha decay products become trapped in the lungs, where they are transmitted to local respiratory tissue. Epidemiologic studies indicate that low-dose long-term exposure is associated with higher risk of lung cancer than high-dose short-term exposure. Radon exposure increases the incidence of all types of lung cancer.

27. D. 100×

Decibels (dB) are a way of describing ratios. This unit of measurement is most often used to describe sound, pressure, and power. The third letter in the decibel abbreviation indicates the type of ratio being described. For example, dBV represents the decibel of voltage. Similarly, dBa represents the loudness of sound perceived by the human ear. To accommodate for large numbers, the decibel scale is based on logarithms, where for every 10 decibels, there is 10 times the sound. For a better understanding, see the table below;

Difference in decibels (dB)	Multiples of ratio
10	10 ×
20	100 ×
30	1000 ×
40	10,000 ×

To solve this problem, one must understand that 100 decibels of sound is 20 decibels higher than 80 decibels of sound. A 20 decibel difference represents a 100 × difference in intensity of sound.

Of note is that a 3 dBa difference in sound represents 2 × louder intensity of sound.

On the dB scale, the smallest audible sound is at 0 dB. This is known as the threshold of sound. Meanwhile, the threshold of pain is typically between 100dBa and 120 dBa.

28. A. Shift 1—8:00 a.m.–6:00 p.m

Shift 2—2:00 p.m.–12:00 a.m.

Shift 3—8:00 p.m.–6:00 a.m.

Shift 4—2:00 a.m.–10:00 a.m.

Fixed-shift work schedules are preferred for workers' well-being, but some employment circumstances do not allow for them. In these cases, rotating-shift work schedules are necessary. When it comes to rotating schedules, the schedule may either be slow rotating or quick rotating. In a slow rotation schedule, the employee holds a specific shift for several weeks (or more) before rotating onto the next shift. In a quicker rotating schedule, the employee will change shifts multiple times weekly or semiweekly.

Numerous studies agree that employees do better in situations where they work in forward shifting rotations compared to random or backward shifting rotations. Forward rotation allows for longer time between shifts, while backward rotation does just the opposite.

Option A is the only answer that does not have a shift moving backward in start time.

29. B. Traveling east 6 time zones

Traveling through different time zones causes a circadian rhythm sleep disorder known as jet lag. Symptoms include fatigue, sleep disturbances, anorexia, reduced cognitive skills, and altered coordination. Eastward travel shortens the circadian cycle, whereas westward travel extends it. Eastward travelers have more problems adjusting because the body has a tendency to resist shortening the cycle.

Prevention of jet lag symptoms can be accomplished by adjusting sleep patterns prior to travel and avoiding light. Taking exogenous melatonin helps travelers to not only prepare for cycle changes, but also to recover from them as well. Resynchronization takes at least 1 day for each time zone traveled westward and slightly longer for each time zone traveled eastward.

Option B is the longest flight east of all the options and would cause the most jet lag.

30. C. Methacholine

Bronchoprovocation testing is a tool used to determine airway hyperresponsiveness. It may be used to evaluate for asthma and assess a worker's response to airway hyperresponsiveness treatment. It is conducted by obtaining a baseline FEV1 and then exposing the patient to a drug that causes bronchospasms. The most commonly used agent is methacholine, a cholinergic drug. As increasing concentrations of methacholine are administered, the FEV1 is monitored. Once the FEV1 falls $\geq 20\%$ from the baseline, the test is considered positive and the concentration is recorded.

31. E. Safety Data Sheet

Formerly known as the material safety data sheet, the safety data sheet (SDS) is a resource that communicates the hazards of chemical products in the workplace. The SDS is a required component of the Hazard Communication Standard (HCS).

32. E. Require all employees to wear earplugs

When implementing workplace safety, the preferred method is to solve the problem through engineering solutions. By engineering a solution, the problem is eliminated at the source. The second preferred method is to implement administrative solutions. When utilized appropriately, administrative controls remove workers from workplace hazards. Asking workers to wear personal protective equipment is the least preferred approach. This is because there is less control of the relationship between workers and the hazardous environment and hence more room for error.

33. **D. Medical Review Officer (MRO)**

MROs are physicians that undergo special training and receive certification to receive, review, interpret, verify, and report drug test results. With increasing evidence that drug free workplaces increase worker productivity and decrease work-related accidents, more companies are mandating drug testing for their employees.

MROs have a comprehensive understanding of substance abuse disorders and substances that may elicit false-positive results. With this knowledge, MROs are able to use clinical judgement to review positive drug test results and determine the odds of whether or not the positive result is due to drug use.

34. **B. Temporary partial disability**

This worker suffered a temporary partial disability due to impairment of her left hand. It is important to distinguish that an impairment refers to functional or anatomical loss, while a disability refers to limitation in performing activities. Disabilities may vary among people with the same impairment. For example, an injured hand would render a chef disabled, while the same injury would not disable a hostess.

There are four types of disabilities:

Temporary total disability—Worker is unable to work temporarily but recovery is expected.

Temporary partial disability—Worker is unable to perform usual duties but is capable of working in modified duty until impairment resolves.

Permanent total disability—Worker is unable to regain full employment in the same position and further treatment will not lead to recovery.

Permanent partial disability—Worker suffers permanent impairment. If possible, worker may work in modified duty.

35. **D. Permanent partial disability**

The four types of disability were explained in Answer #34 (directly above).

In this question, the audiologist has determined that the patient will not regain his hearing and has thus suffered a permanent injury. The miner is disabled because of his hearing ailment, but he is able to work and should be categorized as being partially disabled.

36. B. Eyes

Laser stands for light amplification by the stimulated emission of radiation. It is a nonionizing radiation that may cause thermal or mechanical damage to body tissues. Lasers are classified into four categories according to potential safety hazard, from least to most hazardous.

Eyes are the most sensitive to the effects of lasers. Affected regions of the eye include the cornea, lens and retina.

37. A. Lymphocyte proliferation test

Beryllium exposure can lead to both beryllium sensitization and chronic beryllium disease. A positive beryllium lymphocyte proliferation test without symptoms of lung disease is diagnostic for beryllium sensitization, while a positive test in the presence of granulomatous lung disease is diagnostic for chronic beryllium disease.

38. D. Silica

Chronic silica (aka quartz) inhalation is responsible for silicosis, a restrictive type of pneumoconiosis. Inhalation of silica can result from cutting, grinding, crushing, or drilling materials that contain silica. Once developed, silicosis cases increased risk of lung infection, cancer (especially in smokers), and restrictive lung disease. Symptoms of silicosis may develop many years after the last exposure to silica.

Chest X-rays in patients with silicosis may demonstrate opacities in the upper lobes and hilar lymphadenopathy with peripheral calcifications known as eggshell calcifications.

39. C. Byssinosis

Byssinosis assessment is mandated by OSHA's Cotton Dust Standard, where employees of cotton mills are to be assessed annually and the employer is to intervene if there is any sign of byssinosis.

40. E. Supportive treatment

Metal fume fever is the most common type of inhalation fever and is one of the most common occupational illnesses in welders. It occurs several hours after inhaling metal fumes and metal dusts, most commonly zinc. Symptoms are typically categorized as influenza-like and include thirst, chills, myalgias, shivering, headaches, and nausea. The disease is usually short-lived (resolving in 1–3 days) and self-limited. Diagnostic tools include chest X-rays, PFTs, and arterial blood gasses, all of which are typically within normal limits. Treatment is typically supportive. Clinical judgement must be made to rule out the development of chemical pneumonitis which can lead to fatal complications. If the inhaled metal is cadmium, clinicians should monitor renal function.

Polymer fume fever is another type of inhalation fever and occurs through inhalation of fluorocarbon polymers. Symptoms and treatment are the same as metal fume fever.

41. A. Inhalation hypoxia

There are four recognized types of hypoxia: (1) Histotoxic hypoxia—Results when cells become unable to use oxygen. Examples include cyanide and alcohol; (2) Hypemic hypoxia—Results when blood loses its oxygen carrying capacity. Examples include anemia and carbon monoxide poisoning; (3) Stagnant hypoxia—Results from inadequate blood flow. Examples include congestive heart failure, decompression sickness (DCS), and sustained acceleration forces where blood pools in dependent areas of the body; and (4) Hypoxic hypoxia—Results when there is poor gas exchange in the lungs. Examples include low levels of oxygen in the air and airway obstruction.

Inhalational hypoxia is not a recognized type of hypoxia.

42. C. Dye manufacturing

Aromatic amines are a known carcinogen for bladder cancer. These chemicals are used in the rubber and dye manufacturing industries.

43. Carpal tunnel syndrome is a syndrome resulting from compression of the median nerve. The median nerve innervates the anterior first–third and the lateral portion of the fourth digit in addition the thenar muscles (base of thumb) of the hand.

44. B. RBC cholinesterase

Organophosphate exposure inhibits acetylcholinesterase throughout the body, lowering the level of the enzyme. Because the enzyme is inhibited, endogenous acetylcholine is no longer broken down into choline and acetic acid. When acetylcholine builds up, symptoms occur. Because the cholinesterase variant found in the red blood cell (RBC) is the same as in the nerve synapse, the RBC variant is used as a marker for organophosphate exposure.

45. D. Working during pregnancy increases the overall risk of adverse pregnancy outcomes

Working during pregnancy has not been shown to increase risk for adverse pregnancy outcomes. There are factors that both increase the well-being of pregnant mothers and children, as well as factors that decrease well-being. Every pregnant woman should be screened for potential hazardous occupational risks.

Pregnancy produces physiological changes that effect the way in which the body interacts with toxins. Examples include increased body water leading to expanded volume of distribution, increased ventilation leading to enhanced absorption of volatile compounds, and increased GFR enhancing secretion from the kidneys. These physiologic changes occur at different stages of pregnancy.

The developing fetus also has varying susceptibilities depending on exposure. In the case of radiation, exposures that occur closure to the beginning of pregnancy (2–8 weeks) may lead to structural abnormalities and fetal death, while the same exposure later may lead to functional deficits.

46. D. Psoriatic lesion

Koebner phenomenon is when physical skin pressure precedes a psoriatic lesion at the same site of contact. It is more likely in those with a past medical history of psoriasis.

47. A. Alcohol

Trichloroethylene (TCE) is a common component found in degreasing solvents. Exposure occurs mainly through inhalation. Acute toxicity can lead to central nervous system suppression causing dizziness, drowsiness, nausea, and headaches. Workers exposed to this chemical experience degreaser's flush when they consume alcohol and the two chemicals compete for alcohol dehydrogenase (ADH) causing a disulfiram-like reaction.

48. A. Alcoholism is covered by the ADA

Alcohol addiction is considered a disability and is protected by the Americans with Disabilities Act (ADA) as long as the addicted person is capable of performing essential job tasks. Employers may prohibit use of alcohol in the workplace and have the ability to discipline workers in whom alcohol adversely effects job performance/conduct. Those addicted to illicit drugs are not covered by the ADA. Employers must provide reasonable accommodations to alcoholics but not those addicted to illicit drugs.

Under the ADA, employers are to provide reasonable accommodation to those that are disabled. However, the employer is not obligated to make accommodations that lead to undue hardship (significant difficulty or expense) of business function. Furthermore, an individual must be diagnosed with a disability, not merely regarded as disabled.

The ADA falls under control of the Department of Justice within the Civil Rights Division, and not within the Department of Labor (DOL).

49. B. 22-year-old heroin addict

The ADA protects individuals with qualified disabilities and those associated with individuals with qualified disabilities from discrimination. Qualified disabilities include physical or mental impairments that limit major life activities and those that carry social stigma. The ADA has specific titles to address employment, state and local government activities, public transportation, public accommodations, and telecommunications relay services. The ADA title addressing employment asserts that employers with 15 or more employees may not discriminate against a qualified individual that is capable of performing essential job functions with reasonable accommodations. It affords protection from discrimination in all employment practices (hiring, benefits, job duties, etc.). Employers may not ask about a disability prior to employment but may ask about a candidate's ability to perform specific job functions. The ADA does allow job offers contingent on satisfactory results of a post-offer medical exam as long as all employees vying for the same job must have the exam. Of the employees listed in the question, the only one that does not qualify for protection under the ADA is someone actively addicted to illicit drugs. However, the ADA does provide protection for those recovering from addiction to illicit drugs. The ADA does provide protection for people addicted to alcohol in addition to recovering alcoholics. An employer may discipline an employee whose use of alcohol adversely affects job performance. Pregnancy alone is not considered a disability under the ADA since it is not an impairment caused by a physiological disorder. However, complications arising from pregnancy may constitute impairments that are covered by the ADA. The Pregnancy Discrimination Act prohibits discrimination based on pregnancy status.

The ADA does not cover workers injured from temporary nonchronic conditions such as those with influenza or broken bones (DOL).

50. D. Employee assistance programs are primarily for problems that arise outside of work

Employee assistance programs (EAP) are confidential programs that allow employees to help resolve personal problems that arise outside of work. If an employee is able to resolve personal problems, his/her increased productivity will be beneficial to the employer. Although supervisors may recommend participation, employees must volunteer to participate in an EAP and cannot be coerced into doing so. EAPs are sponsored by the employer and may be conducted in-house or contracted with outside organizations.

51. B. Within 24 hours

Employers must report hospitalizations, amputations, and eye-loss injuries to the OSHA within 24 hours of learning about it. The short duration time allows OSHA to inspect the workplace before the problem has been corrected and interview other employees before they forget important events.

Reports may be submitted to OSHA by filling out the online form at www.OSHA.gov, calling the national toll-free OSHA Hotline, or calling/visiting the local OSHA office.

52. A. 300 Form

As described in OSHA recordkeeping regulation (29 CFR 1904), the OSHA 300 Form (collectively called the 300 log) is a standardized form for the reporting of all injuries and illness that result from employment. The form includes information to classify injuries and note the extent of each case. This mandatory form provides data used for legal, medical, and epidemiological purposes. At the end of the year, employers must create a summary of the year's 300 reports by completing the 300A form.

The Division of Workers' Compensation 25 form (DWC-25) is completed by the treating clinician to summarize a patient's treatment plan and workplace restrictions.

Workplace injury reports may be required by the employer but are not mandated by law.

Under the Hazard Communication Standard, chemical manufacturers and distributors must provide a SDS (formerly known as the MSDS) to explain the hazards of their chemical products to workers and clinicians.

53. B. History of hypertension

The Health Insurance Portability and Accountability Act (HIPAA) introduced sweeping legislation to increase protection of patient privacy and create safeguards of personal health information. However, 45 CFR 164.512(I) states that HIPAA privacy laws are different for workers' compensation insurers, workers' compensation administrative agencies, and employers. Health information relating to a workers' compensation claim may be shared with need-to-know parties and must be the minimum necessary amount of information needed to work through the workers' compensation claim. The employer and workers' compensation insurer do not need to know about a nonrelevant past medical history (such as hypertension) and this information is protected by HIPAA.

54. E. State government

Employees of state and local governments are not covered under the federal OSHA. Most states provide coverage under state programs. Besides state and local government employees, federal OSHA applies to all workers except for the following: Self-employed, immediate family members of farm employers that do not employ outside workers and agencies not afforded occupational regulation by other federal agencies such as the FAA and the Mine Safety and Health Administration.

55. E. 50

An employee at a private organization is covered under FMLA if the company has at least 50 employees in 20 or more workweeks in the current or previous year. Public employees qualify for FMLA regardless of the number of employees.

56. A. 0 weeks

Individual employees covered by FMLA may take up to 12 workweeks of leave in a 12-month period. These weeks are unpaid and may be taken for the following reasons; sickness or illness of employee, birth of a child, or care for a sick direct (sibling, parent, or child) family member. This does not include nephews or nieces.

Employees must observe their employer's requirements for requesting time off. They must also provide enough information for the employer to determine whether the request is reasonable and fits FMLA criteria.

57. B. 8 hour time-weighted average (TWA) of >85 decibels

OSHA has developed guidelines regulating permissible noise exposure for specific time durations as demonstrated in the table below.

Duration per day	Decibels (dB)
8 hours	90
6 hours	92
4 hours	95
2 hours	100
1 hour	105
15 minutes	115

If these levels are exceeded, OSHA mandates that administrative or engineering controls be utilized to lower exposure. If this does not work, personal protective equipment should be provided to employees.

Employees are to institute a hearing conservation program to employees if the 8 hour TWA of noise exposure to employees is 85 dB or higher. Employees are to be notified of the risk and should be allowed to have access to noise measurements. Furthermore, audiometric testing is to be made available to employees that exceed 85 dB per 8 hour TWA at no cost to them. Ideally, a baseline audiogram should be established within 6 months of exposure to this noise. An annual audiogram should then be performed which can then be compared to the baseline.

States and individual employers may design their own standards but may not be less stringent than OSHA's limitations.

58. A. 8 hours

Under OSHA standard number 1904.39, employers are to report the death of an employee within 8 hours of learning about it. If the fatality occurred more than 30 days prior, it does not need to be reported. This regulation does not apply for a motor vehicle accident in a public street or in a public transportation system.

59. E. Length of employment, plus 30 years

According to OSHA standard 1910.1020, medical records of each employee must be retained for the duration of employment, plus 30 years.

60. B. 15

Under the ADA, a private organization with 15 or more employees is required to provide ADA accommodations with the exception for cases of age discrimination (20 or more employees). Other exempt groups include American Indian tribes and bona fide private membership clubs. Partial exemption is given to the following: religious organizations, business on or near an Indian reservation, veterans' preference, and national security.

61. D. Yes: It should be recorded

Because the employee was injured on the employer's premises and was within the course of work activities, the injury is considered workers' compensation.

Being within the building of employment or on the time clock does not determine whether or not an injury should be covered under workers' compensation. The main consideration was whether or not an employee was acting within the scope of her job.

62. D. Maximum medical improvement (MMI)

MMI represents the point where a patient's medical impairment becomes so stable that additional medical intervention will not yield further improvement. If a patient reaches MMI without impairment, they are released from workers' compensation in the same condition they were in prior to the injury. If the worker does not reach 100% recovery, they are found to be impaired. Impairment rating is determined by standardized scales.

Physician input is mandated to perform the following stages of workers' compensation: determination of causality, direction of medical care, determination of work restrictions, MMI determination, and establishing impairment rating.

63. C. Federal Motor Carrier Safety Administration (FMCSA)

Truck drivers in every state are required to maintain Department of Transportation (DOT) certification through the FMCSA. The FMCSA is the branch of the DOT charged to prevent commercial motor vehicle-related fatalities and injuries specifically those involving large trucks and buses. It accomplishes this by developing standards to license commercial truck drivers, disseminating data to improve motor carrier safety, operating safety improvement programs, researching motor carrier safety, and financially supporting safety programs. Health care practitioners must complete training and certification to award certification to drivers. The maximum duration of driver DOT certification is 2 years.

The FMCSA is one of the many divisions within the department of transportation. Other divisions include the National Highway Traffic Safety Administration, Federal Aviation Administration (FAA), Federal Highway Administration, Pipeline and Hazardous Materials Safety Administration, FMCSA, Federal Railroad Administration, Federal Transit Administration, Maritime Administration and the Surface Transportation Board.

The ACGIH is a nonprofit organization composed of independent professionals aimed at defining the science of occupational and environmental health.

The OSHA is a division within the Department of Labor (DOL).

64. B. As small as possible

The volume of the breathing apparatus (hose and mask) increases the functional dead space of a pilot. A larger volume is associated with increased risk of rebreathing exhaled carbon dioxide. Furthermore, a larger dead space means the pilot must work harder to breathe air exchanged in the lungs. The work of breathing increases as the pilot experiences an increase in G-forces.

65. C. Adjusting the seating to sit straight up

Gravity forces (G-forces) can divert blood away from the brain, resulting in loss of consciousness (G-LOC). G-tolerance refers to either how much G-force can be achieved while remaining conscious or how long one can sustain a given gravity force level.

G-tolerance is multifactorial and differs for everyone. It can even vary daily for each individual. Factors such as dehydration and hypoglycemia worsen G-tolerance. G-tolerance can be increased by utilizing measures that maintain blood flow from the heart to the brain.

By reclining a seat, the vertical distance that blood must travel from the heart to the brain is reduced. For this reason, reclining a seat is a measure that helps increase G-tolerance. To the contrary, adjusting a seat to sit straight up does not help increase G-tolerance.

All of the other options listed will help increase G-tolerance. Wearing a G-suit helps maintain pressure within the body that allows blood to continue to reach the brain at higher G-forces. Multiple studies have demonstrated that anaerobic exercise programs can also help with G-tolerance. Using supplemental oxygen protects against hypoxemia which worsens G-tolerance. By saturating the blood with oxygen, there is extra assurance that the brain will receive more oxygen when under the stress of G-forces. Finally, performing a Valsalva maneuver at the right time will increase cardiac output and increase G-tolerance.

66. A. Boyle's law

This patient experienced discomfort from the expansion of gasses in her body due to increased altitude.

Boyle's law states that at constant temperature, the volume that gas occupies is inversely related to the pressure. Commercial airlines are pressurized to a minimum of 8000 ft above sea level. At this altitude, gasses expand roughly 30%. For this reason, it is recommended to use saline to expand endotracheal cuffs in intubated patients that will experience a change in altitude.

Charles' law states that when kept at a constant pressure, the volume of a gas is directly proportional to the temperature. As temperature increases, the volume of the gas increases and vice versa.

Dalton's law states that the total pressure of a gaseous mixture is equal to the sum of pressures applied by each gas in the mixture. This law is also referred to as the law of partial pressures.

Henry's law states that the amount of gas dissolved in a solution is directly proportional to the partial pressure of the gas over the solution. This law explains why nitrogen bubbles dissolve in body tissues of divers, causing DCS.

67. D. Henry's law

The gas laws were explained in Answer #66 (directly above).

68. D. Vision

Human sensory perception is best designed for terrestrial use. When in the air, sensory stimuli (visual, vestibular, and proprioceptive) experience conflicts and deviate from the norm. When the body attempts to put together these deviations, illusions and sensory mismatches occur. These mismatches lead to spatial disorientation. Traveling at great speeds further contributes to disorientation.

Vision is the most important sensory function for maintaining spatial orientation both on the ground and in the air. Vision may provide information about depth, distance, and speed. Most importantly, vision is important to use the horizon as a frame of reference. As opposed to other sensory stimuli, vision is less effected by linear acceleration, angular acceleration and changes in gravity.

Vision has shortcomings of its own. In fact, visual illusions are commonplace in flight. For example, consider a pilot descending toward a narrow runway. A narrow runway may lead a pilot to believe that he/she is at a higher altitude. The opposite happens for wide runways.

69. C. Administration of 100% oxygen

This patient is suffering from DCS, a condition caused by air bubbles forming in body tissues and fluids resulting from ascent from high pressure environments to low pressure environments.

Clues to knowing the answer to this question are that the patient was exposed to high atmospheric pressure while scuba diving underwater, followed by a low pressure atmospheric environment in the airplane. Chest pain, shortness of breath, joint pain, and visual changes are some of the many symptoms of DCS. The most appropriate immediate treatment listed is administration of oxygen. If the rural hospital had more resources available, another consideration would be hyperbaric therapy.

In this case, it would also be appropriate to evaluate for acute coronary syndrome and pneumothorax.

70. C. 3

At 33 ft deep in freshwater at ground level, one extra atmosphere of pressure is present. This is in addition to the one atmosphere that is present at ground level. Every 33 ft deeper into the water produces an extra atmosphere of pressure.

Because the diver in this scenario is 99 ft deep, he is experiencing three extra atmospheres of water in addition to the one atmosphere experienced at ground level.

The increased atmospheric pressure forces nitrogen from inhaled air to dissolve into body tissues and fluids. When the pressure is reduced, the nitrogen becomes less soluble. If the nitrogen solubility decreases too quickly, nitrogen bubbles will form in body tissues. In joint space synovial fluid, this causes pain. In the blood, these bubbles act as emboli. Air bubbles can also wreak havoc when present in cerebrospinal fluid. These phenomena are characteristic of DCS.

To prevent DCS, a diver should not fly immediately after diving. There is no consensus of specific lengths of time recommended before divers drive/fly to higher altitudes. It is known that those engaging in deeper dives should wait longer time prior to flying. An often cited period of time is waiting more than 24 hours to fly after a 10 m (33 ft) dive.

71. B. Intramuscular injection of promethazine

These crew members are suffering from the effects of space motion sickness (SMS). The symptoms of SMS resemble terrestrial motion sickness and occur when microgravity affects the homeostasis of the sensory systems. Microgravity causes fluids to shift cephalad, leading to headaches, back pain, and musculoskeletal changes, symptoms that are components of space adaption syndrome. SMS itself is considered a component of space adaption syndrome.

SMS is considered to be a natural response, affecting most astronauts for the first few days of flight. Symptoms include dyspepsia, headache, drowsiness, nausea, vomiting, pallor, sweating, dizziness, malaise, and loss of initiative. The constellation of SMS symptoms can span a spectrum from being a nuisance all the way to debilitating.

SMS cannot be predicted through ground tests. The best predictor is previous episodes of SMS. Prior to space flight, crew members should be educated on the symptoms of SMS and how to avoid offensive stimulus. To avoid offensive stimulus, astronauts should move slowly, maintain appropriate orientation by not flipping upside-down, staying hydrated, avoiding heat, and avoiding noxious odors. Some astronauts attempt to use prophylactic medications to limit SMS.

Eventually the body adopts to low gravity and normal body physiology returns without pharmaceutical treatment. Moderate and severe symptoms typically resolve by day 2. The preferred treatment for SMS is promethazine administered intramuscularly or in suppository form.

72. B. Isocyanates

Diisocyanates are a wide-ranging group of chemicals frequently called isocyanates. They are common in products such as paints, coatings, adhesives, and sealants. Automobile painters are regularly exposed to isocyanates.

The most common adverse effect of isocyanate exposure is asthma. Isocyanate exposure over a period spanning from months to years leads to sensitization that elicits asthma. Symptoms include cough, shortness of breath, chest tightness, and wheezing. Once sensitization occurs, low exposure to isocyanates can cause symptoms. These symptoms are characteristically temporally associated with isocyanate exposure. Avoidance of exposure improves symptoms.

Outside of asthma, other symptoms of isocyanate exposure include hypersensitivity pneumonitis, dermatitis, and rhinitis.

Carbon monoxide is byproduct of hydrocarbon combustion that decreases hemoglobin's ability to carry oxygen. Symptoms are often described as flu-like. More specifically, symptoms of carbon monoxide inhalation include weakness, dizziness, altered mental status, headache, nausea, and tachypnea

Manganese is common in welders and leads to neurological effects such as a parkinsonian syndrome.

Methacholine is a muscarinic receptor agonist that is commonly used to evaluate asthma. During a methacholine challenge, methacholine is inhaled a spirometry is performed to evaluate lung function. Methacholine is not a chemical that an automobile painter would be exposed to.

Nickel can be inhaled, ingested, or absorbed through the skin. When inhaled in large amounts, it may lead to chemical pneumonitis.

73. E. National Transportation Safety Board (NTSB)

The NTSB is a federal agency that investigates every civil aviation crash and significant crashes in other modes of transportation in the United States. During the investigation, the NTSB determines the probable cause of the crash and makes recommendations to prevent future crashes. Furthermore, the NTSB coordinates resources to assist those impacted by transportation disasters. Although the NTSB was originally founded within the Department of Transportation, it has since been removed and is now an independent federal agency that answers directly to congress.

The FAA may conduct investigations of airline accidents beneath the guide of the NTSB.

74. B. Federal Aviation Administration (FAA)

The FAA is a government agency with law making and enforcement capacity primarily concerned with airline safety. These penalties may be enforced through civil fines or certification action.

The FAA carries the ability to issue certifications and assure qualifications. The FAA certifies the following: airmen, aircrafts, air carriers, air navigation facilities, air agencies, airports, and designees (representatives of administrators). It may require inspections and modifications of already certified airplanes or updates to pilot education. The FAA also manages air traffic control, radio communications, national airports, and the aircraft documentation process.

The FAA will also conduct crash investigation in a subordinate role to the National Transportation Security Board (NTSB). It also investigates near misses in hopes to prevent future crashes.

Relevant to public health, the FAA must consult with the EPA to establish noise standards.

75. D. 25-year-old male with pneumothorax

The only patient with an absolute contraindication to flight is the patient with a pneumothorax. The pressure changes due to altitude have the ability to further shift the structures within the mediastinum, worsening injury. Less absolute restrictions are considered relative contraindications.

It is not recommended for infants less than one-week-old to fly because it is suggested to have a one-week period of observation to assure that the infant is otherwise healthy.

Pressure equalizing (myringotomy) tubes help disperse changes in pressure that are found at increased elevation and do not represent a contraindication to flight.

Patients with active tuberculosis are contraindicated to flight but those with latent tuberculosis infection are not restricted.

Women with uncomplicated pregnancy may fly, but it is discouraged to fly later in pregnancy due to lack of medical resources if delivery were to happen at altitude. Most commercial airlines allow women to fly up until their 36th week of pregnancy.

Individual airlines have the right to refuse transport of anyone they deem a medical risk. However, it is often up to the physician to determine if medical contraindications should keep a person on the ground.

Contraindication to flight

Cardiopulmonary	Baseline hypoxia ($PaO_2 < 70$ mmHg at sea level)
	Coronary artery bypass graft within 14 days
	Decompensated congestive heart failure
	Exacerbation of obstructive/restrictive lung disease
	Myocardial infarction in the prior 2–3 weeks
	Pneumothorax
	Uncontrolled angina
	Uncontrolled dysrhythmia
Neuropsychiatric	Cerebrovascular accident within 10 days
	Uncontrolled psychiatric state/acute psychosis
	Uncontrolled seizure disorder
Pregnancy	>32 weeks for multiple gestation or >36 weeks for single gestation
	Complicated pregnancy
Surgery	ENT, GI, or neurological surgery within 10 days
	Laparoscopic procedure within 5 days
Other	Anemia (<8.5 g/dL)
	Decompression sickness within 3–7 days
	Infant younger than 1 week
	Severe contagious illness
	Sickle cell disease exacerbation within 1 week

76. B. 92%

People with a sea level oxygen saturation less than 92% on room air should receive oxygen supplementation while flying. Furthermore, the Aerospace Medical Association (AsMA) recommends that passengers with a sea level room air $PAO_2 \leq 70$ mmHg receive supplemental oxygen.

77. C. 10th row window seat

Commercial airplanes typically use the laminar airflow pattern as demonstrated below in the airplane cross-section. In this diagram, the curved lines represent currents. These currents are primarily concealed between rows and are less fluid in the front to back direction. Therefore, passengers sitting in the same row have the highest likelihood of sharing respiratory droplets.

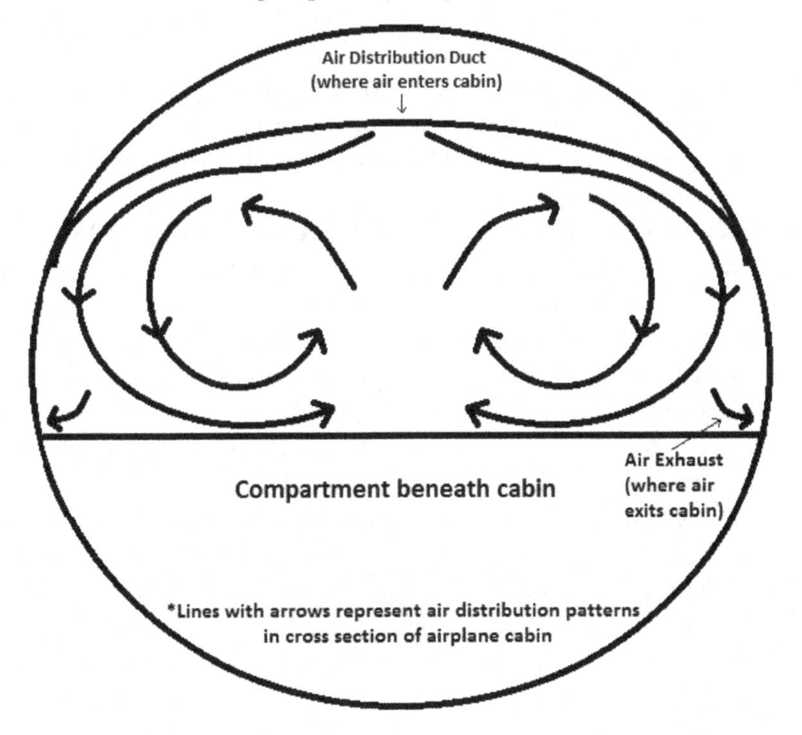

Aircraft engines pull bypass air into the cabin to help maintain pressurization. This air is treated through the AC system and dispersed through a vent in the ceiling. This newly vented air is combined with highly filtered recirculated cabin air. The air is removed from the cabin through outflow valves near the floor. There are more than 20 completed cabin air exchanges per hour, which is much more frequent than office buildings.

The high air exchange rate combined with efficient filtration reduces the risk of exposure to microbes and allergens. Due to the nature of disease transmission and difficulty in keeping track of passengers, it is difficult to study disease transmission. Studies do indicate that disease transmission does occur on planes which present a risk for international spread of disease. Disease transmission may be more likely to take place while the plane is grounded and passengers are boarding or disembarking, when airflow is decreased. During time on the tarmac, airflow to the cabin is reduced and airflow patterns change.

78. C. 8000 ft.

Although most commercial airliners in the United States cruise at an altitude of 30,000–40,000ft, they are required to maintain cabin pressurization to a pressure found at 8000 ft. Many airliners maintain a pressure found at 6000ft.

Depending on the condition of your patient, flying with sickle cell anemia may be contraindicated. Any patient with a resting oxygen saturation lower than 92% at sea level should be advised to fly with supplemental oxygen.

79. E. Occupational Safety and Health Administration (OSHA)

The ACOEM is a professional organization that serves as an academic home for professionals specializing in the fields of occupational and environmental medicine. ACOEM provides resources to advance the cause of preventive medicine and environmental medicine professionals.

The ACGIH is a nonprofit organization composed of independent professionals aimed at defining the science of occupational and environmental health. This organization publishes both threshold Limit Values (TLVs) and biological exposure indices (BEIs) as guidelines to use for well-informed decision-making in the occupational arena.

The EPA is the federal agency tasked with protecting the environment. It is the agency that publishes maximum contaminant levels (MCLs).

The NIOSH is a division within the CDC, which is within the Department of Health and Human Services (DHHS). NIOSH researches occupational problems and provides solutions and recommendations to protect the health of workers. NIOSH creates recommended exposure limits (RELs), which are not enforceable.

The OSHA is a branch within the Department of Labor (DOL) tasked with regulating the work environment to make it safe for employees. It has the power to enforce legislation. If an organization exceeds an OSHA PEL, OSHA may enforce penalties.

80. A. 1

The IARC, part of the WHO, developed the most widely used system to classify human carcinogens. It also performs studies on agents believed to be hazardous to humans and publishes monographs on each agent.

The IARC classification of agents is as follows:

Classification	Explanation
1	Carcinogenic to humans
2a	Probably carcinogenic to humans
	Sufficient evidence of carcinogenicity in animals but not humans
2b	Possibly carcinogenic to humans
	Less than sufficient evidence of carcinogenicity in animals, limited evidence of carcinogenicity in humans
3	Unclassifiable as to carcinogenicity in humans
4	Probably not carcinogenic to humans

81. D. L5–S1

Disk herniation	Nerve	Reflex	Weakness	Numbness
L3–L4	L4	Knee	Knee extension	Medial foot/ankle
L4–L5	L5	None	Dorsiflexion	Hallux
L5–S1	S1	Ankle	Plantar flexion	Lateral foot

BIBLIOGRAPHY

[1] Industrial Hygiene. OSHA training and reference materials library—industrial hygiene, <https://www.osha.gov/dte/library/industrial_hygiene/industrial_hygiene.html>; [accessed 15.10.2016].

[2] Grosse Y, Baan R, Straif K, et al. A review of human carcinogens—part A: pharmaceuticals. Lancet Oncol 2009;10(1):13–14 http://dx.doi.org/10.1016/s1470-2045(08)70286-9

[3] Bonnick SL. Bone densitometry in clinical practice: application and interpretation. 3rd ed Denton, TX: Springer Science & Business Media; 2010. p. 127–9.

[4] Acute radiation syndrome: a fact sheet for physicians. Emergency Preparedness and Response, <https://emergency.cdc.gov/radiation/arsphysicianfactsheet.asp>; [accessed 15.10.2016].

[5] Budd GM. Wet-bulb globe temperature (WBGT)—its history and its limitations. J Sci Med Sport 2008;11(1):20–32. Available from: http://dx.doi.org/10.1016/j.jsams.2007.07.003.

[6] Pinkerton KE, Rom WN. Global climate change and public health. New York: Springer Science & Business Media; 2014. p. 97–8.

[7] Sataloff RT, Sataloff J. Occupational hearing loss. 3rd ed. Boca Raton, FL: CRC Press; 2006. p. 415.

[8] Bansal M. Diseases of ear, nose, and throat. New Delhi, India: JP Medical; 2013. p. 137–48.

[9] Levy BS, Wegman DH, Baron SL, Sokas RK, editors. Occupational and environmental health: recognizing and preventing disease. 6th ed New York: Oxford University Press; 2011.

[10–11] CDC. About NIOSH. The National Institute for Occupational Safety and Health (NIOSH), <http://www.cdc.gov/niosh/about.html>; [accessed 15.10.2016].

[12] Enforcement. Workplace Violence, <https://www.osha.gov/SLTC/workplace-violence/standards.html>; [accessed 15.10.2016].

[13] Baxter PJ, Aw T-C, Cockcroft A, Durrington P, Harrington MJ. Hunter's diseases of occupations. 10th ed. Boca Raton, FL: CRC Press; 2010. p. 225.

[14] Bardal SK, Waechter JE, Martin DS. Applied pharmacology. St. Louis, MO: Elsevier Health Sciences; 2011. p. 122–3.

[15] Baxter PJ, Aw T-C, Cockcroft A, Durrington P, Harrington MJ. Hunter's diseases of occupations. 10th ed. Boca Raton, FL: CRC Press; 2010. p. 173–4.

[16] Smedley J, Dick F, Sadhra S, editors. Oxford handbook of occupational health. 2nd ed Oxford, UK: OUP Oxford; 2013.

[17] Tarlo SM, Cullinan P, Nemery B, editors. Occupational and environmental lung diseases. Hoboken, NJ: Wiley-Blackwell; 2010.

[18] Smedley J, Dick F, Sadhra S, editors. Oxford handbook of occupational health. 2nd ed Oxford, UK: OUP Oxford; 2013.

[19] Costabel U, et al. Interstitial lung disease chronic hypersensitivity pneumonitis. Clin Chest Med 2012;33:151–63.

[20] Reducing exposure to lead and noise at outdoor firing ranges. Workplace solutions from NIOSH, <http://www.cdc.gov/niosh/docs/wp-solutions/2013-104/pdfs/2013-104.pdf>; November 2012 [accessed 16.10.2016]. DHHS (NIOSH) Publication No. 2013 – 104.

[21–22] Levy BS, Wegman DH, Baron SL, Sokas RK, editors. Occupational and environmental health: recognizing and preventing disease.. 6th ed New York: Oxford University Press; 2011.

[23] Levy BS, Wegman DH, Baron SL, Sokas RK, editors. Occupational and environmental health: Recognizing and preventing disease. 6th ed New York: Oxford University Press; 2011.

[24] Seaton A, Agius R. Practical occupational medicine.. 2nd ed. London, England: CRC Press; 2005. p. 109–17.

[25] Wigley FM, Herrick AL, Flavahan NA. Raynaud's phenomenon: a guide to pathogenesis and treatment. New York, NY: Springer; 2015. p. 129–35.

[26] Selinus O, Alloway B, Centeno J, editors. Essentials of medical geology. Revised ed. New York: Springer Science & Business Media; 2013.

[27] Atkinson D, Overton J, Cavagin T. The sound production handbook. 1st ed Cambridge, UK: Psychology Press; 1995. p. 23–5.

[28] Bhattacharya A, McGlothlin JD. Occupational ergonomics: theory and applications. 2nd ed. Boca, Raton, FL: CRC Press; 2011. p. 452–64.

[29] Silverman D, Gendreau M. Medical issues associated with commercial flights. Lancet 2009;373(9680):2067–77 http://dx.doi.org/10.1016/s0140-6736(09)60209-9

[30] Heuer A, Scanlan CL. Wilkins' clinical assessment in respiratory care. 7th ed Maryland Heights, MO: Elsevier Health Sciences; 2014. p. 196.

[31a] Hazard communication standard: safety data sheets—OSHA 3514. OSHA Brief, <https://www.osha.gov/Publications/OSHA3514.html>; [accessed 16.10.2016].

[31b] The Hazard Communication Standard (HCS), [29 CFR 1910.1200(g)].

[32a] Hazard Prevention and Control, <https://www.osha.gov/SLTC/etools/safety-health/comp3.html>; [accessed 16.10.2016].

[32b] Rom WN, Markowitz SB. Environmental and occupational medicine. 4th ed Philadelphia, PA: Lippincott Williams & Wilkins; 2007. p. 902.

[33] Rom WN, Markowitz SB. Environmental and occupational medicine. 4th ed Philadelphia, PA: Lippincott Williams & Wilkins; 2007. p. 85.

[34–35] Guidotti TL, Arnold SM, Lukcso DG, Bender J. Occupational health services: a practical approach. 2nd ed New York: Routledge; 2013. p. 27.

[36] Baxter PJ, Aw T-C, Cockcroft A, Durrington P, Harrington MJ. Hunter's diseases of occupations. 10th ed. Boca Raton, FL: CRC Press; 2010. p. 652.

[37] Balmes JR, Abraham JL, Dweik RA, et al. An official American Thoracic Society statement: diagnosis and management of beryllium sensitivity and chronic beryllium disease. Am J Resp Crit Care Med 2014;190(10):e34–59. Available from: http://dx.doi.org/10.1164/rccm.201409-1722st.

[38] Steenland K, Ward E. Silica: a lung carcinogen. CA—Cancer J Clin 2013;64 (1):63–9. Available from: http://dx.doi.org/10.3322/caac.21214.

[39] Clark K, Cai B, Rao G, Svendsen ER. Byssinosis prevalence and intervention in a US millworker population. C52. Occupational lung disease. Am J Respir Crit Care Med 2011;183:A4805.

[40] Baxter PJ, Aw T-C, Cockcroft A, Durrington P, Harrington MJ. Hunter's diseases of occupations. 10th ed. Boca Raton, FL: CRC Press; 2010. p. 433–7.

[41] Barratt MR, Pool SL. Principles of clinical medicine for space flight. New York: Springer Science & Business Media; 2008. p. 451–2.

[42] Koh D, Takahashi K. Textbook of occupational medicine practice. 3rd ed Hackensack, NJ: World Scientific; 2011. p. 28.

[43] Hegmann K, editor. Occupational medicine practice guidelines. Evaluation and management of common health problems and functional recovery in workers. 3rd ed Elk Grove Village, IL: American College of Occupational and Environmental Medicine (ACOEM); 2011.

[44] Greenberg MI. Occupational, industrial, and environmental toxicology. 2nd ed Philadelphia, PA: Elsevier Health Sciences; 2003. p. 149.

[45] August P. In: Cohen WR, editor. Obstetric medicine: management of medical disorders in pregnancy. 6th ed Shelton, CT: McGraw-Hill Medical; 2014. p. 25–45.

[46] Kanerva L, Elsner P, Wahlberg J, Maibach H. Handbook of occupational dermatology. Heidelberg, Germany: Springer Science & Business Media; 2014. p. 273.

[47] Plant JA, Voulvoulis N, Ragnarsdottir VK, editors. Pollutants, human health and the environment: a risk based approach. 2012th ed. West Sussex, UK: John Wiley; 2012.

[48–49] . Division CR, Rights D. Americans with disabilities act questions and answers, <http://www.ada.gov/qandaeng.htm>; [accessed 16.10.2016].

[50] Fallon FL, Zgodzinski E. Essentials of public health management. 3rd ed Sudbury, MA: Jones & Barlett Learning; 2012. p. 215.

[51] Updates to OSHA's recordkeeping rule: reporting fatalities and severe injuries. OSHA Fact Sheet, <https://www.osha.gov/recordkeeping2014/OSHA3745.pdf>; September 2014 [accessed 16.10.2016].

[52] Work-related injuries and illnesses. OSHA forms for recording, <https://www.osha.gov/recordkeeping/RKform300pkg-fillable-enabled.pdf>; [accessed 16.10.2016].

[53a] When does the privacy rule allow covered entities to disclose information to law enforcement, <http://www.hhs.gov/ocr/privacy/hipaa/faq/disclosures_for_-law_enforcement_purposes/505.html>; [accessed 16.10.2016].

[53b] Sharing information related to mental health, <http://www.hhs.gov/hipaa/for-professionals/special-topics/mental-health/index.html>; [accessed 16.10.2016].

[54] OSHA worker rights and protections, <https://www.osha.gov/workers/index.html>; [accessed 16.10.2016].

[55–56] U.S. Department of Labor Wage and Hour Division. Fact sheet #28: The Family and Medical Leave Act, <https://www.dol.gov/whd/regs/compliance/whdfs28.pdf>; [accessed 16.10.2016].

[57] Occupational noise exposure, 1910.95 (2008).

[58] Updates to OSHA's recordkeeping rule: reporting fatalities and severe injuries, <https://www.osha.gov/recordkeeping2014/OSHA3745.pdf>; September 2014 [accessed 16.10.2016].

[59] Access to employee exposure and medical records, 1910.1020.

[60] Compliance manual chapter 2: threshold issues. EEOC Compliance Manual, <https://www.eeoc.gov/policy/docs/threshold.html#2-III-B-4-b-i>; [accessed 16.10.2016].

[61] OSHA Injury and Illness Recordkeeping: Q & A, <https://www.osha.gov/recordkeeping/faq_search/OSHAFAQ.pdf>; [accessed 16.10.2016].

[62] DeLisa JA, Gans BM, Walsh NE. 4th ed Physical medicine and rehabilitation: principles and practice, vol. 1. Philadelphia, PA: Lippincott Williams & Wilkins; 2005. p. 173–4.

[63a] Federal Motor Carrier Safety Administration, About us, <https://www.fmcsa.dot.gov/mission/about-us>; [accessed 16.10.2016].

[63b] Department of Transportation. About DOT, <https://www.transportation.gov/about>; [accessed 16.10.2016].

[64] Davis JR, Johnson R, Stepanek J. 4th ed Fundamentals of aerospace medicine, 39. Philadelphia, PA: Lippincott Williams & Wilkins; 2008. p. 92.

[65] Seedhouse E, Pulling G. Human responses to high and low gravity. New York, NY: Springer Science & Business Media; 2013. p. 196–207.

[66–67] Davis JR, Johnson R, Stepanek J. Fundamentals of aerospace medicine. 4th ed Philadelphia, PA: Lippincott Williams & Wilkins; 2008. p. 49–50.

[68] Antuñano MJ. Spatial disorientation—visual illusions. Federal Aviation Administration, <https://www.faa.gov/pilots/safety/pilotsafetybrochures/media/spatiald_visillus.pdf>; Updated February 2011 [accessed 16.10.2016].

[69] Hall JE. Pocket companion to Guyton & Hall textbook of medical physiology. 13th ed Philadelphia, PA: Elsevier Health Sciences; 2016. p. 574.

[70] Sanford CA, Pottinger PS, Jong EC. The travel and tropical medicine manual. 5th ed New York, NY: Elsevier; 2017. p. 152–3.

[71] Barratt MR, Pool SL. Principles of clinical medicine for space flight. New York: Springer Science & Business Media; 2008. p. 211–20.

[72] Rom WN, Markowitz SB. Environmental and occupational medicine. 4th ed Philadelphia, PA: Lippincott Williams & Wilkins; 2007. p. 509.

[73–74] Lawrence H. Aviation and the role of government. Dubuque, IA: Kendall/Hunt Publishing Company; 2004. p. 165–70.

[75–76] Silverman D, Gendreau M. Medical issues associated with commercial flights. Lancet 2009;373(9680):2067–77. Available from: http://dx.doi.org/10.1016/s0140-6736(09)60209-9.

[77] Davis JR, Johnson R, Stepanek J. Fundamentals of aerospace medicine. 4th ed Philadelphia, PA: Lippincott Williams & Wilkins; 2008. p. 466.

[78] Campion EW, Nable JV, Tupe CL, Gehle BD, Brady WJ. In-flight medical emergencies during commercial travel. New Engl J Med 2015;373(10):939–45. Available from: http://dx.doi.org/10.1056/nejmra1409213.

[79] Nims D. Basics of industrial hygiene. Danvers, MA: John Wiley; 1999 [chapter 3].

[80] Boslaugh S, editor. Encyclopedia of epidemiology, vol. 1. Thousand Oaks, CA: SAGE Publications; 2008. p. 116–7.

[81] Fiebach NH, Kern DE, Thomas PA, Ziegelstein RC, Barker L, Zieve PD, editors. Barker, Burton, and Zieve's principles of ambulatory medicine. 7th ed Philadelphia: Lippincott Williams & Wilkins; 2006.

CHAPTER SIX

Clinical Preventive Medicine

6.1 CLINICAL PREVENTIVE MEDICINE QUESTIONS

1. Which level(s) of prevention can be achieved with medical nutrition therapy (MNT)?
A. Primary
B. Secondary
C. Tertiary
D. Secondary and tertiary
E. Primary, secondary, and tertiary

2. A 56-year-old woman presents to your office for a routine health evaluation. During discussion, she mentions that her friend was recently diagnosed with breast cancer. Although she does not have any family members that have been diagnosed with breast cancer, she would like to be screened. She has never had a mammogram before and does not have any risk factors that increase her odds of having breast cancer.

What grade recommendation has the US Preventive Services Task Force (USPSTF) established for this patient to receive mammography to screen for breast cancer?
A. A
B. B
C. C
D. D
E. I

3. According to the USPSTF, after completing a negative mammogram, how long should a 56-year-old woman wait until completing her next mammogram?
A. One year
B. Two years
C. Three years
D. Five years

Board Review in Preventive Medicine and Public Health.
DOI: http://dx.doi.org/10.1016/B978-0-12-813778-9.00006-2

E. 10 years

4. A 77-year-old woman comes to your clinic for her annual examination. She reminds you that it has been over 10 years since she had her last colonoscopy. Her overall health is above average and she expects to live >10 years longer. She has a cousin that died of colorectal cancer (CRC), but nobody in her immediate family has been diagnosed with CRC. Furthermore, she has never had an abnormal finding in a colonoscopy.

Which USPSTF recommendation grade applies to this patient receiving a CRC screening?

A. A
B. B
C. C
D. D
E. I

5. A 46-year-old man with diabetes mellitus (DM) presents as a new patient to your clinic. What is the USPSTF recommendation to screen this patient for dyslipidemia?

A. A
B. B
C. C
D. D
E. I

6. A 54-year-old woman with a noncontributory past medical history presents to your clinic for an annual examination. She has no health complaints but would like to be screened for high cholesterol because she has never been screened before. This patient does not have any risk factors for cardiovascular disease (CVD).

What grade does the USPSTF assign for a person in this woman's position regarding cholesterol screening?

A. A
B. B
C. C
D. D
E. I

7. A 53-year-old woman presents to your clinic after witnessing her husband experience a stroke. She is concerned about her own well-being and would like to know if she should start taking a daily aspirin pill to reduce her risk. She does not use tobacco, does not drink alcohol, has no history of hypertension, and her body mass index (BMI) is 29.

What grade does the USPSTF assign for a person in this woman's position regarding the use of aspirin to prevent strokes?

A. A
B. B
C. C
D. D
E. I

8. A 72-year-old man with a family history of colon cancer presents to your office. He tells you that a friend recommended he take an aspirin daily to reduce his risk of developing colon cancer. What grade has the USPSTF published for this patient in regards to taking aspirin to reduce the risk of colon cancer?

A. A
B. B
C. C
D. D
E. I

9. Which of the following is not considered by the USPSTF when publishing a recommendation?

A. Accuracy of screening tests
B. Net benefit of screening
C. Financial costs
D. Net harm of screening
E. Morbidity and mortality of the target disease

10. Which grade is assigned to USPSTF recommendation where there is inconclusive evidence to make a recommendation?

A. B
B. C
C. D
D. E
E. I

11. Which of the following must hold true for the USPSTF to endorse screening for depression in adults?

A. The clinician is specially trained in screening for psychiatric illness
B. The patient must meet specific criteria
C. There must be treatment available

D. A friend or family member must be present with the patient
E. The clinician determines that the patient would actually benefit from therapy

12. At what age does the USPSTF recommend that clinicians begin to screen patients for obesity?
A. Three years
B. Six years
C. 10 years
D. 15 years
E. 18 years

13. A 71 year old asymptomatic white man has asked his primary care doctor to screen his risk of prostate cancer with a prostate-specific antigen (PSA) test. What is the US Preventive Services Task Force's recommendation for this man to receive PSA screening?
A. A
B. B
C. C
D. D
E. I

14. A healthy 28-year-old White man presents to a physician for a well check. He has not seen a doctor for over eight years since he had an upper respiratory infection. He has no significant family history. Which of the following is not recommended?
A. Annual influenza vaccination
B. Blood pressure screen
C. Alcohol screening
D. Testicular cancer screening
E. HIV screening

15. Providing interventions to reduce tobacco use in children and adolescents is a USPSTF grade ______ recommendation.
A. A
B. B
C. C
D. D
E. I

16. As a physician in a state correctional facility, which of the following scenarios is considered to be an ethical practice?
A. Administering punishment by lethal injection to one of your patients

B. Drawing blood to collect genetic evidence for a murder trial on one of your patients
C. Reporting a patient that plans to murder another inmate
D. Performing a cavity search on your patient for the purpose of finding contraband
E. All of the above actions are considered ethical

17. Your patient is a 75-year-old woman that quit smoking 11 years ago after smoking one pack a day for 50 years. She had a negative CT scan of her chest to screen for lung cancer one year ago. Using a CT scan to screen for lung cancer in this patient is a USPSTF grade _____ recommendation?
A. A
B. B
C. C
D. D
E. I

18. A 25-year-old woman comes to your office for cervical cancer screening. She is married and her husband is her only sexual partner of the past five years. Her last cervical cancer screening was three years ago and was negative for abnormal cytology. She read online that the best way to screen for cancer is to have human papillomavirus (HPV) testing performed.

As a primary care physician, what should be your response?
A. The USPSTF supports her having HPV testing and not a Pap smear
B. The USPSTF supports her having HPV testing if she has a Pap smear
C. The USPSTF does not support her having HPV testing at this time
D. The USPSTF does not support her having HPV testing or a Pap smear
E. A combination of two or more of the above answers

19. After watching a show hosted by physicians on daytime television, an asymptomatic 60-year-old man presents to your office requesting a cardiac examination to screen for coronary artery disease (CAD). What grade recommendation is it for clinicians to screen for CAD in asymptomatic patients without increased risk?
A. A
B. B
C. C
D. D
E. I

20. What is the number one cause of death worldwide?
A. Cancer
B. CVD
C. Malaria
D. Trauma
E. Respiratory disease

21. According to the Eighth Joint National Committee (JNC 8), at what age does the target systolic pressure jump from 140 to 150 in those without diabetes or kidney disease?
A. 55
B. 60
C. 65
D. 70
E. 75

22. Which of the following causes the most deaths in the United States annually?
A. Cancer
B. Chronic obstructive pulmonary disease (COPD)
C. Heart disease
D. Stroke
E. Unintentional injuries

23. According to JNC 8, what is the maximum blood pressure for a 75-year-old Black man with DM and renal insufficiency that would not indicate a need for treatment (or change in treatment) for hypertension?
A. 120/80
B. 130/80
C. 140/80
D. 140/90
E. 150/90

24. JNC 8 recommends that Black men and women begin treatment for hypertension with either of which two medications?
A. Calcium channel blockers and ACE inhibitors
B. Calcium channel blocker and thiazide diuretic
C. Thiazide diuretics and ACE inhibitors
D. ACE inhibitors and loop diuretics
E. Thiazide diuretic and beta blocker

25. Approximately what percentage of hypertensive adults in the United States are adequately controlled?
A. 55%
B. 65%

C. 75%
D. 85%
E. 95%

26. What is the leading cause of CVD mortality in the United States?
A. High blood pressure
B. Smoking
C. Poor diet
D. Insufficient physical activity
E. Abnormal serum glucose

27. What is the most common form of hypertension?
A. Cushing's syndrome
B. Essential
C. Pheochromocytoma
D. Primary hyperaldosteronism
E. Renal artery stenosis

28. A primary care doctor has been unsuccessfully treating his hypertensive patient for several years with diet and exercise. The last three readings have been 152/100, 148/108, and 157/104. Which stage of hypertension is this patient classified as having?
A. Normal
B. Prehypertension
C. Stage one hypertension
D. Stage two hypertension
E. Stage three hypertension

29. Which of the following is the modifiable risk factor that carries the highest risk of hypertension?
A. Age
B. Family history
C. High-potassium diet
D. Physical inactivity
E. Sex

30. Which of the following does NOT fit criteria for metabolic syndrome?
A. A man with a waist circumference of 42 inches
B. A man with a serum triglyceride level of 174 mg/dL
C. A man with an HDL cholesterol of 35 mg/dL
D. A man with a blood pressure of 125/80 mmHg
E. A man with a fasting glucose of 115 mg/dL

31. Which of the following groups is burdened with the highest prevalence of asthma?
A. Asian
B. Black
C. Hispanic
D. Pacific Islander
E. White

32. COPD is most commonly caused by which risk factor?
A. α1-Antitrypsin
B. Cigarette exposure
C. Environmental air pollution
D. Lead exposure
E. Personal history of asthma

33. Which organ is transplanted most frequently on an annual basis in the United States?
A. Heart
B. Kidney
C. Liver
D. Lung
E. Pancreas

34. Approximately what percentage of cancers are due to environmental and lifestyle factors?
A. 10%–15%
B. 25%–30%
C. 35%–40%
D. 55%–60%
E. >75%

35. Which is the correct sequence of the multistage process of carcinogenesis?
A. Initiation → Progression → Promotion
B. Initiation → Promotion → Progression
C. Progression → Promotion → Initiation
D. Promotion → Initiation → Progression
E. Promotion → Progression → Initiation

36. What is the approximate lifetime risk of being diagnosed with cancer?
A. 10%
B. 20%
C. 30%

D. 40%
E. 50%

37. What is the approximate lifetime risk of dying from cancer?
A. 5%
B. 10%
C. 15%
D. 20%
E. 25%

38. Which of the following risk factors is responsible for the most cases of bladder cancer?
A. Arsenic exposure
B. Birth defect
C. Occupational exposure
D. Radiation exposure
E. Tobacco smoke

39. Which of the following is a risk factor for breast cancer?
A. Birthing multiple children
B. Early menopause
C. Hormone replacement therapy
D. Late menarche
E. Physical activity

40. Diethylstilbestrol (DES) has been shown to protect against miscarriages. Which adverse effect caused it to be pulled from the market?
A. Bacteremia in the person taking DES due to immunocompromised status
B. Clear-cell adenocarcinoma of the cervix in daughters of mothers that take DES while pregnant
C. Congestive heart failure of the person taking DES
D. Renal cancer of the person taking DES
E. Severe risk of liver failure due to multiple drug interactions with DES

41. Which of the following types of HPV infection is most frequently associated with cervical cancer?
A. 1
B. 2
C. 6
D. 11
E. 16

42. At what age is it no longer recommended that women with a cervix and uterus, without risk factors for cervical cancer, stop receiving Pap smears?

A. 60
B. 65
C. 70
D. 75
E. 80

43. Which of the following risk factors is the number one cause of liver cancer in the United States?

A. Aflatoxin B
B. Alcohol
C. Hepatitis B
D. Hepatitis C
E. Liver flukes

44. Which of the following is a known risk factor for colorectal cancer?

A. Consumption of beef
B. Consumption of chicken
C. Consumption of fish
D. High-fiber diet
E. All of the above

45. Which of the following is not true of stomach cancer?

A. *Helicobacter pylori* presence is a risk factor for stomach cancer
B. Obesity is a risk factor for stomach cancer
C. Salted and preserved foods are protective of stomach cancer
D. Smoking is a risk factor for stomach cancer
E. Stomach cancer is about twice as common in men than women

46. Which of the following is not a risk factor for pancreatic cancer?

A. Age
B. History of chronic pancreatitis
C. DM
D. Family history
E. Low-protein diet

47. Which of the following cancers has the lowest five-year survival rate after diagnosis?

A. Breast cancer
B. Colon cancer
C. Lung cancer

D. Pancreatic cancer
E. Prostate cancer

48. A concerned parent presents to your office with her five-year-old child to show you a bruise on her medial thigh. She read on the internet that it could be a sign of cancer. Which of the following pediatric cancers are most common in children between the ages of one to 14?
A. Leukemia
B. Lymphoma
C. Osteosarcoma
D. Rhabdomyosarcoma
E. Wilms tumor

49. Which type of cancer is the leading cause of death in the United States?
A. Breast
B. Colon
C. Lung
D. Pancreas
E. Prostate

50. After smoking, what is the next most common cause of lung cancer in the United States?
A. Alcohol
B. Asbestos
C. Atmospheric nitrogen
D. Fluorocarbons
E. Radon

51. Which of the following is not a risk factor for cancers of the oral cavity?
A. Alcohol
B. HPV
C. Smokeless Tobacco
D. Smoking Tobacco
E. All of the above are risk factors for cancers of the oral cavity

52. When is it recommended to screen for genetic risk factors of ovarian cancer?
A. ≥ 50 years of age
B. ≥ 50 pack-year smoker
C. Ashkenazi Jewish heritage but no family members with history of cancer

D. Positive history of breast cancer in one sister and ovarian cancer in another sister
E. Answers C and D

53. A 55-year-old woman with a medical history only significant for osteoporosis presents to your clinic. She has had a headache for several days and has read online that it could be ovarian cancer. What grade is the USPSTF position on screening asymptomatic women for the presence of ovarian cancer?
A. A
B. B
C. C
D. D
E. I

54. What type of cancer is staged with a Gleason score?
A. Colon
B. Lung
C. Ovarian
D. Prostate
E. Skin

55. Which type of cancer is most frequently diagnosed in the United States?
A. Lung
B. Breast
C. Skin
D. Prostate
E. Colon

56. Which type of skin cancer accounts for the most skin cancer deaths in the United States annually?
A. Basal cell carcinoma
B. Dermatofibroma
C. Keratoacanthoma
D. Melanoma
E. Squamous cell carcinoma

57. A 75-year-old former roofer presents to your clinic with concerns about a black mole that has been growing on his arm. You decide to take this opportunity to educate him about the ABCs of dysplastic nevi and melanoma detection. What are the ABCs of melanoma detection?
A. Area, bleeding, chronic, dystrophic, elderly
B. Asymmetry, border, color, diameter, enlargement

C. Atopic, barrier, count, definition, erythema
D. Acute, bleeding, congenital, disseminate, epidermal
E. Altered, biopsy, circumscribed, dark, earned

58. As part of a study, a clinician in the United States pulls five patients off the street to screen for testicular cancer. Which of the following people carries the highest risk of testing positive for testicular cancer?
A. 13-year-old Hispanic man with acne
B. 20-year-old White man with HIV
C. 42-year-old Black man with depression
D. 50-year-old Asian man with cataracts
E. 58-year-old mixed race (White and Black) man with diabetes

59. A 24-year-old man presents to your clinic to have his testicles screened for cancer. He has no personal risk factors for testicular cancer and a negative family history for any type of cancer. Which response would be most likely to be endorsed by the USPSTF?
A. Yes. It is appropriate to screen men of all ages for testicular cancer.
B. Yes. It is appropriate to screen men aged 18–24 for testicular cancer.
C. No. It is only appropriate to screen men with a positive history of cancer for testicular cancer.
D. No. It is not appropriate to screen men of any age for testicular cancer.
E. The USPSTF does not have a stance on testicular cancer screening.

60. Which of the following risk factors has been linked with the highest risk of causing thyroid cancer?
A. Alcohol consumption
B. Head/neck radiation
C. Obesity
D. Tobacco use
E. All of the above significantly increase risk of thyroid cancer

61. Which is the most common type of kidney cancer?
A. Renal adenoma
B. Renal cell carcinoma
C. Renal sarcoma
D. Transitional cell carcinoma
E. Wilms tumor

62. Which organization serves as the Organ Procurement and Transplantation Network (OPTN) in the United States?
A. American College of Surgeons (ACS)
B. American Organ Matching Network (AOMN)
C. Coordinated Organ Sharing System (COSS)
D. Federal Organ Rationing System (FORS)
E. United Network for Organ Sharing (UNOS)

63. What is the prevalence of adults in the United States living with DM?
A. 5%
B. 10%
C. 15%
D. 20%
E. 25%

64. What is the most common cause of chronic kidney disease (CKD) in the United States?
A. Cancer
B. DM
C. Shock
D. Hypertension
E. Obesity

65. A mother brings her nine-year-old child to your clinic for evaluation. She believes that his video game habits and poor diet have caused him to become overweight. He is 67 inches tall and weighs 130 pounds. His BMI is measured to be 21 kg/m^2, which places him in the 93rd percentile.
How is this child's weight categorized?
A. Underweight
B. Healthy weight
C. Overweight
D. Obese
E. Not enough information

66. If a patient consumes 200 calories of carbohydrates, how many grams of carbohydrates did he/she eat?
A. 22
B. 29
C. 37
D. 44
E. 50

67. What is the maximum amount of sodium that a person without any risk factors should consume daily?
A. 3000 mg
B. 2300 mg
C. 2000 mg
D. 1500 mg
E. 1000 mg

68. During an appointment, your 68-year-old patient asks how much sodium he should be consuming daily. He has hypertension but no other health conditions.

How much sodium should he be consuming daily?
A. 3000 mg
B. 2300 mg
C. 2000 mg
D. 1500 mg
E. 1000 mg

69. What is the recommended maximum daily limit of dietary cholesterol?
A. 100 mg
B. 150 mg
C. 200 mg
D. 250 mg
E. 300 mg

70. In the national 5-2-1-0 model to combat childhood obesity, the numbers stand for units of measurement for risk factors that contribute to obesity. What does the 0 stand for?
A. Vegetables
B. Screen time
C. Physical activity
D. Sugary drinks
E. Books

71. The DASH diet is used to manage which of the following?
A. Blood pressure
B. Cognitive ability
C. Kidney function
D. Muscle mass
E. Osteoporosis

72. Which of the following risk factors is protective against osteoporosis?
A. Alcohol
B. Caffeine
C. Lack of physical activity
D. Obesity
E. Tobacco

73. A 65-year-old woman presents to your office for treatment after being told by her OBGYN that her dual-energy X-ray absorptiometry (DXA) Scan score was −3.1. What is the name of the ailment that the patient is seeking treatment for?
A. Cervical intraepithelial neoplasia
B. Fibrocystic breasts
C. Menopause
D. Osteopenia
E. Osteoporosis

74. Which type of fracture is most commonly experienced in patients with osteoporosis?
A. Carpal
B. Hip (femur)
C. Rib
D. Tarsal
E. Vertebra

75. A deficiency of which vitamin causes pellagra?
A. Vitamin A
B. Vitamin B
C. Vitamin C
D. Vitamin D
E. Vitamin E

76. What is the most common musculoskeletal ailment worldwide?
A. Ankle sprain
B. Back pain
C. Carpal tunnel syndrome
D. Knee arthritis
E. Rib contusion

77. What is the most common cause of disability among adults in the United States?
A. Arthritis
B. Cancer
C. Diabetes

D. Cardiomyopathy
E. Respiratory problems

78. Which of the following is the most common cause of traumatic brain injury in the United States?
A. Sports injuries
B. Falls
C. Stroke
D. Motor vehicle accidents
E. Domestic violence

79. In the United States, approximately what percentage of adults aged >65 years do not have a single natural tooth remaining?
A. 5%
B. 10%
C. 20%
D. 30%
E. 40%

80. Which of the following conditions is responsible for the most cases of blindness worldwide?
A. Cataracts
B. Diabetes
C. Glaucoma
D. Nutritional deficiency
E. Trachoma

81. A woman who has recently emigrated from rural Mexico discovers that she is having trouble with her night vision. When she presents to her physician for a workup to solve the problem, she reveals to him that she is struggling with alcoholism and had a miscarriage several weeks ago. On physical examination, her eyes are very dry and there is a small ulcer in the temporal side of her left cornea. The physician suggests that she join Alcoholic Anonymous (AA) to help resolve her alcohol addiction. He also suggests that she begin taking vitamin __ to improve her visual acuity. After she starts the vitamin, she notices improvements in her vision and begins to take double the dose, even though she starts to notice cracking of her lips and fingernails. She has a child in the following year and is shocked to see that congenital birth defects are present.

Which vitamin did the patient begin to take?
A. Vitamin A
B. Vitamin B_6
C. Vitamin C

D. Vitamin D
E. Vitamin E

82. In the United States, approximately what percentage of pregnancies are unplanned?
A. 20%
B. 30%
C. 40%
D. 50%
E. 60%

83. Sudden infant death syndrome (SIDS) is defined as the unexplainable death of a child in which age range?
A. 0–6 months
B. 0–12 months
C. 0–18 months
D. 0–24 months
E. 0–30 months

84. For a mother and infant living in the United States, under which condition is it appropriate for a mother to breastfeed her infant through direct (mouth-to-nipple) nursing?
A. Infant has galactosemia
B. Infant has a respiratory infection
C. Mother has active, untreated tuberculosis
D. Mother has HIV, but her viral load is not detectable
E. Mother has active varicella infection of her nipple

85. A 36-week pregnant Black woman presents to your clinic. She is graduating high school in a couple weeks and has heard from her friends that breastfeeding will negatively affect her body. Which of the following is true regarding outcomes of a breastfeeding?
A. Breastfeeding is associated with higher risk of breast cancer
B. Breastfeeding is associated with higher risk of postpartum depression
C. Breastfeeding is associated with a slower return to prepregnancy body weight
D. Breastfeeding is associated with lower risk of ovarian cancer
E. Breastfeeding makes it easier to get pregnant during the time one is breastfeeding

86. Which of the following is not a benefit of breastfeeding for infants?
A. Reduced incidence of asthma
B. Reduced risk of leukemia
C. Reduced risk of gastrointestinal infections

D. Reduction in child, adolescent, and adult obesity
E. All of the above are benefits of breastfeeding

87. In the United States, which of the following is mandated to be enriched with folic acid?
A. Cereal grain
B. Flour
C. Oral contraceptive pills
D. Rice
E. Salt

88. How much folic acid is recommended for daily consumption in women of childbearing age without a history of offspring with a neural tube defect?
A. 0.4 mg
B. 4 mg
C. 40 mg
D. 400 mg
E. 4000 mg

89. What is the leading cause of neonatal mortality in the United States?
A. Asphyxia
B. Diarrheal illness
C. Pneumonia
D. Preterm birth and related complications
E. Sepsis

90. A 15 week pregnant women asks you what constitutes preterm pregnancy.
How many completed weeks of gestation constitutes full-term delivery?
A. 34
B. 35
C. 36
D. 37
E. 38

91. A young married couple presents to your clinic concerned that they are unable to conceive a child. They are hoping that an official diagnosis of infertility will help them receive benefits from their insurance company. Both are overweight and smoke cigarettes.
How long must a couple try to conceive a child until they may be diagnosed with infertility?
A. 6 months
B. 12 months

C. 18 months
D. 24 months
E. 36 months

92. A 4-year-old child that was born in Africa presents to your clinic. After an exhaustive workup for skin changes and developmental delays, she is diagnosed with phenylketonuria (PKU). If she were born in the United States, when should she be screened?
A. Time of birth
B. Two-years old
C. Five-years old
D. Only if clinically indicated
E. Never

93. Which of the following risk factors has not been associated with an increased risk of lung cancer?
A. β-Carotene supplementation in smokers
B. Vitamin E supplements
C. Heavy alcohol consumption
D. Air pollution
E. HIV infection

94. Which organization publishes the Diagnostic and Statistical Manual of Mental Disorders (DSM) guidelines?
A. American Hospital Association
B. American Medical Association
C. American Psychiatric Association
D. American Psychological Association
E. World Health Organization

95. In the United States, which DSM diagnosis carries the highest lifetime prevalence?
A. Anxiety
B. Impulse control disorders
C. Mood disorder
D. Substance abuse disorder
E. None of the above

96. Approximately what percentage of Americans meet the criteria for a mental disorder in their lifetime?
A. 10%
B. 20%
C. 35%
D. 50%
E. 65%

97. Which is the most common type of dementia in the United States?

A. Alcohol
B. Alzheimer's
C. Lewy body
D. Parkinson related
E. Vascular

98. A 31-year-old woman presents to your clinic for amenorrhea. After a series of questioning, she states that she typically drinks "a couple glasses" of wine daily. When prying further, you learn that the glasses she is referring to are actually 32 ounces. She does not feel guilty about her drinking, but does feel that she should cut it down a little bit. When her friends tell her that she should drink less, she gets annoyed by their criticism. She does not think that she is an alcoholic because she never needs to have a drink in the morning as an eye opener. She is employed full time and lives alone.

What is this patient's CAGE score?

A. 1
B. 2
C. 3
D. 4
E. 5

99. Which age group of men has the highest "success" rate of suicide in the United States?

A. 15–24
B. 25–44
C. 45–64
D. 65–74
E. ≥75

100. Which of the following statements comparing suicide in men and women is true?

A. Men attempt suicide more frequently and have a higher "success" rate of suicide
B. Men attempt suicide more frequently and have a lower "success" rate of suicide
C. Men attempt suicide less frequently and have a higher "success" rate of suicide
D. Men attempt suicide less frequently and have a lower "success" rate of suicide
E. None of the above

101. Which of the following regarding autism is false?
A. All children with autism spectrum disorder have intellectual disability
B. Autism may be diagnosed as young as two years of age
C. Genetic variations may cause autism
D. Males are at higher risk of autism than females
E. Vaccines do not cause autism

102. Which of the following is causes the most disability adjusted life years (DALYs) worldwide?
A. Depression
B. Heart disease
C. HIV/AIDS
D. Low back pain
E. Premature birth complications

103. Which of the following causes the most of years lived with disability (YLD) worldwide?
A. Mental illness (including depression)
B. Diabetes
C. Falls
D. Hearing loss
E. Iron-deficiency anemia

104. Approximately what percentage of high school students have used tobacco products in the previous 30 days?
A. 10%
B. 15%
C. 20%
D. 25%
E. 30%

105. Which is the most commonly used form of tobacco used in the past 30 days by high school students?
A. Cigarettes
B. Cigars
C. Electronic cigarettes
D. Hookah pipes
E. Smokeless tobacco

106. Among men $\geq$ 18 years old in the United States, which group has the lowest prevalence of cigarette smokers?
A. Asian
B. Hispanic
C. Non-Hispanic Black

D. Non-Hispanic White
E. American Indian/Alaska Native

107. In the United States, what percentage of cigarette smokers try a cigarette prior to their 18th birthday?
A. 50%
B. 60%
C. 70%
D. 80%
E. 90%

108. Which of the following is true regarding mental illness and tobacco smoking?
A. Adults with mental illness are equally as likely to be smokers as those without mental illness
B. Adults without psychological distress have greater odds of tobacco use than those with distress
C. Cigarette smokers with mental illness smoke more per day than those without mental illness
D. Smoking rates in mentally ill population is decreasing as fast as in the population without mental illness
E. None of the above are true

109. Which of the following is not a risk factor for alcohol abuse?
A. Antisocial personality
B. Cultural background of abuse
C. Parental alcohol abuse
D. High social support
E. All of the above are risk factors for alcohol abuse

110. What is the definition of binge drinking in women?
A. Two drinks in one occasion
B. Three drinks in one occasion
C. Four drinks in one occasion
D. Five drinks in one occasion
E. Six drinks in one occasion

111. Which of the following is NOT one of the "5 A's" used to improve smoking cessation rates?
A. Ask
B. Advise
C. Assess
D. Assist
E. Alter

112. What is the leading method of suicide in men?
- **A.** Drowning
- **B.** Firearms
- **C.** Hanging
- **D.** Poisoning
- **E.** Other

113. The Physical Activity Guidelines Advisory Committee recommends that adults engage in how much physical activity on a weekly basis?
- **A.** 180 min of moderate-intensity activity or 90 min of vigorous-intense activity or a combination of both
- **B.** 150 min of moderate-intensity activity or 75 min of vigorous-intense activity or a combination of both
- **C.** 100 min of moderate-intensity activity or 60 min of vigorous-intense activity or a combination of both
- **D.** 80 min of moderate-intensity activity or 50 min of vigorous-intense activity or a combination of both
- **E.** 50 min of moderate-intensity activity or 30 min of vigorous-intense activity or a combination of both

114. All types of screening are exclusively a form of secondary prevention except for which of the following?
- **A.** Abdominal ultrasound for aortic aneurysm in elderly smokers
- **B.** Colonoscopy for CRC
- **C.** Pap smear for cervical cancer
- **D.** Purified protein derivative (PPD) for tuberculosis
- **E.** Serum cholesterol for atherosclerosis

115. Which of the following is an example of secondary prevention?
- **A.** Immunization against HPV in a person that has not been sexually active
- **B.** Providing fluoride treatment in all children that mainly drink well water
- **C.** Practising healthy diet and exercise
- **D.** Taking a statin drug to maintain appropriate cholesterol
- **E.** Wearing a condom to prevent unwanted pregnancy

116. Which of the following is not considered for a person awaiting organ transplantation?
- **A.** Probability of a successful outcome
- **B.** Reason organ is needed
- **C.** Severity of illness

D. Time on the waiting list
E. All of the above options are considered

117. Which national organization recommends core newborn screening tests to the states?
A. Advisory Committee on Heritable Disorders in Newborns and Children
B. American Academy of Pediatrics
C. Centers for Disease Control and Prevention
D. National Advisory Group for Newborn Disease Prevention
E. USPSTF

118. Which congenital disorder leading to intellectual disability is most likely to be found in an area with iodine deficiency?
A. Congenital hypothyroidism
B. Fragile X syndrome
C. PKU
D. Spina bifida
E. Trisomy 21

119. Which of the following diseases is strongly associated with Ashkenazi Jewish heritage?
A. Congenital hypothyroidism
B. Cystic fibrosis
C. Down syndrome
D. Sickle cell anemia
E. Thalassemia

120. Which of the following parties is most commonly found to be the perpetrator of child abuse?
A. Daycare provider or babysitter
B. Father
C. Friend of family
D. Mother
E. Romantic partner of parent

121. Which of the following is not a core newborn screening test that is recommended by the Advisory Committee on Heritable Disorders in Newborns and Children?
A. Anencephaly
B. Congenital adrenal hyperplasia
C. Cystic fibrosis
D. Primary congenital hypothyroidism
E. Sickle cell anemia

122. Which of the following statements regarding intimate partner violence (IPV) is not true?
- **A.** Roughly one in four women have experienced severe physical violence by an intimate partner
- **B.** Survivors of IPV experience long-term physical and mental health consequences
- **C.** Most perpetrators of rape of a female are unknown by the victim
- **D.** Most victims of sexual assault are under the age of 25
- **E.** Most cases of rape of a man are performed by a man

123. Which of the following is false regarding IPV during pregnancy in the United States?
- **A.** IPV is the leading cause of injuries among US women of child-bearing age
- **B.** IPV in pregnancy is associated with low birth weight
- **C.** Rates of IPV vary across countries
- **D.** Unmarried women are at greater risk of IPV in pregnancy than married women
- **E.** All of the above are true

124. Which of the following groups is the most common perpetrator of elder abuse?
- **A.** Family member
- **B.** Nursing home staff
- **C.** Personal friend
- **D.** Private personal caretaker
- **E.** Other

125. Which age group suffers the highest victimization rate of child maltreatment?
- **A.** 0–2
- **B.** 3–5
- **C.** 6–8
- **D.** 9–11
- **E.** 12–14

126. What is the most common form of child abuse in the United States?
- **A.** Neglect
- **B.** Physical abuse
- **C.** Psychological abuse
- **D.** Sexual abuse
- **E.** None of the above

127. Secondhand tobacco smoke has been shown to increase morbidity and mortality. The largest number of deaths attributed to secondhand smoke is directly caused by which disorder?

A. Asthma
B. Ischemic heart disease
C. Lower respiratory infection
D. Lung cancer
E. SIDS

128. Lack of which of the following vitamins is associated with rickets and osteomalacia?

A. Vitamin A
B. Vitamin B
C. Vitamin C
D. Vitamin D
E. Vitamin E

129. A diet rich in potassium is associated with which of the following?

A. Hyperglycemia
B. Increased bone loss
C. Increased risk of stroke
D. Lowering of blood pressure
E. Risk of kidney stone development

130. Which of the following is closest to the basal energy expenditure?

A. Estimated energy requirements
B. Nonexercise activity of thermogenesis
C. Resting energy expenditure
D. Thermic effect of food
E. Total energy expenditure

131. According to the Dietary Guidelines for Americans, what percentage of calories consumed should come from fats in healthy adults?

A. $<5\%$
B. 5%–10%
C. 10%–15%
D. 15%–20%
E. $>20\%$

132. What total percentage of total daily calories should come from consumption of transfats?

A. $<1\%$
B. 1%–5%
C. 5%–10%

D. 10%–20%
E. 20%–30%

133. Which of the following foods is routinely fortified with iodine?
A. Bread
B. Cereal
C. Drinking water
D. Milk
E. Salt

134. Which of the following nutrients can cause night blindness when deficient and has been fortified into milk?
A. Calcium
B. Iodine
C. Iron
D. Vitamin A
E. Vitamin B

135. Which of the following is not true of a disease management program?
A. A disease management program may focus on prevention of comorbid conditions
B. It is a cost-effective way to manage disease
C. Patients experience better healthcare outcomes
D. Coordinated care is provided amongst different practitioners
E. All of the above are true

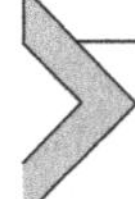

6.2 CLINICAL PREVENTIVE MEDICINE ANSWERS

1. **E. Primary, secondary, and tertiary**
   MNT is a flexible form of individually tailored therapy that achieves nutrition-related goals by utilizing a coordinated team approach consisting of patients, nutritionists, clinicians, and other ancillary care providers. MNT plays a role in primary, secondary, and tertiary prevention. It is most commonly used for the treatment of diabetes and CVD.

   Primary prevention is accomplished by delaying the onset of disease. Secondary prevention attempts to avoid complications once signs of the disease begin to manifest and tertiary prevention attempts to control complications of disease. It is never too late for prevention. Studies have demonstrated that MNT can lower Hemoglobin A1C as much as 2% and reduce LDL cholesterol by 15–25 mg/dL.
2. **B. B**
   The USPSTF gives a grade B recommendation for this woman to receive mammography to screen for breast cancer. For more information on mammography recommendations, please see the following table:

Target	USPSTF recommendation to use mammography screening for breast cancer[a]
Women age 40–49	Grade C
Women age 50–74	Grade B
Women age ≥75	Grade I

[a]USPSTF recommends biennial mammography.

3. **B. Two years**
   The USPSTF recommends that women aged 50–74 without elevated risk of breast cancer receive a screening mammogram every two years.
4. **C. C**
   Due to this patient's age and lack of significant risk factors, the USPSTF gives a grade C recommendation for routine screening colonoscopy, or any other CRC screening method. The patient's physician should make an individual decision of whether or not to have CRC screening and which type is most appropriate.

For more information on USPSTF CRC screening recommendations, see the following table:

Target	USPSTF recommendation for colorectal cancer screening[a]
Adults age 50–75	Grade A
Adults age 76–85	Grade C

[a]Colorectal cancer screening includes:
– Screening Colonoscopy every 10 years
or
– Flexible sigmoidoscopy every five years with fecal occult blood test every three years
or
– Fecal occult blood test every year

In 2008, the USPSTF published positions regarding CRC screening in adults >85 years of age (grade D), use of computed tomography colonography (grade I), and use of fecal DNA testing (grade I), but these recommendations were removed in a 2016 revision.

Numerous professional organizations have established their own recommendations for CRC screening. Each has established guidelines for people at normal risk which vary from those at increased risk.

The American Cancer Society (ACS) recommends that those at average risk of CRC should initiate screening at age 50. Flexible sigmoidoscopy, double-contrast barium enema, and CT colonography may be performed every five years. Colonoscopies may be performed every 10 years. If any of the five-year tests are positive, a colonoscopy should be performed. Those at risk requiring earlier testing include those with a personal history of CRC /adenomatous polyps, inflammatory bowel disease, family history of CRC /polyps, and those with a hereditary syndrome that increases risk if CRC.

The American College of Gastroenterology states that using a screening colonoscopy beginning at age 50 in average risk individuals is still the preferred CRC screening strategy. This group believes that Black men and women should initiate CRC screening at age 45. Other endorsed CRC screening tests include flexible sigmoidoscopy (every five to 10 years), CT colonography (every five years), and the fecal immunochemical test for blood (FIT). Similar to recommendations set forth by other groups, the American College of Gastroenterology states that close relatives of those with CRC should begin CRC screening at age 40 or 10 years younger than the age of diagnosis of the affected relative.

The American College of Physicians recommends that CRC screening should begin at age 50 in average risk individuals and cease at age 75 in those with a life expectancy of less than 10 years. The college also recommends that screening in high-risk individuals should start at age 40 or 10 years younger than the age of CRC diagnosis in a closely related family member. Other acceptable screening tests in high-risk individuals include FIT testing, flexible sigmoidoscopy, and optical colonoscopy.

5. A. A

Due to his age and history of DM, the USPSTF guidelines strongly recommend that this patient be screened for dyslipidemia.

The following table summarizes the USPSTF recommendations to screening for lipid disorders

Target group	USPSTF recommendation to screen for lipid disorders
Women not at increased risk of CHD[a]	Grade C
Women age 20–45 at increased risk for CHD	Grade B
Women age ≥ 45 at increased risk for CHD	Grade A
Men age 20–35 not at increased risk of CHD	Grade C
Men age 20–35 at increased risk for CHD	Grade B
Men age ≥ 35 with and without increased risk for CHD	Grade A

[a]CHD = coronary heart disease.

6. C. C

The USPSTF holds a grade C recommendation to screen for dyslipidemia in men aged 20–35 or in women aged 20 and older that are not at increased risk for coronary heart disease. USPSTF recommendations for lipid disorders were outlined in answer #5 (directly above).

Multiple different organizations have weighed in with recommendations for lipid screening.

The American Heart Association (AHA) encourages patients to have their physicians screen for lipids, especially those with risk factors. The AHA has stated that an overall medical care plan should include recording serum cholesterol, blood pressure, BMI, and fasting blood sugar. Furthermore, the AHA believes that all screening test results should be interpreted by that person's physician, stating that people who interpret their own lab results may become scared or confused.

Another expert panel that has created cholesterol screen recommendation is the Expert Panel on Integrated Guidelines for Cardiovascular Health and Risk Reduction in Children and Adolescents. In their summary report, they endorse routine screening in children.

7. **B. B**

The USPSTF has lumped together the use of aspirin for prevention of both CVD and CRC. The recommendations are as follows:

Recommendation for aspirin to prevent CVD and CRC

Target population	USPSTF grade
Age <50	I
Age 50–59	B[a]
Age 60–69	C[a]
Age ≥70	I

[a]If ≥10% risk of CVD in 10 years and ≥10 year life expectancy and not at increased risk of bleeding.

Most cases of CRC arise from adenomatous polyps. There is ample evidence that aspirin and/or NSAIDs taken in high doses reduces the incidence of adenomatous polyps.

8. **E. I**

Answer #7 (directly above) describes recommendations for when to use aspirin to prevent CVD and CRC.

9. **C. Financial cost**

The USPSTF is an independent entity sponsored and supported by the Agency for Healthcare Research and Quality (AHRQ) that makes recommendations on preventive services to primary care clinicians. Recommendations include screening tests, individual patient counseling, and preventive therapies. Each recommendation is intended for use by primary care providers as opposed to specialists. The USPSTF is composed of 16 volunteer members including professors, medical directors, chief health officers, clinicians.

In creating a recommendation, the USPSTF moves through the steps of creating a research plan, developing an evidence review, creating a recommendation statement, and finally disseminating the recommendation. In reviewing the evidence, the panel extensively analyzes existing peer-reviewed evidence. Considerations are made to the benefits versus the harms of the screening test, accuracy of the test, and the severity of the disease being examined. Although opportunity costs (time clinicians spend with patients, etc.) may be considered in the task force's recommendations, financial costs are not.

10. E. I

The USPSTF publishes recommendations to help guide decision making for primary care providers. The five guidelines are as follows:

A. There is high certainty that the net benefit is substantial. These services are recommended.

B. There is high certainty that the net benefit is moderate or there is moderate certainty that the net benefit is moderate to substantial. These services are recommended.

C. There is at least moderate certainty that the net benefit is small. These services should be selectively offered.

D. There is moderate or high certainty that the service has no net benefit or that the harms outweigh the benefits. These services are not recommended.

I- There is insufficient evidence to assess the balance of benefits and harms. This is neither a recommendation for or against screening.

There is no grade E recommendation.

11. C. There must be treatment available

The USPSTF gives a grade B recommendation to screen for depression in those aged 18 and older when staff-assisted depression care supports are in place to assure accurate diagnosis, effective treatment, and follow-up.

Target	USPSTF recommendation for screening for depression[a]
Population ≥18 years old	B
Population 12–18 years old	B
Population ≤11 years old	I
Pregnant women	B
Postpartum women	B

[a]Screening for depression should only be implemented when there are appropriate measures in place to provide effective treatment.

In general, screening should be used selectively when therapy is not available.

Screening may be an important public health tool when the treatment is not available. Take for example the HIV outbreak of the 1980s. Although a full understanding of the disease and effective medication did not exist, public health officials were able to partially control the outbreak by screening for those with the disease and reducing transmission.

12. B. Six years old

The USPSTF recommends that clinicians screen children aged six years and older for obesity. These children should be offered comprehensive, intensive behavioral interventions to promote appropriate weight.

13. D. D

The US Preventive Services Task Force recommends against routine PSA screening for prostate cancer in men older than 70 years old. This is because the harms of a positive test greatly outweigh the benefits of finding prostate cancer. These harms include false positives, diagnosis in men with nonaggressive forms of prostate cancer, and complications in men that have prostate cancer. Specific complications include anxiety, depression and impotence and/or incontinence from surgery.

The USPSTF used to recommend against screening for other age groups as well, but this stance is softening due to emerging evidence of benefits. It is now a grade C recommendation to use PSA to screen for prostate cancer men aged 55-69. In both Black people and men with a positive family history of prostate cancer, the benefits of screening may be greater than in other groups.

14. D. Testicular cancer screening

All of the services are recommended for this patient except for testicular cancer screening, which has been given a grade D recommendation by the USPSTF.

15. B. B

Providing interventions to reduce tobacco use in children and adolescents has received a grade B recommendation from the USPSTF.

16. C. Reporting a patient that plans to murder another inmate

Prisoners are entitled to equitable and patient-centered care that is respectful to their individual needs. To accomplish this, the patient–physician relationship should not change just because the patient has been incarcerated.

As a clinician in a prison, values of patient confidentiality, informed consent, privacy, nonmaleficence (do no harm) and beneficence are no less important. Use of medical skills to obtain evidence that is used to prosecute or punish patients (even if they are inmates) is unethical and will inevitably damage the doctor–patient relationship. If this relationship is damaged, a patient may withhold valuable information about his or her health.

In circumstances of pending harm to others, it is acceptable for a physician to breech their commitment of privacy by notifying proper authorities (answer C). Guidelines such as those set forth by the National Commission on Correctional Health Care recommend that correctional facilities hire outside help if they require a medical professional to perform a service that is outside of the standard doctor–patient relationship. These services include administering lethal injection, collecting evidence for prosecution, and performing cavity searches.

17. B. B

The USPSTF gives a grade B recommendation for annual lung cancer screening with a low-dose CT scan for adults aged 55–80 that have >30 pack-year history of smoking and discontinued smoking within 15 years. This only applies to those that are willing to undergo treatment and corrective surgery with a reasonable lifespan.

18. C. The USPSTF does not support her having HPV testing at this time

The USPSTF, American College of Obstetricians and Gynecologists, and American Cancer Society all have the same stance on cervical cancer screening. These recommendations (listed below) do not advise screening for HPV before age 30.

USPSTF grade	Cervical cancer screening recommendation
A	Screening in those aged 21–65 with cytology every three years or Screening in those aged 30–65 with cytology and HPV testing every five years
D	Screen for HPV alone or in conjunction with cytology in patients <30 years old
D	Screen for cervical cancer in patients <21 years old
D	Screen patients >65 years old that have had adequate prior screening and are not at high risk of cervical cancer
D	Screen patients that have had a hysterectomy with removal of the cervix and do not have a history of cervical intraepithelial neoplasia (CIN) 2/3 or cervical cancer

19. D. D

The USPSTF recommends against screening with resting or exercise electrocardiography (ECG) for the prediction of CHD events in asymptomatic adults at low risk for CHD events.

20. **B. Cardiovascular disease**
According to the WHO, ischemic heart disease and stroke are the top two leading causes of death worldwide.
21. **B. 60**
According to the JNC 8 hypertension management guidelines, the target blood pressure for patients without diabetes or kidney disease increases from $\frac{140}{90} \rightarrow \frac{150}{90}$ at age 60.
22. **C. Heart disease**
Starting from the most frequent cause of death and listed in descending order, the top five most common causes of death in the United States are the following:
– Heart disease
– Cancer
– Chronic lower respiratory disease
– Unintentional injuries
– Stroke
23. **E. 150/90**
For those age $\geq$60 years old, regardless of diabetes/kidney status and race, the recommendation is to maintain a systolic blood pressure <150 and a diastolic blood pressure <90.
24. **B. Calcium channel blocker and thiazide diuretic**
JNC 8 recommends that blood pressure management for Black patients either be initiated on a calcium channel blocker and/or a thiazide diuretic. Non-Black patients may be started on any combination of calcium channel blockers, thiazide diuretics, ACE inhibitors, and/or angiotensin receptor blockers.
25. **A. 55%**
According to the AHA, only 54% of hypertensive adults in the United States are adequately controlled. Meanwhile, 76.5% are treated for hypertension and 82.7% are aware that they have hypertension. This means that 17.3% of adults with hypertension are undiagnosed.
26. **A. High blood pressure**
In the United States, nearly 33% of adults $\geq$20 years of age have hypertension with Black individuals boasting the highest numbers. Roughly 40% of CVD mortality can be attributed to high blood pressure. In descending order, the population attributable fractions for CVD mortality are as follows: smoking (13.7%), poor diet

(13.2%), insufficient physical activity (11.9%), and abnormal serum glucose (8.8%).

27. B. Essential

Essential hypertension is defined as elevated blood pressure with no identifiable cause. The elevated pressure likely results from a combination of genetic and environmental factors. Essential hypertension is estimated to account for as many as 90% of American adults living with hypertension.

28. D. Stage two hypertension

This patient should be classified as having stage two hypertension. The JNC classifications for adults with hypertension is summarized in the below table.

Classification of hypertension	Systolic blood pressure	Diastolic blood pressure
Normal blood pressure	<120	<80
Prehypertension	120–139	80–89
Stage one hypertension	140–159	90–99
Stage two hypertension	≥160	≥100

29. D. Physical inactivity

Of the risk factors listed for hypertension, only high-potassium diet and physical activity are modifiable risk factors. Of these two options, physical inactivity increases risk of hypertension, while potassium is protective of hypertension. Age, family history, and [male] sex are all nonmodifiable risk factors for hypertension.

30. D. A man with a blood pressure of 125/80 mmHg

Metabolic syndrome is defined as having three or more of the following criteria.

Criteria for metabolic syndrome

	Men	Women
Waist circumference	>102 cm (>40 in)	>88 cm (>35 in)
Triglycerides	≥150 mg/dL	≥150 mg/dL
HDL cholesterol	<40 mg/dL	<50 mg/dL
Blood pressure	≥130/≥85 mmHg	≥130/≥85 mmHg
Fasting glucose	≥110 mg/dL	≥110 mg/dL

31. B. Black

Asthma is an inflammatory respiratory condition that occurs from both genetic and environmental causes. It is more common in children than adults and in the Black population more than any other group. Furthermore, those with a lower socioeconomic status (SES) experience a higher prevalence of asthma. When uncontrolled, asthma may cause morbidity, decreased well-being, absenteeism, and significant expenses.

Environmental exposures include both indoor and outdoor air pollutants. Indoor asthma triggers include tobacco smoke, dust mites, animal dander, and mold. Outdoor triggers include criteria air pollutants and products of hydrocarbon combustion, such as automobile exhaust. There is a direct correlation between housing proximity to traffic and prevalence of asthma.

Treatment of asthma should take a multidisciplinary approach that includes improving air quality, removing exposure to environmental irritants, addressing disparities, increasing access and quality of care, and conducting asthma research. Many of these actions are taken by the Center for Disease Control and Prevention's National Asthma Control Program (NACP). The NACP maintains extensive asthma-related records and funds private and public organizations to improve asthma surveillance and educate both health professionals and the general public about asthma prevention.

32. B. Cigarette exposure

COPD is the primary contributor to the classification of chronic lower respiratory diseases, one of the top five leading causes of mortality in the United States. It affects over 5% of the American population and is undiagnosed in roughly 50% of those that meet criteria for COPD. It is more common in the following demographics: women, lower level of education, lower income, not working (unemployed, retired, disabled), those no longer with their spouse (divorced, widowed, separated), current smokers, history of asthma, and elderly.

Diagnosis is made through recording the first second of forced expiratory volume (FEV1) and forced vital capacity (FVC) via spirometry, with a postbronchodilator FEV_1/FVC of less than 70% being the threshold for diagnosis. Once diagnosed, percentage of predicted FEV_1 may be used for staging. Mild COPD is a predicted $FEV_1 > 80\%$. Moderated COPD is 50%–80%, severe is 30%–50%, and very severe is $<30\%$.

Cigarette exposure causes >90% of COPD cases. There is a genetic component to cigarette smokers developing COPD, as only 20%–30% of regular smokers develop COPD. Another genetic risk factor is α_1-antitrypsin deficiency. Lead exposure and air pollution may also cause COPD. Smoking cessation is the most effective treatment to reduce COPD progression. In some cases, spirometry may show partial reversibility in COPD.

33. B. Kidney

Kidney transplants far outnumber all other transplants performed in the United States.

In descending order, the three most common causes of kidney transplants in the United States are diabetes, hypertension, and glomerulonephritis.

34. E. >75%

Cancer results from multiple mutations that change the growth and death dynamics of a cell. Estimates vary but a general consensus is that >75% of all cancers are due to environmental and lifestyle factors. These factors include smoking, diet, alcohol, and infectious disease. The rest is due to genetics.

35. B. Initiation → Promotion → Progression

There are several theories for the origin of carcinogenesis. The Multistage Process of Carcinogenesis Theory (also known as the Initiation–Promotion–Progression Theory) includes the following chronological phases: initiation, promotion, progression, invasion, and metastasis. This theory states that an initiating factor for cancer must meet with a promoting factor to progress. Eventually, the cancer no longer needs the promoting factor and can progress on its own. Invasion and metastasis are often grouped into the progression stage.

In the initiation stage, the cellular genome undergoes genetic mutations that predispose it to neoplastic development. The DNA sequences that contribute to these changes are called oncogenes.

In the promotion stage, the initiated cell undergoes further proliferation toward a cancer cell under the influence of a promoting factor. Promoting factors include inflammation and hormones.

In the progression stage, the process continues through further proliferation, often times into malignant subpopulations. The change becomes irreversible.

36. D. 40%

The overall lifetime risk of being diagnosed with cancer is 40%. This number includes a lifetime risk of 43% in men and 37% in women. As this is only the recognized diagnosis of cancer, the true number is likely higher. Cancer is frequently discovered incidentally postmortem. Prostate cancer is the most frequently diagnosed cancer in men, while breast cancer is the most frequently diagnosed cancer in women. Lung cancer is the most frequently diagnosed cancer overall and causes the most deaths of any type of cancer.

37. D. 20%

The overall lifetime risk of dying from cancer is approximately 21%. This number includes a lifetime risk of 23% in men and 19% in women. Cancer is the second-leading cause of death in the United States, behind CVD.

Lung cancer is the leading cause of cancer-related death in men and women. Prostate cancer is the second leading cause of cancer-related death in men and breast cancer is the second leading cause of cancer-related death in women.

Prostate cancer is the most common cancer in men and breast cancer is the most common cancer in women.

38. E. Tobacco smoke

Smoking is one of the most important risk factors associated with bladder cancer. Incidence for this disease is nearly four times higher in smokers than nonsmokers. In fact, half of all cases of bladder cancer can be attributed to smoking. Other important risk factors for bladder cancer include gender (incidence is four times higher in men) and older age.

Arsenic exposure (typically in drinking water), urinary birth defects, occupational exposures (paint, dye, rubber, leather, aluminum), and radiation exposure are other risk factors associated with bladder cancer.

39. C. Hormone replacement therapy

Worldwide, breast cancer is the most prevalent cancer in women. Risk factors for breast cancer result from unopposed hormones and include hormone replacement therapy, early menarche, late menopause, nulliparity, late age at birth of first child, and oral contraceptives. Alcohol consumption, physical inactivity, obesity, and genetics are other risk factors.

40. B. Clear-cell adenocarcinoma of the cervix in daughters of mothers that take DES while pregnant

DES is a synthetic estrogen which in the past was routinely given to women to prevent miscarriages. It was later found through epidemiological investigation that DES exposure during pregnancy is association with reproductive tract changes in offspring. In girls, it may lead to a T-shaped uterus, poor pregnancy outcomes (premature birth, miscarriages, and ectopic pregnancy), fertility problems, breast cancer, and clear-cell adenocarcinoma of the uterus, cervix, and vagina. In boys, DES exposure in utero is associated with cryptorchidism, hypospadias, and hypoplastic testis. It has also been demonstrated that DES exposure during pregnancy may cause effects in grandchildren such as hypospadias.

The effects of DES exposure during pregnancy are a great example of epigenetics, in which information is passed on to offspring outside of DNA.

41. E. 16

There are over 120 officially recognized types of HPV with many more classifications pending and a total expected count over 200. HPV is extremely common, with a prevalence between 50%–80% of sexually active adolescents. An intact cell-mediated immune response is capable of clearing 70% of HPV infections in the first year and 90% in two years. Meanwhile, HPV infection can present as productive, subclinical, or latent infections in the skin and/or mucosa.

Different HPV strains are associated with different pathological changes. HPV types one, two, and four are associated with plantar warts, while types two, four, and seven are associated with common warts. Meanwhile, types six, 11, 40, 42, and several others are associated with anogenital warts, with types six and 11 most commonly implicated.

On the cancer front, HPV types 16, 18, 31, 33, 45, 51, and 52 are associated with anogenital cancers. More specifically, HPV 16 and 18 account for 50% and 20% cases of anogenital cancers, respectively. More than 99% of cervical cancers contain HPV DNA. Types 16 and 18 are also the most common causes of oropharyngeal carcinoma with type 16 being even more of a causative factor for oral cancer than anogenital cancer. HPV 16 is also responsible for about one out of three cases of anal intraepithelial neoplasia (AIN).

42. B. 65

For women at low risk of cervical cancer that are greater than 65 years of age with an adequate history of prior screening, the USPSTF recommends against further cervical cancer screening. In this population, the risks and harms associated with screening for cervical cancer are small.

Screening in those older than 65 may be appropriate in high-risk individuals or that that have never been adequately screened for cervical cancer. The ACS, American Society for Colposcopy and Cervical Pathology (ASCCP), and the American Society of Clinical Pathologists (ASCP) all recommend three consecutive negative cytology results or two consecutive negative HPV tests within 10 years prior to discontinuing cervical cancer screening. In addition, they state that routine screening should continue for 20 years after resolution (medical or spontaneous) of a high-grade pre-cancerous lesion, even if the patient is greater than 65 years old.

In women that have had a hysterectomy with removal of the cervix for reasons other than cervical cancer, there is little benefit to continue screening for cervical cancer.

43. B. Alcohol

Alcohol is the number one cause of liver cancer in the United States. Worldwide, however, alcohol is a small risk factor compared to both hepatitis B and hepatitis C. Aflatoxin B and liver flukes are other known risk factors in the development of liver cancer. Often times, all of these risk factors interact synergistically.

44. A. Consumption of beef

CRC is the 3rd most commonly diagnosed cancer in males and second in females worldwide. These numbers are lowest in Africa and increasing in countries undergoing industrialization which previously enjoyed low incidence of CRC. In the United States, it is the third most common cancer in women and men with lung cancer and breast cancer being more common in women and with lung cancer and prostate cancer being more common in men. Deaths from CRC are decreasing in the United States due to advancements in education, detection, and treatment.

Risk factors for CRC include consumption of red meat, processed meat, and alcohol. Obesity, sedentary lifestyle, and tobacco use are also associated with CRC. Meanwhile, a high-fiber diet may be protective of CRC.

45. C. Salted and preserved food are protective of stomach cancer

Stomach cancer risk factors include *H. pylori*, obesity, tobacco use, and foods preserved with salt or pickling. Rates are highest in Eastern Asia, Eastern Europe, and South America and lowest in North America and Africa. As availability of refrigeration and fresh fruits/vegetables increases, the need for food preservation via pickling and salting decreases, leading to a lower risk of stomach cancer. Prevalence is typically twice as high in men as women.

46. E. Low-protein diet

Risk factors for pancreatic cancer include increasing age, history of chronic pancreatitis, diabetes mellitus, high fat diet, family history, obesity, male sex, tobacco smoking, diets high in meat and low in vegetables, and occupational exposures (hydrocarbons and nickel). Cigarette smoking and family history are the most dominant risk factors. A low-protein diet is not a recognized risk factor for pancreatic cancer.

47. D. Pancreatic Cancer

The five-year survival rate of those with pancreatic cancer is among the lowest of all cancers, hovering from 4%–7%. This is in large part due to late diagnosis, as the cancer is typically clinically silent until metastasis occurs. Surgical resection is the main treatment and is much less successful after metastasis occurs. By the time the disease is discovered, roughly 80%–85% of patients are no longer candidates for surgery. Presenting symptoms of pancreatic cancer typically include abdominal pain, back pain, jaundice, and weight loss.

After pancreatic cancer, lung cancer has the next lowest five-year survival rate, followed in increasing order by colon cancer, breast cancer, and prostate cancer.

48. A. Leukemia

Despite dramatic improvements in pediatric cancer mortality over the past several decades, cancer is the second leading cause of death in children aged one to 14, only behind accidents. Accounting for nearly 30% of pediatric cancer cases, leukemia is the most common type of cancer diagnosed in children between the ages of one to 14. The cancers that follow in descending order are: Central nervous system tumors (26%), lymphoma (8%), neuroblastoma (6%), Wilms tumor (5%), rhabdomyosarcoma (3%), and osteosarcoma (2%).

The majority (91%) of leukemia cases are found in patients that are ≥20 years old. In this population, chronic lymphocytic leukemia is the most common. However, the most common type of leukemia in children is acute lymphocytic leukemia, accounting for more than three out of four cases.

49. C. Lung

Lung cancer is the leading cause of cancer mortality and the second most prevalent form of cancer in both men and women in the United States. The most common cancer in men is prostate cancer and the most common cancer in women is breast cancer.

50. E. Radon

Radon is the second leading cause of lung cancer, ahead of secondhand smoke, occupational exposures, asbestos, and air pollution. Radon is a gas consisting of alpha particles resulting from the decay of uranium in the soil. The alpha particles are blocked by the skin but when the gas is inhaled, the alpha particles can then damage the lungs.

51. E. All of the above are risk factors for cancers of the oral cavity

Oral cancer may be caused by excessive alcohol consumption, HPV, smokeless tobacco, and smoking tobacco. Although these risk factors have synergistic effects, tobacco is the largest cause of oral cancers worldwide. Regions that are experiencing an increase in tobacco use are seeing increased number of oral cancer cases, where regions that have seen a decline in tobacco use are also enjoying a decrease in oral cancer cases. Other causes of oral cancer may include mouthwashes that contain alcohol, poor oral hygiene, and low SES. The majority of oral cancers are squamous cell carcinomas.

52. D. Positive history of breast cancer in one sister and ovarian cancer in another sister

Ovarian cancer causes more deaths in the United States than any other gynecological cancer. Incidence and mortality increase with age and it is most common in White women.

The USPSTF does not recommend routine screening for asymptomatic women without a significant family history (grade D). This is because screening with serum Cancer Antigen 125 (CA-125) and transvaginal ultrasound yields risks that outweigh benefits.

Risk factors for ovarian cancer include genetics, race, age, nulliparity, and endometriosis. Meanwhile, low-fat, high-fiber diets may lower risk of ovarian cancer.

The USPSTF recommends genetic screening for risk of ovarian cancer based off family history. If family history does not indicate increased risk of mutations in BRCA1 and BRCA2 genes, the USPSTF assigns grade D recommendation for ovarian cancer genetic screening. If there is a positive family history suggesting BRCA1 or BRCA2 gene mutation, the USPSTF gives a grade B recommendation for ovarian cancer genetic screening.

The following criteria suggests BRCA1 or BRCA2 gene mutation for those without Ashkenazi Jewish heritage:

- ≥ Two first or second degree relatives with ovarian cancer
- ≥ Two first or second degree relatives with a combination of breast/ ovarian cancer
- ≥ One male relative with breast cancer
- ≥ One first degree relative with bilateral breast cancer
- ≥ Three first or second degree relatives with breast cancer

The following criteria suggests BRCA1 or BRCA2 gene mutation for those with Ashkenazi Jewish heritage:

- ≥ One first degree relative with breast or ovarian cancer
- ≥ Two second degree relatives with a combination of breast/ ovarian cancer on same side of family

USPSTF recommends that those with positive screening results receive genetic counseling. If indicated, they may then consider prophylactic surgery to remove any combination of female reproductive organs.

53. D. D

Please see the explanation for question #52 (directly above)

54. D. Prostate

The Gleason score is the most widespread technique for grading prostate cancer. It can be used as a guideline for prognosis of survival and is completed after taking a biopsy from two separate sites of the prostate. Each sample is scored between one and five. The two scores are then combined, creating a range of two through 10. The higher the score, the worse the prognosis.

55. C. Skin

Skin cancer is by far the most frequently diagnosed cancer in the United States. It is estimated that there are roughly twice as many cases of skin cancer annually as there are cases of all other cancer combined. squamous cell carcinomas (SCCs) and basal cell carcinomas (BCCs) are the most prevalent form of skin cancer but they are not required to be reported to cancer registries.

56. D. Melanoma

BCCs and SCCs make up the vast majority of skin cancers. While Melanomas account for under 2% of all skin cancers, they are responsible for the majority of skin cancer deaths. Risk factors for melanomas include light skin, sun sensitivity, sun exposure (occupational or recreational), family history of melanoma, numerous (>50) moles, immunosuppression, and a history of skin cancer.

Keratoacanthoma is a type of SCC. Meanwhile, dermatofibromas are not a type of skin cancer.

57. B. Asymmetry, border, color, diameter, enlargement

The ABCs of melanoma detection are asymmetry, border, color, diameter, and enlargement. Clinicians and patients should be aware of skin lesions that have an asymmetric shape, irregular border, changing or dark color, large diameter (at least >5 mm) and recent enlargement.

58. B. 20-year-old White man with HIV

Of the five options, only one carries risk factors for testicular cancer. Testicular cancer incidence peaks between the ages of 15–35. Other risk factors for testicular cancer include history of cryptorchidism, history of orchitis, hypospadias, HIV infection, Down syndrome, Klinefelter syndrome, gonadal dysgenesis, Caucasian heritage, and a family history of testicular cancer.

59. D. No. It is not appropriate to screen men at any age for testicular cancer.

The USPSTF gives a D recommendation for clinicians to screen for testicular cancer or teach patients to check for it themselves. Regardless of the stage that the cancer is discovered, over 90% of all newly diagnosed cases will be cured. For this reason, discovering the disease early (via screening) is unlikely to offer meaningful health benefits, especially considering the low incidence of testicular cancer. Potential harms from screening tests include false-positive results leading to anxiety and complications from related procedures.

60. **B. Head/neck radiation**
Radiation to the head and neck, especially at a young age, is a strong risk factor for thyroid cancer. Meanwhile, tobacco smoke, alcohol use, and obesity have not been shown to be risk factors for thyroid cancer. In fact, several studies have found these factors to be protective of thyroid cancer.

61. **B. Renal cell carcinoma**
Renal cell carcinomas are by far the most common form of cancer of the kidney, accounting for >85% of renal cancers. This disease most often manifests in people between ages 40–60 and has a two: one male to female predominance. Risk factors include familial history, tobacco use, cystic kidneys resulting from end-stage renal disease, and occupational exposures.

62. **E. United Network for Organ Sharing (UNOS)**
The OPTN is a public–private partnership designed to match organ donors, organ recipients, and organ transplantation professionals in the United States. Goals of OPTN include increasing access to organ transplants and improving transplant outcomes. The UNOS is a private nonprofit organization that is contracted through the Health Resources and Services Administration (HRSA) to operate as the sole OPTN in the United States.

63. **B. 10%**
Roughly one out of every 10 adults in the United States are living with DM with >90% having type two DM. Men with DM live on average 7.5 years less and women with DM live on average 8.2 years less.

64. **B. Diabetes mellitus**
Roughly 10% of adults in the United States suffer from varying levels of CKD. Risk of developing CKD increases with age. DM is the leading cause of CKD in the United States. Roughly one out of every three adults in the United States with diabetes have developed CKD. Hypertension is the second leading cause of CKD in the United States. Other risk factors include CVD, obesity, dyslipidemia, and lupus.

65. C. Overweight

This question highlights how differently children and adolescents are classified as obese compared to adults. In children and adolescents, the BMI, calculated in kg/m^2, is taken into consideration for the age. A BMI of 21 for an adult would be considered healthy weight, but it places a nine-year-old boy in the 93rd percentile.

The table below identifies how to calculate the weight category of children, adolescents and adults.

Category	Children and adolescent BMI percentile	Adult BMI
Underweight	$<$ 5th percentile	$<$ 18.5
Healthy weight	5th–85th percentile	18.5–24.9
Overweight	85th–95th percentile	25.0–29.9
Obese	$>$ 95th percentile	$>$ 30

66. E. 50

The following table depicts calories per gram of four sources that provide energy.

Source	Calories per gram
Carbohydrate	Four
Protein	Four
Alcohol	Seven
Fat	Nine

If someone consumes 200 calories of carbohydrates and each gram of carbohydrate contains four calories, that would mean that they consumed $200/4 = 50$ grams of carbohydrates. The same numbers would hold true for proteins. Meanwhile, 200 calories of alcohol would contain 28.5 (200/7) grams of alcohol and 200 calories of fat would contain 22.2 (200/9) grams of fat.

In the United States, most carbohydrates are consumed in the form of starch.

67. B. 2300 mg

The US Department of Health and Human Services (DHHS) and US Department of Agriculture (USDA) recommend that Americans consume no more than 2300 mg of sodium daily. However, African Americans (of any age), persons ≥ 51 years old, diabetics, persons with high blood pressure, or persons with kidney disease should have no more than 1500 mg of sodium daily.

These numbers represent the tolerable upper intake level (UL). The UL is the level that there is likely no risk of adverse effects for most individuals in the population. On the lower end of the spectrum is the adequate intake (AI) level which is the amount of a nutrient that is necessary to maintain normal biological function in healthy individuals. The sodium AI for persons aged nine to 50 years old is 1500 mg per day. Meanwhile, sodium AI is 1300 mg per day for persons aged 51–70 years old and 100 mg per day in people ≥71 years old.

As a whole, Americans average higher than the UL. The majority of sodium ingestion comes from salt added during food processing. A small percentage of sodium comes from salt added during the cooking process or at the table.

68. D. 1500 mg

It is recommended by the Department of Health and Human Services (DHHS) and the Department of Agriculture that people ≥51 years old should have no more than 1500 mg of sodium daily.

Sodium recommendations were explained in answer #67 (directly above).

69. E. 300 mg

According to the US Department of Agriculture and the US Department of Health and Human Services (DHHS), the recommended maximum daily limit of cholesterol is 300 mg.

Eggs provide the major source of cholesterol in the United States.

70. D. Sugary drinks

5-2-1-0 is a nationally recognized childhood obesity prevention program. It stands for daily quotas of at least five servings of fruits and vegetables, two or less hours of screen time, one or more hours of physical activity, and 0 sugary drinks.

71. A. Blood pressure

The Dietary Approach to Stop Hypertension (DASH) study was a landmark study that helped prove a scientific correlation between diet and hypertension. Many clinicians now recommend the DASH diet to manage blood pressure in hypertensive patients. It includes consumption of fruits, vegetables, and low-fat dairy products all while reducing intake of saturated fats and limiting sodium intake.

72. D. Obesity

Obesity is the only risk factor listed that is protective of osteoporosis. Wolff's law states that bone will adapt to the stresses that it is placed under. In other words, osteoblastic activity will increase to produce stronger bone when placed under more weight.

Alcohol, caffeine, physical inactivity, and tobacco are all risk factors that contribute to osteoporosis.

73. E. Osteoporosis

DXA (formerly DEXA) scans are a tool used to create a T score to evaluate the mineral density and structural integrity of bones. Osteoporosis is defined as bone mineral density at least 2.5 standard deviations below peak bone mass. A precursor of osteoporosis is osteopenia, with a bone mass of 1.0 to 2.5 standard deviations mean peak mass.

DXA scans must be evaluated with clinical context. Bone loss does not occur at the same pace throughout the body. For example, the bone density of a patient's heal may be within normal limits, while the bone density of the same person's lumbar spine may reveal osteoporosis. Meanwhile, vertebral compression fractures (VCFs) falsely increase bone density on DXA Scan.

Bone loss naturally occurs in the normal aging process. It may be accelerated by alcohol, smoking, caffeine, beverages high in phosphate (soda), high-protein diet, steroids, phenytoin, heparin, and lack of vitamin D. Other risk factors for osteoporosis include family history of osteoporosis, low body weight, early menopause, immobilization, chemotherapy, and hypogonadism.

74. E. Vertebra

Vertebral compression fractures (VCFs) are the most common type of fracture resulting from osteoporosis. VCFs are more common with increasing age. One VCF is predictive of future VCFs, where having one VCF will increase the risk of a second by five-fold. Nonosteoporotic causes of VCFs include malignancy and trauma.

75. B. Vitamin B

Consequences of vitamin deficiencies

Vitamin	Consequences of deficiency
Vitamin A	Night blindness, keratomalacia
Vitamin B_1 (thiamine)	Beriberi, Wernicke–Korsakoff syndrome
Vitamin B_2 (riboflavin)	

(continued)

(Continued)
Consequences of vitamin deficiencies

Vitamin	Consequences of deficiency
	Angular stomatitis, glossitis, cheilosis, seborrheic dermatitis
Vitamin B_3 (niacin)	Pellagra
Vitamin B_6 (pyridoxine)	Peripheral neuropathy, glossitis, cheilosis
Vitamin B_9 (folic acid)	Megaloblastic anemia, thrombocytopenia, fatigue
Vitamin B_{12} (cobalamin)	Megaloblastic anemia, peripheral neuropathy
Vitamin C deficiency	Scurvy
Vitamin D deficiency	Rickets, osteoporosis
Vitamin E deficiency	Abetalipoproteinemia, hemolytic anemia, ataxia, neuropathy

76. B. Back pain

Back pain is the most common musculoskeletal ailment worldwide. Roughly 80% of the world's population will experience low back pain at one point or another. In the United States alone, around 20% of the populace experiences back pain annually. Cervical and lumbar pain is the most common type of back pain with thoracic spine pain being relatively rare.

Most cases of back pain are nonspecific and the etiology is not discovered during treatment. Literature shows that etiology of back pain varies with age. Sprains and strains are the most common causes of pain in younger patients. Additionally, onset of back pain in younger patients may be from developmental problems, including scoliosis, spondylolysis, and spondylolisthesis. Older patients more commonly advance to ailments associated with degeneration, including arthritis and spinal stenosis.

77. A. Arthritis

Arthritis is the leading cause of disability in the United States. Roughly 23% of adults aged ≥ 18 years have self-reported doctor-diagnosed arthritis. This percentage jumps to 50% in the ≥ 65 year old population, a group that will continue to grow.

Arthritis is more common in adults with multiple comorbidities.

The leading causes of disability in the United States order are listed as follows in descending order:

- Arthritis
- Back or spine problems

- Heart ailment
- Lung ailment
- Mental ailment
- Diabetes
- Hearing ailment
- Stiffness of limbs
- Vision ailment
- Stroke
- Cancer

78. B. Falls

Falls are the most frequent cause of traumatic brain injury in the United States.

79. C. 20%

Roughly 19% of adults age 65 and older are considered edentulous, the term used for someone without a single natural remaining tooth. Adults older than 75 years of age are twice as likely (26%) to be edentulous than adults aged 65–74 (13%). Non-Hispanic Black adults carry the distinction of having the highest risk of being edentulous in the ≥65 age group. The ≥65 population has the highest prevalence of periodontitis in the United States. Of the ≥65 adults with permanent teeth, roughly 96% have dental caries.

Dental health in the ≥65 year old population is particularly problematic, with only 30% of US patients in this age range having dental insurance. Medicare typically covers dental services for oral cancer and dental trauma. Thanks to private insurance coverage, there has been a steady increase in the number of patients ≥65 year old with dental insurance.

80. A. Cataracts

Cataracts is the leading cause of blindness worldwide and is second to uncorrected refractive errors in global causes of visual impairment. Risk factors for cataracts include alcohol abuse, diabetes, UV radiation, smoking, and genetics. Cataracts is much more prevalent in developing nations than developed nations. Additionally, it is much more prevalent in people ≥50 years old. A short surgical procedure can permanently cure cataracts, therefore it is categorized as a reversible type of visual impairment.

Nearly 80% of all visual impairments may be reversed or prevented. Visual impairment is a global burden that disproportionately affects the poor and elderly. In fact, roughly 90% of people living with visual impairment reside in low-income settings. Greater than 80% of people living with blindness are older than 50 years old.

The WHO has been appointed by the United Nations to coordinate international health efforts. This responsibility extends to vision services, where the WHO works to strengthen national and local efforts to provide eye services through public health and direct patient care. WHO's global initiatives to improve vision include "Vision 2020: The Right to Sight", a program to eliminate avoidable blindness and "Universal Eye Health: A Global Action Plan" to improve public health measures to reduce avoidable visual impairment.

81. A. Vitamin A

This vignette contains multiple clues to indicate that the patient began to take vitamin A. Initially, she discovers night blindness, which is a common finding in vitamin A deficiency. Next, it is revealed that she abuses alcohol, a risk factor for vitamin A deficiency. Her recent pregnancy also predisposed her to vitamin A deficiency. Corneal ulceration is a frequent finding in vitamin A deficiency. Vitamin A toxicity may lead to cracking of the lips and fingernails. Finally, after taking large doses of vitamin A, she experiences the teratogenic effects of vitamin A toxicity.

Vitamin A deficiency leads to xerophthalmia and keratomalacia, conditions characterized by night blindness and drying of the cornea. This dryness may lead to corneal ulcers and even areas of decreased vision (Bitot spots) on the cornea. It is more common in malnourished countries, where there is lack of access to vitamin A-rich foods: Liver, dairy, fruits, and vegetables. In fact, it is the leading cause of pediatric blindness worldwide. It is also more common in alcoholics and pregnant women. Vitamin A levels in the blood can be monitored by serum retinol levels.

82. D. 50%

Unintended pregnancies include unwanted and mistimed pregnancies. In the United States, unintended pregnancies are responsible for roughly 50% of all pregnancies and >33% of all live births. This is amongst the highest rates of all industrialized countries. When not planning to become pregnant, a woman may not be in preferred childbearing health. Studies demonstrate that unintended pregnancy causes an elevated risk of adverse social, economic, and health outcomes for the mother and child. Specific factors associated with unintended pregnancy include delayed prenatal care, reduced breast feeding, smoking while pregnant, and poor health outcomes in childhood. Children born from unexpectedly pregnant teenagers are more likely to be preterm and born at lower birth weight.

Risk factors for unintended pregnancy include younger age of the woman, lack of education, low income, Black race, and cohabitating outside of marriage. Since the inception of the Healthy People program, an emphasis has been placed on reducing disparities amongst groups. One strong emphasis is teenage pregnancy. Four out of five teenage mothers became pregnant unintentionally. This account for slightly over 20% of unintended births.

83. B. 0–12 months

SIDS is defined as the death of a child under one year old that can't be explained, despite a complete history, examination of environment, and autopsy. In the United States, it is the number one cause of death in infants aged one month through 12 months. It is classified as a type of sudden unexplained infant death (SUID).

84. B. Infant has a respiratory infection

An infant with an infection not is contraindicated to breastfeed. In fact, breastmilk contains numerous components that boost the infant's immune system and help fight infection. Infants should not consume breastmilk from women that have brucellosis or human T-cell lymphotrophic virus. Additionally, mothers should not directly breastfeed infants if they have active untreated tuberculosis, H1N1 influenza, varicella, or a herpetic lesion of the breast. In these cases, pumping and subsequent bottle feeding is preferred. Mothers with HIV in the United States (and all other industrialized countries) are discouraged from breastfeeding. To the contrary, breastfeeding is encouraged in HIV-positive mothers in developing nations because the benefits are thought to outweigh the risks.

The only infant condition that contraindicates breastfeeding is that the infant has galactosemia.

In cases where a mother is unable to provide breastmilk, it is often recommended that the infant receive pasteurized donor breastmilk.

85. D. Breastfeeding is associated with lower risk of ovarian cancer

The only correct answer to this question is answer D. Cumulative time lactating decreases the risk of both breast and ovarian cancer.

Furthermore, breastfeeding is associated with a decreased occurrence of postpartum depression. As breastfeeding consumes a lot of calories and requires an increased amount of work by the body, it has been shown to help women return to prepregnancy weight quicker than not breastfeeding. While breastfeeding, it becomes much more difficult to become pregnant because of the lactational amenorrhea that occurs due to hormonal changes. Additionally, women that breastfeed have more rapid involution of the uterus and decreased postpartum blood loss.

Younger mothers initially attempt to breastfeed less frequently than older mothers. Meanwhile, Black women initially attempt breastfeeding less frequently than any other group.

86. E. All of the above are benefits of breastfeeding

The list below highlights some of the numerous benefits of breastfeeding compared with children that do not receive breastmilk. Research suggests that many of these benefits are in a dose-response relationship according to the duration of breastfeeding. The American Academy of Pediatrics (AAP) recommends exclusive breastfeeding for the first six months. At that point, breastfeeding should occur in conjunction with the introduction of complimentary foods until the first birthday. Introducing complimentary foods while consuming breastmilk has been shown to reduce food allergies, presumably due to the immunoprotective components of human milk.

Breastfed babies have lower risk of the following ailments than nonbreastfed babies:

- Asthma
- Atopic dermatitis
- Eczema
- Gastrointestinal infections
- Leukemia
- Obesity (childhood, adolescent, and adult)
- Otitis media
- Respiratory infections
- SIDS

87. A. Cereal grain

In 1998, the United States mandated folic acid fortification of enriched cereal grain products. Cereal grain products must contain 140 μg of folic acid per every 100 g.

Although the FDA has approved oral contraceptive pills fortified with folic acid, it is not a requirement that all oral contraceptive pills contain it.

88. A. 0.4 mg

The CDC and the USPSTF recommend that women capable of becoming pregnant should take at least 400 micrograms (0.4 mg) of folic acid daily to prevent neural tube defects (NTDs). In women that have experienced a pregnancy resulting in a neural tube defect, it is recommended that they take 4 mg of folic acid daily if attempting to have another child. The two most common NTDs are spina bifida and anencephaly.

89. D. Preterm birth and related complications

Preterm birth and related complications are the leading cause of neonatal mortality, accounting for nearly 50% of neonatal deaths. Preterm birth includes both spontaneous preterm birth and indicated preterm birth, where delivery is induced due to medical indications.

See question #90 (directly below) for explanation of what constitutes preterm.

Preterm birth is typically the outcome of multifactorial etiologies with each pathway toward prematurity being influenced by gene–environment interactions. Some factors that may lead to preterm birth include stress, maternal infection, cervical insufficiency, uterus overdistention, placental thrombosis/ischemia, maternal stress, pesticide exposure, air pollution, and most frequently from smoking. There are disparities in the rates of preterm birth amongst different racial and ethnic socioeconomic groups.

Preterm infants suffer from a myriad of health ailments that may expand all the way into adulthood. Studies have demonstrated that preterm infants suffer from the following: Temperature instability, respiratory distress, infections, apnea, hypoglycemia, seizures, kernicterus, feeding problems, necrotizing enterocolitis, leukomalacia, vision impairments, hearing impairments, cardiovascular complications, and rehospitalizations.

90. D. 37

Beginning the 38th week of pregnancy, the pregnancy is considered to be term. Therefore, babies born before the 37th completed week of pregnancy are considered to be preterm. Preterm encompasses extremely preterm (under 28 weeks), very preterm (between 28–32 weeks), and moderately preterm (32–36 weeks).

91. B. 12 months

Infertility is defined as the inability to become pregnant despite having unprotected sexual intercourse with the same partner at least once monthly for 12 consecutive months. Infertility is considered to be a component of impaired fecundity, in which a woman has trouble becoming pregnant and carrying a pregnancy to live birth.

The true number and cause of infertility and impaired fecundity is underreported. With this in mind, it is estimated that between 6%–7% of married women are considered infertile and 12% have impaired fecundity. Furthermore, it is estimated that nearly one in three cases of infertility are due to the female, one in three cases are due to the male, and 1/3 cases are unexplained or due to a combination of problems between both partners.

92. A. Time of birth

According to the Advisory Committee on Heritable Disorders in Newborns and Children's (ACHDNC) recommendations that were approved by the Department of Health and Human Services (DHHS), PKU is one of the core recommendations that should be screened at birth. By identifying those with PKU early on, dietary modifications may be made and complications of the disease can be averted later on in life.

In descending order, the four most common conditions diagnosed by routine screening in newborns are hearing loss, primary congenital hypothyroidism, cystic fibrosis, and sickle cell disease.

93. B. Vitamin E supplements

Of the five choices, vitamin E supplementation is the only factor that has not been associated with a higher risk of lung cancer. This was demonstrated through multiple studies that failed to meet their primary end point of declaring vitamin E as a protective factor for lung cancer.

β-Carotene supplementation has been shown to be a risk factor for lung cancer in patients that smoke but not in nonsmokers.

94. C. American Psychiatric Association

The DSM criteria is published by the American Psychiatric Association to provide standardized classification of mental disorders. It is used by virtually all healthcare providers in both clinical and nonclinical settings. It is important in the field of public health due to its ability to classify and track the psychiatric well-being of the American public.

95. A. Anxiety

In a widely publicized 2005 study, lifetime prevalence of anxiety was found to be 28.8%. This was the highest of all other mental disorders. It was followed by impulse control disorders (24.8%), mood disorders (20.8%), and substance use disorders (14.6%). These prevalences still hold true today.

96. D. 50%

In a widely publicized study, 46.4% of Americans experienced a lifetime prevalence of a mental disorder. This study also revealed that 27.7% of respondents have had two or more psychiatric disorders in their lifetime. The prevalence of both of these figures is higher in older populations.

97. B. Alzheimer's

Alzheimer's dementia is the most common type of dementia, accounting for more than half of all dementia cases. Dementia is defined as a decline in more than one cognitive capability that causes impairment in function but not in alertness or attention. It is a syndrome rather than a distinct illness. The DSM5 proposed that the word dementia be replaced with "neurocognitive disorder".

Dementia is primarily a disease of the elderly, with a prevalence between 9%–13% in persons older than 65 years. Suggested risk factors include physical inactivity, depression disorder, smoking, hypertension, obesity, low educational attainment, decreased cognitive activity, diabetes, head trauma, and family history of dementia.

98. B. 2

The CAGE Questionnaire is one of many tools used to subjectively describe a person's alcohol use. It stands for the following:

- Do you feel the need to cut down on your drinking?
- Do you feel annoyed by criticism of your drinking?
- Do you feel guilty about your drinking?
- Have you ever had a first drink in the morning to act as an eye opener and steady your nerves?

Each question answered "yes" receives a score of one, while each question answered "no" receives a score of zero. A total score of two or greater is reflects significant alcohol problems.

99. E. ≥75

The highest rate of men that successfully commit suicide are aged 75 or older. Meanwhile, the age group with the highest success rate of suicide in women is 45–64 years old.

100. C. Men attempt suicide less frequently and have a higher "success" rate of suicide.

Suicide in the United States claims more lives annually than homicides. It is a public health challenge with a well-known epidemiology. While women attempt suicide four times more often than men, men are four times more likely to successfully commit suicide than women. Women typically try to commit suicide in nonviolent ways, such as overdosing on pills. Meanwhile, men typically attempt suicide in more violent ways, such as firearms, jumping from high places, and hanging.

101. A. All children with autism spectrum disorder have intellectual disability

Autism is wide-spanning spectrum of developmental disabilities characterized by social and communication impairment that causes varying degrees of physical limitations. Individuals with autism characteristically display restricted interests and repetitive behaviors. Due to the wide spectrum of what is included in the diagnosis of autism, estimates of prevalence vary. One commonly cited number by the CDC is one in 68 children.

The American Academy of Pediatrics recommends that children should be screened for autism spectrum disorder (ASD) at ages 18 and 24 months. Increasing screening is a national Healthy People objective.

Patients with autism may live independent lives. Between 30% through 55% of those with ASD have comorbid intellectual disability (ID). It is unknown whether ID and ASD are correlated because they share a genetic basis or that the cognitive disability exposes limitations in social skills.

The etiology of ASD is heavily researched. Environmental factors may interplay with genetic factors to produce autism. Multiple studies have implicated that genetic structural variations cause autism. This variations may occur through epigenetics (the passage of genetic information not encoded in chromosomes) or through chromosomal genetics. Genetic involvement is further shown in the observation that males are over four times more likely than females to develop ASD.

There are numerous associations between potential risk factors and development of ASD. Genetics were previously mentioned. During pregnancy, risk factors include infection during the first trimester, exposure to organophosphates, gestational diabetes, maternal bleeding, and maternal medication use (valproate increases ASD risk eight times). Perinatal risk factors include umbilical cord complications, fetal distress, birth injury, low birth weight, prematurity, low APGAR score, hyperbilirubinemia, meconium aspiration, and ABO or Rh incompatibility. Children born from mothers born abroad are also at increased risk of being diagnosed with autism.

No potential risk factor for ASD has garnered as much attention as vaccinations. Multiple studies have demonstrated that vaccinations are not correlated with autism.

102. B. Heart disease

Disability adjusted life years (DALYs) are a measure that combines YLD and time lost to premature death. By using this metric, nonfatal illness can be considered along with mortality in weighing in on the severity of disease in measures such as cost-effectiveness analysis. One DALY may be considered a year of healthy life lost and a value of zero represents perfect health.

$$\text{DALY} = \text{YLD} + \text{YLL},$$

where:

YLD = Years lived with disability

YLL = Years of life lost

There is no universal consensus of the number of DALYs worldwide. According to the WHO's Global Burden of Disease, the top ten worldwide list of DALYs is as follows:

1. Ischemic heart disease
2. Lower respiratory infections
3. Stroke
4. Diarrheal disease
5. HIV/AIDS
6. Malaria
7. Low back pain
8. Premature birth complications
9. COPD
10. Road injury

As developing countries continue to experience the shift from infectious to chronic diseases, this list will continue to evolve.

103. A. Mental illness (including depression)

According to the WHO, the top ten worldwide causes of YLDs is as follows:

1. Unipolar depressive disorders
2. Back and neck pain
3. Iron-deficiency anemia
4. COPD
5. Anxiety disorders
6. DM
7. Hearing loss
8. Falls
9. Migraines
10. Osteoarthritis

104. D. 25%

Tobacco addiction typically begins during youth and young adulthood. In 2015, 25.3% of all high school students reported current use of a tobacco product within the 30 days prior to being interviewed. On an average, boys smoked more than girls, while White students smoked slightly more than all other groups. Among White students, Hispanic students and students classified as other, electronic cigarettes were the most commonly used form of tobacco. Among Black students, cigar use was most common.

105. C. Electronic cigarettes

Electronic Nicotine Delivery Systems (ENDS), including electronic cigarettes have become the most commonly used tobacco products among high school students, with roughly 16% using one in the past 30 days. Cigarettes (9.3%) and cigars (8.6%) smoking are on the decline, but still represent the second and third most common form of tobacco use in high school students.

106. A. Asian

In the United States, nearly 20.4% of men ≥18 years old smokes cigarettes. The prevalence of cigarette smoking in men is listed in ascending order below:

- Asian
- Hispanic
- Non-Hispanic Black
- Non-Hispanic White
- American Indian/Alaska native

107. E. 90%

Nearly 90% of cigarette smokers first light up prior to their 18th birthday and 99% of cigarette smokers first tried smoking by their 26th birthday. Although the number of cigarette smokers is decreasing in middle and high school students, the number of electronic cigarette users is increasing as is the number of hookah users.

108. C. Cigarette smokers with mental illness smoke more per day than those without mental illness

The only correct option is that cigarette smokers with mental illness smoke more cigarettes daily than those without mental illness. Those with mental illness are more likely to be smokers than those without mental illness. Adults experiencing psychological distress have greater odds of tobacco use than those without distress. Furthermore, while the smoking rate in the population without mental illness is decreasing, rates remain the same in the population with mental illness.

109. D. High social support

Having a supportive social network is a protective factor for alcohol abuse. Alcohol abuse is found more frequently in those with antisocial personality and anxiety. Excessive alcohol use is more common in specific cultures. There is a genetic component of alcohol abuse.

110. C. Four drinks in one occasion

Binge drinking is defined as consuming five or more drinks in one occasion for men or four or more drinks in one occasion for women.

111. E. Alter

The 5 A's of smoking cessation are listed in the table below.

Action	Description
Ask	Identify those that use tobacco
Advise	Urge to quit and explain benefits
Assess	Determine willingness to quit
Assist	Provide resources and support to quit
Arrange	Schedule follow up contact.

112. B. Firearms

Firearms are the leading method of suicide in men, accounting for roughly 55% of all successful suicides. To the contrary, poisoning (including medication overdose) is the most common means of suicide in women, accounting for nearly one in three successful suicides.

113. B. 150 min of moderate-intensity activity or 75 min of vigorous-intense activity or a combination of both.

The Department of Health and Human Service's Physical Activity Guidelines Advisory Committee recommends that adults engage in 150 min of moderate-intensity activity or 75 min of vigorous-intense activity weekly or equivalent combinations of both. In addition, the committee states that adults should participate in muscle-strengthening activities involving the major muscle groups on at least two days weekly. About half of adults in the United States achieve these guidelines. Children and adolescents are encouraged to partake in at least 60 min of physical activity daily. Meanwhile, older adults should exercise in accordance with their physical abilities and if possible perform exercises that will maintain or improve balance if they have a risk of falling.

The Physical Activity Guidelines Advisory Committee recommendations are also endorsed by the US National Physical Activity Plan Alliance (NPAPA), a nonprofit partially government funded group tasked with developing the US National Physical Activity Plan (NPAP). The goal of NPAP is to facilitate action to help society address the American pandemic of physical inactivity. It does so by providing targeted strategies to obtain the goals set forth by the physical activity guidelines in varied societal environments.

114. B. Colonoscopy for CRC

Secondary prevention includes all screening programs, including the five potential answers to the question. By removing polyps before they develop into cancer, colonoscopies achieve both primary and secondary prevention.

115. D. Taking a statin drug to maintain appropriate cholesterol

Outside of option D which is an example of secondary prevention, all of the other options are examples of primary prevention.

Primary prevention activities aim to prevent a health ailment before it initiates. Activities of primary prevention include health education, environmental modification, and individually specific protective measures. Examples of health education include the acquisition of knowledge leading to practicing a healthy diet, exercising, and avoiding recreational drug use. Environmental modification achieves primary prevention by engineering or administering practices that reduce injuries and diseases from falls, fires, automobile accidents, dirty water, and polluted air. Specific protective measures that achieve primary prevention include condom use to prevent unwanted pregnancy and vaccination to prevent disease.

Secondary prevention includes the early detection and treatment of disease. It includes all screening tests and treatment that prevents development of a disease. For example, a screening test may detect elevated cholesterol and then early treatment reduces risk of stroke.

Tertiary prevention occurs once the disease has occurred. In the example of a stroke, tertiary prevention takes place once the stroke has occurred to prevent future complications and work toward rehabilitation.

116. E. All of the above options are considered

There is an insufficient supply of viable human organs for transplantation to meet the ever-growing demand. Therefore, explicit rationing is implemented to determine who receives organs. The UNOS is the national system for matching organ donors and recipients. UNOS maintains the OPTN's policies which prioritize recipients on a point scale according to probability of successful transplantation outcome, severity of illness, and time on the waiting list. Additionally, there is criteria stating contraindications to transplantation, including the reason the organ is needed. For example, alcoholics with cirrhosis are typically not considered candidates for liver transplantation until a specified period of alcohol abstinence has been achieved.

117. A. Advisory Committee on Heritable Disorders in Newborns and Children

Healthy appearing infants may be born with unseen disorders in which pathologic development may be averted if screened and addressed soon enough. As a form of primary prevention, each state mandates their own universal newborn screening tests to approve the health outcome of affected children. The process of newborn screening is a coordinated effort consisting of education, diagnosis, treatment, follow-up, management, and program evaluation.

Because each state mandates their own screening panel, there is variation of the number and type of mandatory screening tests between states. For this reason, the Advisory Committee on

Heritable Disorders in Newborns and Children was founded within the Health Resources and Services Administration (HRSA) of the Department of Health and Human Services (DHHS). The purpose of the Advisory Committee on Heritable Disorders in Newborns and Children is to recommend universal screening tests to the states to consider when fashioning their own state mandates.

The Advisory Committee on Heritable Disorders in Newborns and Children maintains a list of both core disorders and secondary disorders based on scientific evidence and expert opinion. These tests identify diseases at an early time (24–48 hours after birth) at which it would not be clinically detected. The tests boast high sensitivity and specificity and are accompanied by test algorithms. The tests have available efficacious treatments with clear benefits of early intervention that prevent morbidity and mortality.

The core conditions are listed in the table below.

Core conditions recommended newborn screening panel, as recommended by the Advisory Committee on Heritable Disorders in Newborns and Children

Organic acid conditions	Propionic acidemia, methylmalonic acidemia (methylmalonyl-CoA mutase), methylmalonic acidemia (cobalamin disorders), isovaleric acidemia, 3-methylcrotonyl-CoA carboxylase deficiency, 3-hydroxy-3-methylglutaric aciduria, holocarboxylase synthetase deficiency, β-ketothiolase deficiency, glutaric acidemia type 1
Fatty acid oxidation disorders	Carnitine uptake defect/carnitine transport defect, medium-chain acyl-CoA dehydrogenase deficiency, very long-chain acyl-CoA dehydrogenase deficiency, long-chain L-3 hydroxyacyl-CoA dehydrogenase deficiency, trifunctional protein deficiency
Amino acid disorders	Argininosuccinic aciduria, citrullinemia type 1, maple syrup urine disease, homocystinuria, phenylketonuria, tyrosinemia type 1
Endocrine disorder	Primary congenital hypothyroidism, congenital adrenal hyperplasia
Hemoglobin disorder	S,S disease, S, β-thalassemia, S,C disease
Other disorders	Biotinidase deficiency, critical congenital heart disease, cystic fibrosis, classic galactosemia, glycogen storage disease type II (Pompe), hearing loss, severe combined immunodeficiencies

118. A. Congenital hypothyroidism

Congenital hypothyroidism is insufficient thyroid hormone production in newborns. It may result from defects in thyroid development, errors in thyroid hormone synthesis, thyroid hormone receptor defects, radiation exposure, and iodine deficiency (known as cretinism). Newborns with congenital hypothyroidism appear normal for the first several months before developing physical and mental abnormalities. Intellectual disability may be profound. All newborns should be screened for hypothyroidism at birth. Due to an initial TSH surge after birth, the preferred screening test is T4. If treated within the first month of life, the physical and mental impairments may be averted.

Fragile X syndrome is the most commonly inherited form of mental retardation. It is more pronounced in males as their lone X chromosome is affected.

PKU is another condition that should be routinely screened in newborns. It is an inherited disorder that causes phenylalanine to build up in the body if certain foods are consumed. High levels of phenylalanine may lead to intellectual disability, social problems, and changes in the skin/hair. By screening early and avoiding specific foods, effects of PKU may be averted.

Spina bifida is a type of neural tube defect. Depending on the severity, it may also lead to physical and intellectual disabilities. Causes include genetics, lack of folic acid in the mother's diet, and epilepsy medication taken during pregnancy.

Trisomy 21 is also known as Down syndrome. It is the most common chromosomal abnormality. Intellectual disability typically results from trisomy 21.

119. B. Cystic fibrosis

This question addresses the need for prenatal screening. Carrier screening is used to identify those that carry genes of a disease of interest in a specific population. The American Congress of Obstetricians and Gynecologists (ACOG), as well as other organizations, recommend carrier screening for selected disorders.

Screening tests should be voluntary and obtained confidentially with informed consent. Both parents may be screened, although it is also acceptable to screen the partner at risk first. If the partner at risk is found to be a disease carrier, then the companion should be screened.

Roughly 90% of those with Jewish ancestry in the United States are Ashkenazi. ACOG recommends that if one or both of the future parents is Ashkenazi, they should be screened for a select panel of diseases. These diseases include Tay–Sachs, cystic fibrosis, Canavan disease, familial dysautonomia, mucolipidosis IV, Niemann–Pick disease type A, Fanconi anemia type C, Bloom syndrome, and Gaucher disease.

If the partners are found to be disease carriers, genetic counseling is to be offered. Counseling should include education of the disease and relative risk that children will be symptomatic. Furthermore, options such as donor gametes (sperm and ova) and adoption should be discussed.

Outside of cystic fibrosis, none of the other options listed are more common in Ashkenazi Jews than the general population.

120. D. Mother

Roughly 37% of child abuse cases are perpetrated exclusively by the victim's mother. Meanwhile, 19% of child abuse of child abuse cases are perpetrated exclusively by the victim's father. Mother and father combinations account for 19% of child abuse cases. Daycare providers only account for 0.4%, family friends account for 0.2%, and romantic partners of parents account for 2.7% of child abuse cases.

121. A. Anencephaly

Healthy appearing infants may be born with unseen disorders in which pathologic development may be averted if screened and addressed soon enough. As a form of primary prevention, each state mandates their own universal newborn screening tests to approve the health outcome of affected children. The purpose of the Advisory Committee on Heritable Disorders in Newborns and Children is to recommend universal screening tests to the states to consider when fashioning their own state mandates.

These tests identify diseases at an early time (24–48 hours after birth) at which it would not be clinically detected. The tests boast high sensitivity and specificity. All of the tests have efficacious treatments available with clear benefits of early intervention that prevent morbidity and mortality.

Anencephaly does not meet primary criteria to be on the recommended newborn screening tests by the Advisory Committee on Heritable Disorders in Newborns and Children. This is because

Anencephaly is easily clinically detected and there is no effective medical intervention.

122. C. Most perpetrators of rape of a female are unknown by the victim

IPV occurs when there is threatened or completed physical, sexual, or psychological abuse by a current or past intimate partner. This includes stalking. Roughly one in four women and one in seven men have endured severe physical violence by an intimate partner. Moreover, approximately one in five women and one in 71 men have been raped or attempted to be raped. More than half of female rape victims were victimized by intimate partners. The majority of nonintimate perpetrators of rape are personal acquaintances.

IPV and rape are most common in the under 25 age group. Roughly 80% of victims of rape experienced their first rape by the age of 25.

Survivors of IPV may suffer long-term mental health consequences including depression, anxiety, insomnia, low self-esteem, substance abuse, and suicidal ideations. Additionally, victims of IPV may endure long-term physical manifestations of gastrointestinal illness, gynecological complications, chronic pain, and poor overall physical health.

Male victims of rape and unwanted sexual involvement report primarily male perpetrators. Half of all cases of stalking a male is performed by a male. Nonsexual IPV against men is chiefly committed by women.

123. E. All of the above are true

IPV, including IPV of pregnancy, constitutes a significant public health problem worldwide. Rates of pregnancy IPV differ across countries. Due to the privacy of domestic violence and differing severities of abuse, cases are widely underreported and rates are difficult to calculate. The US Department of Justice estimates that over a lifetime roughly 52% of women experience IPV. It is especially common in younger people as IPV is a leading cause of injuries and death of women of childbearing age in the United States.

Studies show mixed results on whether IPV increases or decreases during pregnancy. One certainty is that lower SES is a risk factor for IPV in all groups. Income and education levels are the two most significant risk factors for pregnancy IPV. Other risk factors for pregnancy IPV include younger age, nonmarried status, and belonging to a minority group. Native Americans and Blacks have the highest incidence of pregnancy IPV.

IPV during pregnancy has been associated with preterm birth and low birth weight. However, it is not clear how much this is confounded by other factors such as substance abuse and lower SES.

124. A. Family member

Elder abuse may occur as neglect, physical abuse, financial abuse, sexual abuse, and verbal/psychological abuse. Most of the time, elder abuse is caused by family members with adult children being the most common offender.

Due to increased clinician education and recent legislation requiring reporting of elder abuse, the number of elder abuse cases has risen. However, the current numbers are still felt to be widely underreported.

Elders that endure abuse have a high risk of death compared to those that had not been abused. Health problems resulting from elder abuse include depression, anxiety, chronic pain, hypertension, and cardiomyopathy.

125. A. 0–2

The youngest children are the most vulnerable to maltreatment. More than quarter victims of child abuse are less than three years old. More specifically, children younger than one year old have the highest rate of victimization. Although White children make up the highest number of child abuse victims in the United States (~44% of all child abuse cases), Black children suffer the highest rate of victimization per race (>14/1000).

126. A. Neglect

Neglect accounts for nearly 80% of all cases of child abuse.

127. B. Ischemic heart disease

The annual worldwide mortality attributed to secondhand smoke has been estimated to be as high as 1% of all deaths. Although this number is debated by some, one thing that is not debated is that there is no safe level of exposure to tobacco smoke. Environmental tobacco smoke (ETS) from cigarettes and cigars is caused by both secondhand and thirdhand smoke.

The largest number of estimated deaths attributed to secondhand smoke is due to ischemic heart disease. In descending order following ischemic heart disease, the next most attributed deaths resulting from secondhand smoke come from lower respiratory infections, asthma, and lung cancer.

The International Agency for Research on Cancer (IARC) has stated that banning smoking in the workplace reduces acute coronary events by 10%–20% after the first year of the ban. Considering that tobacco smoke is one of the most common indoor pollutants worldwide, there are many workers that would benefit from workplace smoking bans. Smokefree policies not only reduce cigarette consumption, they aid increasing rates of tobacco cessation and contribute to the effort to denormalize smoking. Worldwide, children suffer from the highest exposure to tobacco smoke, most commonly at home.

128. D. Vitamin D

Vitamin D has the ability to increase serum calcium. Deficiency of this key vitamin may result in rickets or osteomalacia. It is synthesized by the body after being exposed to sunlight and may also be consumed in the diet. Many foods, including milk, yogurt, orange juice and soy beverages, are fortified with vitamin D. The recommended daily allowance (RDA) in children and adults is 600 IU, while adults 70 years and older should consume at least 800 IU daily.

129. D. Lowering of blood pressure

High dietary potassium blunts the adverse effects of a diet high in sodium. Because of this, high-potassium diets are associated with lowering of blood pressure and decreased risk of heart attack and stroke. Other benefits include decreased bone loss and decreased risk of kidney stone formation. The adequate intake (AI) of potassium is 4700 mg per day.

130. C. Resting energy expenditure (REE)

Understanding nutrition and calorie balance is important in fighting the obesity epidemic. Americans as a whole are in a calorie imbalance, where more calories are being stored than spent.

Total energy expenditure (TEE) reflects the sum of all energy spent by the body. TEE includes resting energy expenditure, the thermic effect of food and energy used for physical activity:

TEE = resting energy expenditure + thermic effect of food + energy used for physical activity

Basal metabolic rate (BMR) represents the energy that the body uses at complete rest. It includes functions such as heartbeat, muscle tone, and respiration. Resting metabolic rate (RMR) is similar to

BMR, with the difference being that RMR is easier to measure diagnostically than BMR because BMR requires very stringent testing requirements. Basal energy expenditure (BEE) and resting energy expenditure (REE) are the BMR and RMR, respectively, extrapolated to 24 hours. The REE/BEE represents the majority of TEE. It increases with muscle mass.

Energy used for physical activity represents the second most common source of TEE. In addition to physical activity such as running, it also includes activities such as fidgeting. This classification also includes nonexercise activity of thermogenesis (NEAT) which is heat produced through nonintentional physical activity.

The thermal effect of food represents the lowest contributor to TEE. It is the heat and energy that is released during the digestive process.

Estimated energy requirements represents the energy intake that will maintain appropriate energy balance in a healthy individual.

131. E. >20%

Calories from fat consumption are essential in the human diet. According to the Dietary Guidelines for Americans, published every five years in a joint effort from the US Department of Agriculture (USDA) and Department of Health and Human Services (DHHS), between 20%–35% of daily calories consumed by adults should be from fats. Recommended caloric fat intakes are 30%–40% in children aged one to three and 25%–35% in those aged four to 18 years.

Dietary fats can further be broken down to include monounsaturated, polyunsaturated, saturated, and transfatty acids. Monosaturated fats are typically liquid at room temperature. These fats are present in olive oil, canola oil, nuts, and avocados. Consumption of monosaturated fats is associated with a more favorable cardiovascular profile. Polysaturated fats are found more frequently in butters and animal meats. They are typically solid at room temperature and yield a less favorable cardiovascular profile. It is widely recommended that less than 7%–10% of calories consumed should be from saturated fatty acids. Most Americans consume too much saturated fats and not enough unsaturated fats. The human body is capable of producing its own saturated fatty acids for physiological and structural functions and there is no dietary requirement for saturated fatty acid consumption.

132. A. <1%

Transfatty acids may be found naturally or as a product of food processing. They are stable for longer periods than other fatty acids and therefore have longer shelf lives at the store. They are not essential to the body and have been associated with an especially unfavorable cardiovascular profile. It is recommended to consume as few calories from transfatty acids as possible.

Fats were further explained in answer #131 (directly above).

133. E. Salt

Adding nutrients to food is a form of primary prevention that addresses population-wide nutrient deficiencies. There are three scenarios in which nutrients are added to food: In order to restore nutrients lost during food processing, to increase the level already present, and to add nutrients that were not present in the first place (fortification). In the United States, food enrichment and fortification is governed by the FDA.

Lack of dietary iodine causes hypothyroidism and associated disorders such as cretinism. By adding iodine to salt, public health officials have been able to reduce the incidence of hypothyroidism.

The table below shows examples of nutrients that have been added to foods.

Nutrient	Food that nutrient is added to
Calcium	Orange juice
Iodine	Salt
Iron	Wheat flour, milk
Omega fatty acids	Eggs
Vitamin A	Margarine, milk, oil, sugar flour
Vitamin B (includes folic acid)	Bread, cereals, wheat flour
Vitamin D	Milk, orange juice, soy milk

134. D. Vitamin A

Nutrient enrichment and fortification was explained in answer #133 (directly above).

Of the options listed, only vitamin A causes night blindness.

135. E. All of the above are true.

Disease management is a coordinated multidiscipline approach to healthcare of a specific disease, in addition to prevention and sometimes care of related comorbid conditions. By pooling healthcare resources and defragmenting care across the spectrum of the disease,

disease management is a cost-effective way to manage disease and increase health outcomes. Challenges to each disease management program include correctly identifying patients that would benefit from the program and determining the most effective way to treat the disease.

BIBLIOGRAPHY

[1] Bantle JP, Wylie-Rosett J, Albright AL, Apovian CM, Clark NG, Franz MJ, et al. Nutrition recommendations and interventions for diabetes: a position statement of the American Diabetes Association. Diabetes Care 2008;31 (Suppl 1):S61–8.

[2–3] Final update summary: Breast cancer: Screening. <https://www.uspreventiveservicestaskforce.org/Page/Document/UpdateSummaryFinal/breast-cancer-screening1?ds = 1&s = breast> [accessed 20.10.2016].

[4a] Final update summary: Colorectal cancer: Screening. <https://www.uspreventiveservicestaskforce.org/Page/Document/UpdateSummaryFinal/colorectal-cancer-screening2> [accessed 24.10.2016].

[4b] American Cancer Society recommendations for colorectal cancer early detection. American Cancer Society. <http://www.cancer.org/cancer/colonandrectumcancer/moreinformation/colonandrectumcancerearlydetection/colorectal-cancer-early-detection-acs-recommendations> [accessed 20.10.2016].

[4c] Rex DK, Johnson DA, Anderson JC, Schoenfeld PS, Burke CA, Inadomi JM. American College of Gastroenterology guidelines for colorectal cancer screening 2008. Am J Gastroenterol 2009;104(3):739–50.

[4d] Qaseem A, Denberg T, Hopkins R, et al. Screening for colorectal cancer: A guidance statement from the American college of physicians. Ann Intern Med 2012;156(5):378.

[5–6a] Final update summary: Lipid disorders in adults (cholesterol, Dyslipidemia): Screening. <http://www.uspreventiveservicestaskforce.org/Page/Document/UpdateSummaryFinal/lipid-disorders-in-adults-cholesterol-dyslipidemia-screening> [accessed 24.10.2016].

[6b] Position Statement- Public cholesterol screening (adults and children). American Heart Association. <http://www.heart.org/HEARTORG/Conditions/Cholesterol/SymptomsDiagnosisMonitoringofHighCholesterol/Public-Cholesterol-Screening-Adults- and-Children_UCM_305617_Article.jsp#.Vua5j0_SmUk> [accessed 24.10.2016].

[7–8] Final update summary: Aspirin use to prevent cardiovascular disease and Colorectal cancer: Preventive medication. <http://www.uspreventiveservicestaskforce.org/Page/Document/UpdateSummaryFinal/aspirin-to-prevent-cardiovascular-disease-and-cancer?ds = 1&s = aspirin> [accessed 24.10.2016].

[9] Methods and processes. United States Preventive Services Task Force (USPSTF). <http://www.uspreventiveservicestaskforce.org/Page/Name/methods-and-processes> [accessed 24.10.2016].

[10] Understanding how the USPSTF works: USPSTF 101. <https://www.uspreventiveservicestaskforce.org/Page/Name/understanding-how-the-uspstf-works> [accessed 24.10.2016].

[11] Final update summary: Depression in adults: Screening. <https://www.uspreventiveservicestaskforce.org/Page/Document/UpdateSummaryFinal/depression-in-adults-screening1?ds = 1&s = depression> [accessed 24.10.2016].

[12] Final update summary: Obesity in children and adolescents: Screening. <http://www.uspreventiveservicestaskforce.org/Page/Topic/recommendation-summary/obesity-in-children-and-adolescents-screening> [accessed 24.10.2016].

[13] USPSTF Draft Prostate Screening Recommendations. <https://screeningforprostatecancer.org/> [accessed 14.04.2017].

[14] Final update summary: Testicular cancer: Screening. <http://www.uspreventiveservicestaskforce.org/Page/Topic/recommendation-summary/testicular-cancer-screening> [accessed 24.10.2016].

[15] Final recommendation statement: Tobacco use in children and adolescents: Primary care interventions. <http://www.uspreventiveservicestaskforce.org/Page/Document/RecommendationStatementFinal/tobacco-use-in-children-and-adolescents-primary-care-interventions#consider> [accessed 24.10.2016].

[16] Favier K. Humane health care for prisoners: ethical and legal challenges. Praeger: Santa Barbara, CA; 2017 [chapter 1].

[17] Final update summary: Lung cancer: Screening. <https://www.uspreventiveservicestaskforce.org/Page/Document/UpdateSummaryFinal/lung-cancer-screening> [accessed 24.10.2016].

[18] Final update summary: Cervical cancer: Screening. <https://www.uspreventiveservicestaskforce.org/Page/Document/UpdateSummaryFinal/cervical-cancer-screening?ds = 1&s = hpv> [accessed 24.10.2016].

[19] Final update summary: Coronary heart disease: Screening with Electrocardiography. <https://www.uspreventiveservicestaskforce.org/Page/Document/UpdateSummaryFinal/coronary-heart-disease-screening-with-electrocardiography> [accessed 24.10.2016].

[20] WHO. The top 10 causes of death. World Health Organization. <http://www.who.int/mediacentre/factsheets/fs310/en/> [accessed 24.10.2016].

[21] James PA, Oparil S, Carter BL, et al. Evidence-based guideline for the management of high blood pressure in adults. JAMA 2014;311(5):507.

[22] CDC. Leading causes of death. National Center for Health Statistics. <http://www.cdc.gov/nchs/fastats/leading-causes-of-death.htm> [accessed 24.10.2016].

[23–24] James PA, Oparil S, Carter BL, et al. Evidence-based guideline for the management of high blood pressure in adults. JAMA 2014;311(5):507.

[25–26] Mozaffarian D, Benjamin EJ, Go AS, Arnett DK, Blaha MJ, Cushman M, de Ferranti S, DesprésJ-P; on behalf of the American Heart Association Statistics Committee and Stroke Statistics Subcommittee. Heart disease and stroke statistics—2015 update: A report from the American Heart Association [published online ahead of print December 17, 2014]. Circulation. doi: 10.1161/CIR.0000000000000152.

[27] Fiebach NH, Kern DE, Thomas PA, Ziegelstein RC, Barker L, Zieve PD, editors. Barker, Burton, and Zieve's principles of ambulatory medicine. 7th ed Philadelphia: Lippincott Williams and Wilkins; 2006.

[28] James PA, Oparil S, Carter BL, et al. Evidence-based guideline for the management of high blood pressure in adults. JAMA 2014;311(5):507.

[29] Bope ET, Kellerman RD. Conn's current therapy: 2014. Philadelphia, PA: Saunders; 2014. p. 445.

[30] National Cholesterol Education Program (NCEP) Expert Panel on Detection, Evaluation, and Treatment of High Blood Cholesterol in Adults (Adult Treatment Panel III). Third report of the National Cholesterol Education Program (NCEP) expert panel on detection, evaluation, and

treatment of high blood cholesterol in adults (Adult Treatment Panel III) final report. Circulation 2002;106(25):3143–421.

[31] Fielding JE, Teutsch SM, Caldwell S. Public health practice: What works. New York, NY: OUP USA. Copyright; 2013. p. 134–8.

[32a] Bope ET, Kellerman RD. Conn's current therapy: 2014. Philadelphia, PA: Saunders; 2014. p. 385–90.

[32b] Centers for Disease Control and Prevention (CDC). Chronic obstructive pulmonary disease among adults—United States, 2011. MMWR. 2012;61:938–943.

[33a] 2015 UNOS annual report. United Network for Organ Sharing. <https://www.unos.org/about/annual-report/> [accessed 24.10.2016].

[33b] Matas AJ, Smith JM, Skeans MA, et al. OPTN/SRTR 2012 annual data report: Kidney. Am J Transplant 2014;14(Suppl 1):11–44.

[34] Anand P, Kunnumakara AB, Sundaram C, et al. Cancer is a preventable disease that requires major lifestyle changes. Pharm Res 2008;25(9):2097–116.

[35] Braun CA, Anderson CM. Pathophysiology: Functional alterations in human health. Philadelphia, PA: Lippincott Williams & Wilkins; 2007. p. 156–7.

[36–37] Lifetime Risk (Percent) of Being Diagnosed with Cancer by Site and Race/Ethnicity, Both Sexes, 18 SEER Areas, 2009-2011. SEER Cancer Statistics Review, National Cancer Institute. <http://seer.cancer.gov/archive/csr/1975_2011/results_merged/topic_lifetime_risk_diagnosis.pdf> [accessed 25.10.2016].

[38] Cancer Facts & Figures 2015. American Cancer Society. <http://www.cancer.org/acs/groups/content/@editorial/documents/document/acspc-044552.pdf>. Published 2015 [accessed 25.10.2016].

[39] Jemal A, Bray F, Center MM, Ferlay J, Ward E, Forman D. Global cancer statistics. CA Cancer J Clin 2011;61(2):69–90.

[40] Tollefsbol T. Transgenerational epigenetics, evidence and debate, 129–130. Waltham, MA: Elsevier; 2014. p. 312.

[41] Cubie HA. Diseases associated with human papillomavirus infection. Virology 2013;445(1–2):21–34.

[42] Final recommendation statement: Cervical cancer: Screening. <https://www.uspreventiveservicestaskforce.org/Page/Document/RecommendationStatementFinal/cervical-cancer-screening#consider> [accessed 25.10.2016].

[43–45] Jemal A, Bray F, Center MM, Ferlay J, Ward E, Forman D. Global cancer statistics. CA Cancer J Clin 2011;61(2):69–90.

[46–47] Vincent A, Herman J, Schulick R, Hruban RH, Goggins M. Pancreatic cancer. Lancet 2011;378(9791):607–20 <http://www.sciencedirect.com/science/article/pii/S0140673610623070> [accessed 25.10.2016]

[48] Cancer Facts & Figures 2015. American Cancer Society. <http://www.cancer.org/acs/groups/content/@editorial/documents/document/acspc-044552.pdf>. Published 2015 [accessed 25.10.2016].

[49] U.S. Cancer Statistics Working Group. United States Cancer Statistics: 1999–2013 Incidence and Mortality Web-based Report. Atlanta (GA): Department of Health and Human Services, Centers for Disease Control and Prevention, and National Cancer Institute; 2016. Available at: <http://www.cdc.gov/uscs>.

[50] Cancer Facts & Figures 2015. American Cancer Society. <http://www.cancer.org/acs/groups/content/@editorial/documents/document/acspc-044552.pdf> [accessed 25.10.2016] Published 2015.

[51] Jemal A, Bray F, Center MM, Ferlay J, Ward E, Forman D. Global cancer statistics. CA Cancer J Clin 2011;61(2):69–90.

[52–53] Bope ET, Kellerman RD. Conn's current therapy: 2014. Philadelphia, PA: Saunders; 2014. p. 1015–6.

[54] Loeffler AG, Hart MN. Introduction to human disease: Pathophysiology for health professionals. 6th ed Burlington, MA: Jones & Bartlett; 2015. p. 252.

[55–56] Cancer Facts & Figures 2015. American Cancer Society. <http://www.cancer.org/acs/groups/content/@editorial/documents/document/acspc-044552.pdf> [accessed 25.10.2016] Published 2015.

[57] Wolff K, Johnson R, Saavedra A. Fitzpatrick's color atlas and synopsis of clinical dermatology. 7th ed. New York, NY: McGraw-Hill Education/Medical; 2013. p. 254–5.

[58] Faber E, Ferrara L, Slyer J. Primary care: An interprofessional perspective. 2nd ed New York, NY: Springer Publishing Co; 2015. p. 947.

[59] Final update summary: Testicular cancer: Screening. <https://www.uspreventiveservicestaskforce.org/Page/Document/UpdateSummaryFinal/testicular-cancer-screening> [accessed 25.10.2016].

[60] Stansifer KJ, Guynan JF, Wachal BM, Smith RB. Modifiable risk factors and thyroid cancer. Otolaryngol Head Neck Surg 2014;152(3):432–7.

[61] Bope ET, Kellerman RD. Conn's current therapy: 2014. Philadelphia, PA: Saunders; 2014. p. 896–7.

[62] About the OPTN. Organ Procurement and Transplantation Network. <https://optn.transplant.hrsa.gov/governance/about-the-optn/> [accessed 28.10.2016].

[63] Mozaffarian D, Benjamin EJ, Go AS, Arnett DK, Blaha MJ, Cushman M, de Ferranti S, on behalf of the American Heart Association Statistics Committee and Stroke Statistics Subcommittee. Heart disease and stroke statistics—2015 update: A report from the American Heart Association. Circulation. 2015;131:e29–e322.

[64] Centers for Disease Control and Prevention (CDC). National Chronic Kidney Disease Fact Sheet: General Information and National Estimates on Chronic Kidney Disease in the United States, 2014. Atlanta, GA: US Department of Health and Human Services, Centers for Disease Control and Prevention; 2014.

[65–68] U.S. Department of Agriculture and U.S. Department of Health and Human Services. Dietary Guidelines for Americans, 2010. 7th Edition, Washington, DC: U.S. Government Printing Office, December 2010.

[69] U.S. Department of Agriculture and U.S. Department of Health and Human Services. Dietary Guidelines for Americans, 2010. 7th Edition, Washington, DC: U.S. Government Printing Office, December 2010.

[70] Childhood obesity prevention program. <http://www.letsgo.org/> [accessed 28.10.2016].

[71] Nnakwe N. Illinois State University. Community nutrition: Planning health promotion and disease prevention, 24. Burlington, MA: Jones & Bartlett Learning; 2013. p. 255.

[72] Lenzi A, Migliaccio S, Donini LM. Multidisciplinary approach to obesity: From assessment to treatment. Switzerland: Springer; 2015. p. 83.

[73–74] Fiebach NH, Kern DE, Thomas PA, Ziegelstein RC, Barker L, Zieve PD, editors. Barker, Burton, and Zieve's principles of ambulatory medicine. 7th ed Philadelphia: Lippincott Williams and Wilkins; 2006.

[75] Escott-Stump S. Nutrition and diagnosis-related care. 6th ed Baltimore, MD: Lippincott Williams & Wilkins; 2008. p. 115.

[76] Bope ET, Kellerman RD. Conn's current therapy: 2014. Philadelphia, PA: Saunders; 2014. p. 48–52.
[77a] Barbour KE, Helmick CG, Theis KA, et al. Centers for Disease Control and Prevention (CDC). Prevalence of doctor-diagnosed arthritis and arthritis-attributable activity limitation-United States, 2010-2012. MMWR Morb Mortal Wkly Rep 2013;62(44):869–73.
[77b] Centers for Disease Control and Prevention (CDC). Prevalence and most common causes of disability among adults--United States, 2005. MMWR Morb Mortal Wkly Rep 2009: 58:421–426.
[78] Sterling DA, O'Connor JA, Bonadies J. Geriatric falls: Injury severity is high and disproportionate to mechanism. J Trauma 2001;50(1):116–19.
[79a] Dye BA, Thornton-Evans G, Li X, Iafolla TJ. Dental caries and tooth loss in adults in the United States, 2011–2012 no 197 NCHS data brief. Hyattsville, MD: National Center for Health Statistics; 2015
[79b] Manski RJ, Brown E. No. 17 Dental use, expenses, private dental coverage and changes, 1996 and 2004. MEPS chartbook. Rockville MD: Agency for Healthcare Research and Quality; 2007
[79c] Nasseh K, Vujicic M. Dental care utilization continues to decline among working-age adults, increases among the elderly, stable among children. Health Policy Resources Center Research Brief. American Dental Association. October 2013. Available from: <http://www.ada.org/sections/professionalResources/pdfs/HPRCBrief_1013_2.pdf>.
[80] WHO. Visual impairment and blindness. World Health Organization. <http://www.who.int/mediacentre/factsheets/fs282/en/> [accessed 28.10.2016].
[81] Merson MH, Black RE, Mills A. International public health: Diseases, programs, systems, and policies. 2nd ed Sudbury, MA: Jones & Bartlett Learning; 2005. p. 203–11.
[82] Mosher WD, Jones J, Abma JC. no 55 Intended and unintended births in the United States: 1982–2010. National health statistics reports. Hyattsville, MD: National Center for Health Statistics; 2012.
[83] CDC. About SUID and SIDS. Sudden Unexpected Infant Death and Sudden Infant Death Syndrome. <http://www.cdc.gov/sids/aboutsuidandsids.htm> [accessed 28.10.2016].
[84–86] Breastfeeding and the use of human milk. Pediatrics. 2012;129(3): e827–e841. doi:10.1542/peds.2011-3552.
[87–88] Williams J, Mai CT, Mulinare J, et al. Centers for Disease Control and Prevention (CDC). Updated estimates of neural tube defects prevented by mandatory folic acid fortification—United States, 1995–2011. MMWR Morb Mortal Wkly Rep 2015;64(01):1–5.
[89–90] Simmons LE, Rubens CE, Darmstadt GL, Gravett MG. Preventing preterm birth and neonatal mortality: Exploring the epidemiology, causes, and interventions. Semin Perinatol 2010;34(6):408–15.
[91] Chandra A, Copen CE, Stephen EH. no 67 Infertility and impaired fecundity in the United States, 1982–2010: Data from the National Survey of Family Growth. National health statistics reports. Hyattsville, MD: National Center for Health Statistics; 2013.
[92] Howell RR, Terry S, Tait VF, et al. Centers for Disease Control and Prevention (CDC). CDC grand rounds: Newborn screening and improved outcomes. MMWR Morb Mortal Wkly Rep 2012;61:390–3.
[93] Chapman S. Public health advocacy and tobacco control: Making smoking history. Malden, MA: Blackwell Publishing; 2007. p. 118–9.

[94] About DSM-5. American Psychiatric Association. <http://www.dsm5.org/about/Pages/Default.aspx> [accessed 29.10.2016].

[95–96] Kessler RC, Berglund P, Demler O, Jin R, Merikangas KR, Walters EE. Lifetime prevalence and age-of-onset distributions of DSM-IV disorders in the national comorbidity survey replication. Arch Gen Psychiatry 2005;62(7):768.

[97] Rabins P, Blass D. In the clinic. Dementia. Ann Intern Med 2014;161:3.

[98] Ewing JA. Detecting alcoholism. The CAGE questionnaire. JAMA 1984;252(14):1905–7.

[99] Curtin SC, Warner M, Hedegaard H. no 241 Increase in suicide in the United States, 1999–2014. NCHS data brief. Hyattsville, MD: National Center for Health Statistics; 2016.

[100] Suicide. CDC 2015 Facts at a Glance. <https://www.cdc.gov/violenceprevention/pdf/suicide-datasheet-a.pdf> [accessed 29.10.2016].

[101] Chaste P, Leboyer M. Autism risk factors: Genes, environment, and gene-environment interactions. Dialogues Clin Neurosci 2012;14(3):281–92.

[102–103] WHO methods and data sources for global burden of disease estimates 2000-2011. Global Health Estimates Technical Paper WHO/HIS/HSI/GHE/2013.4. Available at <http://www.who.int/healthinfo/statistics/GlobalDALYmethods_2000_2011.pdf>.

[104–105] Singh T, Arrazola RA, Corey CG, Husten CG, Neff LJ, Homa DM, King BA. Centers for Disease Control and Prevention (CDC). Tobacco use among middle and high school students – United States, 2011–2015. Morb Mortal Wkly Rep 2016;65(14):361–7 Accessible online: <https://www.cdc.gov/mmwr/volumes/65/wr/mm6514a1.htm?utm_source = Youth.gov&utm_medium = Announcements&utm_campaign = Reports-and-Resources>.

[106] Mozaffarian D, Benjamin EJ, Go AS, Arnett DK, Blaha MJ, Cushman M, de Ferranti S, DesprésJ-P, Fullerton HJ, Howard VJ, Huffman MD, Judd SE, KisselaBM, Lackland DT, Lichtman JH, Lisabeth LD, Liu S, Mackey RH, Matchar DB, McGuire DK, Mohler ER 3rd, Moy CS, Muntner P, Mussolino ME, Nasir K, Neumar RW, Nichol G, Palaniappan L, Pandey DK, Reeves MJ, Rodriguez CJ, Sorlie PD, Stein J, Towfighi A, Turan TN, Virani SS, Willey JZ, Woo D, Yeh RW, Turner MB; on behalf of the American Heart Association Statistics Committee and Stroke Statistics Subcommittee. Heart disease and stroke statistics—2015 update: a report from the American Heart Association [published online ahead of print December 17, 2014]. Circulation. doi: 10.1161/CIR.0000000000000152.

[107] U.S. Department of Health and Human Services. The Health Consequences of Smoking: 50 Years of Progress. A Report of the Surgeon General. Atlanta, GA: U.S. Department of Health and Human Services, Centers for Disease Control and Prevention, National Center for Chronic Disease Prevention and Health Promotion, Office on Smoking and Health, 2014. Printed with corrections, January 2014.

[108] Steinberg ML, Williams JM, Li Y. Poor mental health and reduced decline in smoking prevalence. Am J Prev Med 2015;49(3):362–9.

[109] Poikolainen K. Risk factors for alcohol dependence: A case-control study. Original Article 2000;35(2):190–6.

[110] Stahre M, Roeber J, Kanny D, Brewer RD, Zhang X. Contribution of excessive alcohol consumption to deaths and years of potential life lost in the United States. Prev Chronic Dis 2014;11.

[111] Five Major Steps to intervention (the "5 A's"). Agency for Healthcare Research and Quality. <http://www.ahrq.gov/professionals/clinicians-providers/guidelines-recommendations/tobacco/5steps.html> [accessed 29.10.2016].

[112] Suicide. CDC 2015 Facts at a Glance. <https://www.cdc.gov/violenceprevention/pdf/suicide-datasheet-a.pdf>[accessed 29.10.2016].

[113a] Physical Activity Guidelines Advisory Committee. Physical Activity Guidelines Advisory Committee Report, 2008. Washington, DC: U.S. Department of Health and Human Services, 2008.

[113b] Kraus WE, Bittner V, Appel L, et al. The national physical activity plan: A call to action from the American heart association: A science advisory from the American heart association. Circulation 2015;131(21):1932–40.

[114–115] Katz DL, Elmore JG, Wild DMG, Lucan SC. Jekel's epidemiology, biostatistics, preventive medicine, and public health. 4th ed Philadelphia, PA: Saunders; 2014. p. 176.

[116] Bodenheimer T, Grumbach K. Understanding health policy: A clinical approach. 6th ed New York, NY, United States: MacGraw Hill; 2012. p. 159.

[117a] American College of Medical Genetics Newborn Screening Expert Group. Newborn screening: Toward a uniform screening panel and system--executive summary. Pediatrics 2006;117:S296–307.

[117b] Recommended uniform screening panel. Advisory Committee on Heritable Disorders in Newborns and Children. <http://www.hrsa.gov/advisorycommittees/mchbadvisory/heritabledisorders/recommendedpanel/> [accessed 29.10.2016].

[118] Hay WW, Levin MJ, Deterding RR. Current diagnosis and treatment pediatrics. 22nd ed New York, NY: McGraw-Hill Education; 2014. p. 1065–7.

[119] Gabbe SG, Niebyl JR, Simpson JL, et al. Obstetrics: Normal and problem pregnancies. 7th ed Philadelphia, PA: Elsevier Health Sciences; 2017. p. 206–7.

[120] U.S. Department of Health and Human Services, Administration for Children and Families, Administration on Children, Youth and Families, Children's Bureau. (2012). Child Maltreatment 2011. Available from <http://www.acf.hhs.gov/programs/cb/research-data-technology/statistics-research/child-maltreatment>.

[121a] American College of Medical Genetics Newborn Screening Expert Group. Newborn screening: Toward a uniform screening panel and system—executive summary. Pediatrics 2006;117: S296–S307.

[121b] Recommended uniform screening panel. Advisory Committee on Heritable Disorders in Newborns and Children. <http://www.hrsa.gov/advisorycommittees/mchbadvisory/heritabledisorders/recommendedpanel/> [accessed 29.10.2016].

[122] Black, M.C., Basile, K.C., Breiding, M.J., Smith, S.G., Walters, M.L., Merrick, M.T., Chen, J., & Stevens, M.R. (2011). The National Intimate Partner and Sexual Violence Survey (NISVS): 2010 Summary Report. Atlanta, GA: National Center for Injury Prevention and Control, Centers for Disease Control and Prevention.

[123] Bailey B. Partner violence during pregnancy: Prevalence, effects, screening, and management. Int J Womens Health 2010;2:183–97.

[124] Amstadter AB, Cisler JM, McCauley JL, Hernandez MA, Muzzy W, Acierno R. Do incident and perpetrator characteristics of elder mistreatment differ by gender of the victim? Results from the national elder mistreatment study. J Elder Abuse Negl 2010;23(1):43–57.

[125–126] U.S. Department of Health and Human Services, Administration for Children and Families, Administration on Children, Youth and Families, Children's Bureau. (2012). Child Maltreatment 2011. Available from <http://www.acf.hhs.gov/programs/cb/research-data-technology/statistics-research/child-maltreatment>.

[127] Öberg M, Jaakkola MS, Woodward A, Peruga A, Prüss-Ustün A. Worldwide burden of disease from exposure to second-hand smoke: A retrospective analysis of data from 192 countries. Lancet 2011;377(9760):139–46 http://dx.doi.org/10.1016/s0140-6736(10)61388-8.
[128–129] U.S. Department of Agriculture and U.S. Department of Health and Human Services. Dietary Guidelines for Americans, 2010. 7th Edition, Washington, DC: U.S. Government Printing Office, December 2010.
[130] Insel P, Ross D, Bernstein M, McMahon K. Discovering nutrition. 5th ed Burlington, MA: Jones & Bartlett Learning; 2015 [chapter 8].
[131–132] U.S. Department of Agriculture and U.S. Department of Health and Human Services. Dietary Guidelines for Americans, 2010. 7th Edition, Washington, DC: U.S. Government Printing Office, December 2010.
[133–134] Preedy VR, Srirajaskanthan R, Patel VB. Handbook of food Fortification and health: From concepts to public, 1. New York, NY: Springer Science & Business Media; 2013. p. 15–29.
[135] Task Force on Community Preventive Services. Recommendations for healthcare system and self-management education interventions to reduce morbidity and mortality from diabetes. Am J Prev Med 2002;22(4S):10–14.

CHAPTER SEVEN

Infectious Disease

7.1 INFECTIOUS DISEASE QUESTIONS

1. Which of the following agents of infectious disease are preventable with an injected live vaccine?
   A. Polio
   B. Influenza
   C. Yellow fever
   D. Hepatitis A
   E. Rabies
2. Which genus of mosquito is the main carrier of malaria?
   A. Aedes
   B. Anopheles
   C. Culex
   D. Hemagogus
   E. Ochlerotatus
3. Which of the following diseases has the World Health Organization (WHO) targeted for eradication?
   A. Dracunculiasis
   B. Gonorrhea
   C. Influenza
   D. Plague
   E. Sudden infant death syndrome
4. Which of the following diseases was the first to be successfully eradicated?
   A. Filariasis
   B. Plague
   C. Poliomyelitis
   D. Smallpox
   E. Tetanus

Board Review in Preventive Medicine and Public Health.
DOI: http://dx.doi.org/10.1016/B978-0-12-813778-9.00007-4

5. Which of the following infectious diseases causes enlargement of the parotid glands?
   A. Measles
   B. Mumps
   C. Rubella
   D. Rubeola
   E. Varicella
6. Which of the following animals is an amplifier host of *West Nile virus* (*WNV*)?
   A. Birds
   B. Horses
   C. Crocodiles
   D. Frogs
   E. Leeches
7. Which of the following can be a host for influenza B?
   A. Birds
   B. Pigs
   C. Horses
   D. Bats
   E. None of the above
8. What is the primary mode of hepatitis A transmission?
   A. Blood borne
   B. Fecal–oral
   C. Respiratory droplets
   D. Sexual
   E. Vertical
9. Despite the warning signs at the all-you-can-eat buffet, a 68-year-old man with Crohn's disease and cirrhosis decided to eat raw oysters. Three days later, he presents to a clinic complaining of chills, sweats, severe abdominal pain, and diarrhea. On physical examination, he had some tender erythematous lesions on his leg that appeared infected.

   Upon further questioning, the man revealed that he went to the buffet with six friends and the four that ate the oysters also developed similar symptoms.

   After collecting stool samples and rendering treatment, the physician then called the epidemiologist at the local health department to report his findings.

   What is the most likely culprit of this patient's symptoms?
   A. *Campylobacter*

B. *Giardia*
C. Hepatitis C
D. Histoplasmosis
E. *Vibrio*

10. Which of the following is true of a ring vaccination strategy?
    A. It builds strong herd immunity
    B. It is best for international travelers
    C. It occurs while an outbreak is occurring
    D. It is best used for common illnesses
    E. All of the above
11. Which route of HIV transmission is categorized as vertical?
    A. Blood transfusion
    B. Breastfeeding
    C. Intravenous drug use
    D. Sexual intercourse
    E. Spread from the bite of an insect
12. Which of the following animals carries the highest risk of transferring toxoplasmosis infection to pregnant women?
    A. Bird
    B. Cat
    C. Dog
    D. Mouse
    E. Turtle
13. Which of the following types of organisms can cause hemolytic uremic syndrome (HUS)?
    A. *Campylobacter*
    B. *Escherichia coli*
    C. *Giardia*
    D. *Salmonella*
    E. *Staphylococcus*
14. Which of the following types of bacteria may produce shiga toxin?
    A. *Campylobacter*
    B. *Escherichia coli*
    C. *Giardia*
    D. *Salmonella*
    E. *Staphylococcus*
15. Worldwide, what is the most common cause of severe gastroenteritis in children under five years old?
    A. *Campylobacter*

B. *Giardia*
C. *Norovirus*
D. *Rotavirus*
E. *Salmonella*

16. Which of the following medications does the US Preventive Services Task Force (USPSTF) recommend to be prophylactically administered after birth to prevent complications conjunctival gonorrhea?
A. Amoxicillin
B. Ceftriaxone
C. Erythromycin
D. Silver Sulfadiazine
E. Trimethoprim–sulfamethoxazole

17. What is the appropriate way to read a tuberculin skin test (TST)?
A. Measure the diameter of the erythema in the direction that yields the highest number
B. Measure the diameter of the erythema from one lateral border to the other
C. Measure the diameter of the erythema from the proximal border to the distal border
D. Measure the diameter of the induration in the direction that yields the highest number
E. Measure the diameter of the induration from one lateral border to the other

18. Which of the following TSTs would be considered positive in a person that has never experienced any induration on a TST?
A. A 4 mm induration in an HIV + patient
B. A 6 mm induration of a healthy person that has a spouse with active TB
C. A 7 mm induration of a healthy adult immigrant from Haiti
D. A 9 mm induration of a prisoner in the United States
E. A 13 mm induration on a medical student without any clinical experience or other exposures to tuberculosis (TB)

19. Which of the following drugs should be taken with pyridoxine (vitamin B6)?
A. Ethambutol
B. Isoniazid
C. Pyrazinamide
D. Rifampin
E. Streptomycin

20. When administering a standard four visit, two-step TB purified protein derivative (PPD), how much time is it recommended to wait before giving the second TST, assuming that the first test is negative?
A. 72 hours to one week
B. 1–3 weeks
C. 4–8 weeks
D. 8–10 weeks
E. 20–40 weeks

21. As a public health officer in a local health department, you are participating in a TB outbreak investigation. How long does it take a TST to convert to positive in someone that was previously negative after exposure to TB?
A. Immediately
B. 1–2 weeks
C. 2–8 weeks
D. 10–20 weeks
E. 20–40 weeks

22. Which antibiotic is the drug of choice for pertussis infection?
A. Amoxicillin
B. Ciprofloxacin
C. Clindamycin
D. Erythromycin
E. Trimethoprim–sulfamethoxazole

23. After the initial day of treatment for latent TB infection (LTBI) with isoniazid, your patient develops an allergic reaction. You then decide to treat the infection with daily rifampin monotherapy. For how long should treatment be provided?
A. Four months
B. Five months
C. Six months
D. Seven months
E. Nine months

24. A young woman presents to your clinic complaining of dysuria. She is not sexually active. What is the most common organism responsible for urinary tract infections (UTI) in both men and women?
A. Candida
B. *Escherichia coli*
C. *Klebsiella pneumoniae*
D. *Proteus mirabilis*
E. *Staphylococcus aureus*

25. Which of the following is the leading cause of mortality from opportunistic infections in people with HIV worldwide?
  A. Candidiasis
  B. Cytomegalovirus
  C. Histoplasmosis
  D. Toxoplasmosis
  E. Tuberculosis

26. Roughly how what percentage of those infected with acute hepatitis C develop chronic hepatitis C infection?
  A. 20%
  B. 40%
  C. 60%
  D. 80%
  E. 100%

27. Which of the following is the leading contributor to new hepatitis C infections in the United States?
  A. Hemodialysis
  B. Heterosexual intercourse with hepatitis C positive partner
  C. Injection drug use
  D. Reception of blood transfusion products
  E. Transmission in pregnancy

28. While working in the epidemiology division of the local health department, you get a call from a local ER physician. He frantically explains that a young girl just came through the door after suffering deep trauma from the bite of a stray dog. After biting the girl, the dog ran off.

The ER physician asks you for guidance on how to treat this previously unimmunized girl. What is the most appropriate response?
  A. Keep a close eye on the patient for signs of rabies, but do not immunize yet
  B. Give a rabies shot immediately
  C. Give a human rabies immunoglobulin (HRIG) shot immediately
  D. Give both a rabies shot and HRIG shot immediately
  E. Capture the stray dog and give it a rabies immunoglobulin shot

29. Which of the following serious complications of varicella is the most common?
  A. Coagulopathy
  B. Encephalitis
  C. Pneumonia
  D. Seizures
  E. Sepsis

30. How many varicella (chicken pox) vaccinations are recommended by the Advisory Committee on Immunization Practices (ACIP) for healthy children?

A. Zero
B. One
C. Two
D. Three
E. Four

31. A park ranger in the Tennessee wilderness presents to your clinic with a rash covering his entire body. He complains of fatigue, headache, and myalgias. The rash began on his ankles and wrists but then spread centrally and toward his palms and soles. When asked about his job, he states that he frequently finds ticks on his skin but always pulls them off when he checks over his skin before going to bed.

What is the appropriate antibiotic used to treat this tick-borne illness?

A. Ceftriaxone
B. Doxycycline
C. Levofloxacin
D. Mupirocin
E. Trimethoprim–sulfamethoxazole

32. Which of the following patients is most likely to have histoplasmosis?

A. Chicken farmer in Cincinnati, Ohio
B. Engineer in Cape Canaveral, Florida
C. Fisherman in Anchorage, Alaska
D. Physician in London, England
E. Taxi driver in Las Vegas, Nevada

33. A church group went on a hiking trip through a wooded area in Delaware. Several days after the trip, a young woman noticed a tick behind her right knee. She immediately brushed it off with a washcloth. Several days later, she developed a rash on her back. It was flat with a red spot inside of a larger red circle. There was no open lesion or discharge. She immediately scheduled a dermatology appointment but by the time she was able to go to her appointment eight days later, the rash was gone.

What is the appropriate treatment for this patient?

A. No treatment is necessary because the rash is gone
B. Hydrocortisone ointment
C. Ketoconazole cream

D. Oral trimethoprim–sulfamethoxazole
E. Oral doxycycline

34. Trichinosis can result from consumption of which of the following?
A. Apple juice
B. Chicken
C. Fish
D. Pork
E. Sprouts

35. A young woman presents to your primary care clinic excited to announce that she has recently learned that she is pregnant. You discourage her from eating hot dogs, deli meats, unpasteurized milk, and soft cheeses. Which microorganism is responsible for these dietary restrictions?
A. *Escherichia coli*
B. *Listeria monocytogenes*
C. *Plasmodium falciparum*
D. *Pseudomonas aeruginosa*
E. *Staphylococcus aureus*

36. A four-year-old girl and her mother report to your clinic complaining of diarrhea two weeks after returning from an emergent unplanned month-long trip to India. Upon questioning, they also complain of fatigue, fever, malaise, headache, and a loss of appetite. After seeing the result of their positive stool cultures, you inform them that they could have reduced their risk of contracting this ailment if they prophylactically took the typhoid vaccine.

Which bacteria was confirmed in the stool culture?
A. *Escherichia coli*
B. *Giardia*
C. *Salmonella*
D. *Shigella*
E. *Vibrio*

37. Which of the following organisms is responsible for causing lymphogranuloma venereum?
A. *Chlamydia trachomatis*
B. *Haemophilus ducreyi*
C. Human papillomavirus
D. *Neisseria gonorrhoeae*
E. *Treponema pallidum*

38. What is the causative organism of the sexually transmitted infection with the highest annual incidence in the United States?

A. *Chlamydia trachomatis*
B. Human immunodeficiency virus
C. Human papillomavirus
D. *Neisseria gonorrhoeae*
E. *Trichomonas vaginalis*

39. A 23-year-old man presents to your clinic after his sexual partner notified him that she tested positive for chlamydia. His past medical history is noncontributory but he has an allergy to doxycycline. He wants presumptive treatment for chlamydia but none of the other diseases that are commonly associated with it. Which medication is most appropriate?

A. Azithromycin
B. Ceftriaxone
C. Ciprofloxacin
D. Clindamycin
E. Penicillin

40. A 50-year-old woman presents to your clinic for follow-up of a lab report confirming positive rapid plasma reagin (RPR). She is asymptomatic and contributes this positive RPR status to an unprotected sexual encounter several weeks prior. Her RPR from two months prior was nonreactive.

You decide to treat this patient for syphilis. Your clinic has prepacked Benzathine penicillin G (Bicillin L-A) in 1,200,000 unit syringes. What is the appropriate treatment?

A. One syringe injected intramuscularly
B. Two syringes injected (in same visit) intramuscularly
C. One syringe injected weekly for two weeks
D. One syringe injected intramuscularly weekly for three weeks
E. Two syringes injected (in same visit) intramuscularly weekly for three weeks

41. What is the treatment of choice for a pregnant patient diagnosed with syphilis that has a penicillin allergy?

A. Desensitization and treatment with penicillin
B. Levaquin
C. Rifampin
D. Tetracycline
E. Trimethoprim–sulfamethoxazole

42. Which of the following series of oral polio vaccine (OPV) and/or inactivated polio vaccine (IPV) carries the highest risk of causing vaccine-associated paralytic polio (VAPP)?

A. OPV, OPV, OPV
B. IPV, OPV, OPV
C. IPV, IPV, OPV
D. OPV, IPV, IPV
E. All of the above

43. A 32-year-old American missionary presents to a travel clinic in preparation for an upcoming trip to India and Sri Lanka. Because cholera is endemic in these countries, the physician on staff recommends his patient receive the oral cholera immunization. The patient should take this immunization at least how many days prior to travel?

A. Same day
B. 10 days
C. 30 days
D. 60 days
E. There is no cholera immunization

44. A 29-year-old woman that has never received a single vaccination presents to your clinic. She is 20-weeks pregnant and is now requesting to be caught up on every vaccine that she missed. You respond by saying that not all vaccinations are appropriate to her situation. Which of the following vaccinations is contraindicated in pregnancy?

A. Tetanus diphtheria and pertussis (Tdap)
B. Pneumococcal
C. Measles, mumps, rubella (MMR)
D. Haemophilus influenzae type B (Hib)
E. Hepatitis A

45. The rotavirus vaccination has been frequently associated with which adverse event?

A. Autism
B. Eczema
C. Intussusception
D. Leukemia
E. Russel Silvers syndrome

46. The first dose of which immunization is typically recommended immediately after birth?

A. Hepatitis A
B. Hepatitis B

C. Polio
D. Pneumococcal
E. Varicella

47. According to the ACIP which of the following previously unvaccinated patients should be immediately given the Hib vaccine?
A. Ten-year-old boy about to have his tonsils and adenoids surgically removed
B. Six-year-old boy presenting to the ER with a fractured femur from child abuse
C. Eight-year-old girl with a history of frequent episodes of otitis media
D. Seven-year-old girl recently diagnosed with sickle cell disease
E. All of the above

48. Which type of vaccination is used in the United States to prevent cervical cancer?
A. Capsular polysaccharide
B. Double stranded DNA
C. Live-attenuated
D. Virus-like particle
E. Whole cell

49. What type of vaccine is the hepatitis B vaccination?
A. Live
B. Recombinant
C. DNA
D. Whole cell
E. Toxoid

50. The pneumococcal conjugate vaccine (PCV13) is recommended to be given in a four-dose series. Which of the following ages is not one of the recommended dates of administration in a child that has begun the PCV13 series from the beginning?
A. Two months
B. Four months
C. Six months
D. 12 months
E. 24 months

51. A healthy 76-year-old woman presents to your primary care clinic. She is unsure of whether or not she has ever received a pneumococcal vaccination. Which pneumococcal vaccination(s) is/are recommended by the ACIP?
A. None, she has already received the appropriate pneumococcal vaccination

B. PPSV23 vaccination booster only
C. PCV13 vaccination only
D. PCV13 vaccination followed by PPSV23 in 6–12 months
E. PPSV23 vaccination followed by PCV13 in 6–12 months

52. Which of the following countries requires travelers to provide proof they have received a yellow fever vaccination?
A. Canada
B. Italy
C. Nigeria
D. Poland
E. Russia

53. How often is a booster of the yellow fever vaccine recommended in a healthy adult?
A. Never
B. One year
C. 10 years
D. 20 years
E. 30 years

54. Under which scenario does the CDC and the ACIP recommended that tetanus immunoglobulin (TIG) be administered to someone with an acute injury?
A. A minor wound with a history of one Tdap vaccinations
B. A minor wound with a history of two Tdap vaccinations
C. A major wound with a history of two Tdap vaccinations
D. A major wound with a history of three Tdap vaccinations
E. All of the above

55. A 23-year-old man presents to your clinic after lacerating his hand on a knife while cutting an avocado at home. The cut is clean and superficial. He had a Tdap shot 11 years ago. Prior that, he had completed the recommended childhood vaccination diphtheria, tetanus, acellular pertussis (DTaP) series. Which of the following answers is most appropriate regarding the tetanus and diphtheria toxoids (Td) vaccine?
A. There is no indication to give the Td vaccine
B. Give the Td vaccine now
C. Monitor for three weeks and give the Td vaccine if there are signs of tetanus
D. Give TIG only
E. Give tetanus toxoid (TT) vaccination only

56. What type of vaccination is tetanus?
  A. Conjugated
  B. DNA
  C. Live virus
  D. Toxoid
  E. Whole cell

57. Your patient in the clinic is there to seek medical attention for his lacerated foot after stepping on a rusty knife. The triage nurse followed your clinic's standard protocol and administered the Td vaccination. When you finally see the patient and get to suture the wound, he tells you that his wife is 38-weeks pregnant. What should you do to ensure that he does not transmit whooping cough to his future newborn child?
  A. Tell him that the shot he received will protect against whooping cough
  B. Give him the Tdap vaccination now
  C. Tell him to come back in two weeks for the Tdap vaccination
  D. Tell him to come back in four weeks for the Tdap vaccination
  E. Administer TIG immediately, then give Tdap vaccination.

58. What is the Healthy People 2020 goal for percentage of healthcare workers to receive annual influenza vaccination?
  A. 100%
  B. 90%
  C. 80%
  D. 70%
  E. 60%

59. When is the Tdap vaccination recommended in pregnant women?
  A. During every pregnancy
  B. Td, not Tdap is recommended during the first pregnancy
  C. During a woman's first pregnancy
  D. Not recommended if woman has completed her Tdap series
  E. Two weeks after every pregnancy

60. Approximately how long does it take for the majority of people to build immunity after receiving an inactivated virus influenza vaccination?
  A. Immediately
  B. Two days
  C. Four days
  D. Six days
  E. Two weeks

61. While working in a primary care clinic, your patient presents with his mother complaining of one day of fever and myalgias. A rapid influenza test confirms that he has influenza B. After treating your patient, his mother turns to you and asks if there is any medical intervention to protect her against influenza. Which intervention is most appropriate to reduce the chances of becoming ill with influenza in your patient's mother?

A. Azithromycin
B. Begin wearing a surgical mask
C. Influenza vaccination
D. Oseltamivir
E. Supportive care

62. At what age is it recommended that children begin to receive the annual influenza vaccination?

A. Zero months
B. Six months
C. 12 months
D. 18 months
E. 24 months

63. After getting the first hepatitis A vaccine, how long is the recommended wait period until the second vaccine is administered?

A. Six months
B. 24 months
C. 36 months
D. 48 months
E. 60 months

64. Which vaccination is responsible for the drastically reduced incidence of epiglottitis in the United States?

A. DTaP
B. Hib
C. MMR
D. Pneumococcal conjugate
E. *RV*

65. Due to state legislation to expand scope of practice for pharmacists, a pharmacy manager makes the decision to start carrying vaccinations. Which of the following is false of vaccine information statement (VIS)?

A. A VIS describes the benefits and risks of the vaccine
B. A VIS must be given prior to each dose of a multi-dose series
C. Each vaccine has a specific VIS

- **D.** VISs are produced by the vaccine manufacturer
- **E.** Vaccine providers are mandated to give a VISs to vaccine recipients

66. A pregnant woman presents to your clinic to receive prophylaxis against malaria for her upcoming vacation. She does not have any medication allergies or a significant past medical history. Which of the following medications is most appropriate for her to take for malaria prophylaxis, assuming she is traveling to an area without strains of drug resistant malaria?
- **A.** Atovaquone–proguanil (Malarone)
- **B.** Chloroquine
- **C.** Doxycycline
- **D.** Primaquine
- **E.** None of the above

67. Which type of birth defect is most closely associated with the Zika virus?
- **A.** Cleft palate
- **B.** Congenital adrenal hypoplasia
- **C.** Microcephaly
- **D.** Cardiac septal defects
- **E.** Thyromegaly

68. Chikungunya is categorized as which type of infection?
- **A.** DNA virus
- **B.** RNA virus
- **C.** Gram-negative bacteria
- **D.** Gram-positive bacteria
- **E.** Protozoa

69. Which of the following is not a sign or symptom of Dengue?
- **A.** Dysuria
- **B.** Fever
- **C.** Retroorbital pain
- **D.** Bone pain
- **E.** Macular skin rash

70. What is the recommended age to administer the herpes zoster vaccine to a person without elevated risk of a shingles outbreak?
- **A.** One year old
- **B.** 12 years old
- **C.** 45 years old
- **D.** 60 years old
- **E.** 65 years old

71. While working abroad in Africa, you become aware of a schistosomiasis outbreak. Which of the following organisms is a host for the microbe responsible for this disease?
A. Bird
B. Dog
C. Goat
D. Snail
E. Tsetse Fly

72. Which maternal infection is the most common cause of congenital deafness in the United States?
A. *Cytomegalovirus*
B. Group B *streptococcus*
C. Measles
D. Syphilis
E. Toxoplasmosis

73. Which organism causes blindness resulting from trachoma?
A. *Borrelia*
B. *Chlamydia*
C. *Cytomegalovirus*
D. River fluke
E. *Vibrio*

74. What type of vaccination is the Bacille Calmette–Guérin (BCG) vaccine?
A. DNA
B. Whole cell
C. Toxoid
D. Inactivated
E. Live

75. A patient presents to your clinic concerned that he shared an elevator with a person that was found to be positive for bacterial meningitis. Your patient is now requesting post-exposure prophylaxis. Records indicate he is already up to date on his vaccinations. Which treatment is most appropriate?
A. Ceftriaxone
B. Doxycycline
C. Hib vaccination
D. Ciprofloxacin
E. Nothing

76. When you arrive at your clinic at 8:00 a.m., a seven-year-old boy and his mother are waiting. The boy is complaining of severe sore throat for one day. He is afebrile. A throat swab tests positive for *streptococcus* and antibiotics are prescribed. The mother asks when her son can return to school.

What is the appropriate response?

A. Immediately
B. As soon as he takes the antibiotics
C. 24 hours after taking antibiotics
D. 72 hours after taking antibiotics
E. One week

77. A patient's mother calls your office to see if it is okay to send her daughter to school. Her daughter has been out of school for three days since you saw her for watery diarrhea. At that appointment, you diagnosed her with gastroenteritis and did not prescribe any antibiotics. Her last episode of diarrhea was two days ago and she had a solid bowel movement yesterday.

When can she return back to school?

A. Immediately
B. One day
C. Two days
D. Five days
E. When she provides a negative stool sample

78. Which of the following potentially results from antibiotic resistance?

A. Inability to perform medical actions (e.g., chemotherapy)
B. Increased mortality rates
C. Increased treatment cost
D. Longer duration of illness
E. All of the above

79. Which of the following does not contribute to antibiotic resistance?

A. Animal exposure to antibiotics
B. Nonprescription antibiotic use
C. Prescription antibiotic use
D. Use of diagnostic tests
E. Wastewater treatment facilities

80. Which organism should be screened for rectovaginal colonization in pregnant women between gestational weeks 35–37?

A. *Escherichia coli*
B. GBS
C. *Rubella*
D. Syphilis
E. Zika

81. Which two causes of death are most responsible worldwide for mortality of those aged one to 59 months?
 A. Injury and malaria
 B. Injury and diarrhea
 C. Malaria and diarrhea
 D. Malaria and pneumonia
 E. Diarrhea and pneumonia

82. What is the most frequent route of HIV transmission worldwide?
 A. Oral sex
 B. Heterosexual vaginal intercourse
 C. Homosexual anal intercourse
 D. Intravenous drug use
 E. Mother-to-child transmission

83. You have been deployed to act as part of a medical team to help manage a natural disaster in a foreign country. After a couple days, an epidemic of diarrhea occurs. The diarrheal infection is unlike anything you have seen in the United States. Patients are saying that their diarrhea looks like rice water. Those with the diarrhea are suffering from complications of hypoglycemia, dehydration, and renal failure. Which bacteria is most likely responsible for this epidemic?
 A. *Mycobacteria*
 B. *Norovirus*
 C. *Shiga toxin-producing Escherichia coli*
 D. *Staphylococcus*
 E. *Vibrio cholerae*

84. Which type of bacteria is capable of causing typhoid fever, enteric fever, and gastroenteritis?
 A. *Escherichia coli*
 B. *Giardia*
 C. *Salmonella*
 D. *Shigella*
 E. *Vibrio*

85. Which of the following types of fleas is responsible for transmitting Plague (*Yersinia pestis*)?
 A. Rat flea (*Xenopsylla cheapis*)
 B. Dog and cat flea (*Ctenocephalides sp.*)
 C. Human flea (*Pulex irritants*)
 D. All of the above
 E. None of the above

86. Which form of botulism is most common in the United States?
- **A.** Foodborne
- **B.** Infantile
- **C.** Inhaled
- **D.** Radiation
- **E.** Wound

87. Up to what age is the TST nearly always preferable to the interferon-gamma release assay (IGRA)?
- **A.** One year old
- **B.** Two years old
- **C.** Three years old
- **D.** Four years old
- **E.** Five years old

88. Which of the following is true of IGRA?
- **A.** IGRA should not be taken within four weeks of any vaccination
- **B.** IGRA may be taken regardless of time of any vaccination
- **C.** IGRA should only be taken at same time as attenuated vaccination but never live vaccination
- **D.** IGRA should be taken during time of live vaccination or >4 weeks after live vaccination
- **E.** IGRA provides a boost of the BCG vaccination

89. A hopeful nursing student presents to the local urgent care clinic for a physical evaluation prior to school enrollment. Part of the enrollment requirement mandate that she be up to date with recommended vaccinations (influenza, MMR, Tdap) and be screened for TB. She needs to present all of her documentation to her school within two weeks.

If she receives a varicella vaccination and the clinician follows the proper guidelines, which of the following would she not be able to have completed by her two week deadline?
- **A.** Influenza
- **B.** MMR
- **C.** Tdap
- **D.** Two-step PPD
- **E.** She will be able to complete all of these school requirements within two weeks

90. A patient plans to travel to Saudi Arabia to participate in the Hajj, an annual Islamic gathering of greater than 2,000,000 people. Which vaccination is required by the Saudi government to participate in this event?
- **A.** Hib

B. Influenza
C. Meningococcal
D. Pneumococcal
E. Shingles

91. Which of the following diseases carries the highest risk of transmission from a needlestick injury if the original source of the needlestick is a carrier?
A. Hepatitis A
B. Hepatitis B
C. Hepatitis C
D. HIV
E. B and C carry the same risk

92. While working at an occupational medicine clinic, you have a patient that works as a dental hygienist that was stuck by a sharp dental instrument that had just been used on a patient with hepatitis B. Your patient is not sure whether or not he had ever received the hepatitis B vaccination series.

Regarding hepatitis B, what is the recommended initial treatment for your patient?
A. Wait for the needle to test positive to determine if medical intervention is necessary
B. Give antiviral medication
C. Give hepatitis B immunoglobulin
D. Initiate the hepatitis B vaccination series
E. Both answers C and D

93. Which of the following describes a vaccination fractional dosing strategy?
A. Administering a reduced amount of the recommended dose of a vaccine
B. Lengthening the amount of time between doses in a vaccination series
C. Reducing the age at which a vaccination is recommended to be administered
D. Reducing the amount of vaccines administered in a single day
E. Reducing the length of time between doses in a vaccination series

94. What is the name of the system that notifies health departments of recent immigrants that carry infectious diseases?
A. Electronic Disease Notification (EDN)

B. Healthy Immigrant Assurance (HIA)
C. Immigration Screening System (ISS)
D. Infectious Migration Network (IMN)
E. Keep Infections Out (KIO)

95. Which type of vaccine is an example of an inactivated whole cell vaccine?
A. Diphtheria
B. Haemophilus influenzae
C. Hepatitis A
D. Hepatitis B
E. Pertussis

96. Which of the following is a conjugate vaccine?
A. Haemophilus influenzae
B. Hepatitis B
C. Rabies
D. Shingles
E. Tetanus

97. Which of the following pathogens is responsible for causing Hansen's disease (leprosy)?
A. *Aspergillus*
B. *Mycobacterium*
C. *Neisseria*
D. *Orthomyxovidae*
E. *Staphylococcus*

98. What percentage of the population needs to be vaccinated to achieve herd immunity?
A. 60%
B. 70%
C. 80%
D. 90%
E. Depends on multiple factors

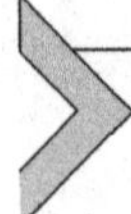

7.2 INFECTIOUS DISEASE ANSWERS

1. **C. Yellow fever**
   The polio, influenza, and yellow fever vaccinations all contain live viruses. However, yellow fever is the only live virus listed that is also injected. The live version of the polio vaccine is taken orally. The live version of the influenza vaccine is intranasal. Both the hepatitis A and rabies vaccines are inactivated viral vaccinations.
2. **B. Anopheles**
   There are over 3000 species of mosquitos grouped into 41 genera.
   Refer to the table below for diseases transmitted by some of the most common species of mosquitos.

Mosquito Genus	Diseases transmitted
Aedes	Dengue, chikungunya, La crosse encephalitis, West Nile virus, yellow fever, Zika
Anopheles	Malaria, West Nile virus, orungo virus
Culex	St. Louis encephalitis, West Nile virus, western equine encephalitis, Japanese encephalitis

3. **A. Dracunculiasis**
   The WHO has been appointed by the United Nations to coordinate international health efforts. Dracunculiasis, Yaws and poliomyelitis are official disease eradication targets of the WHO. The WHO has also endorsed a program to eliminate Filariasis. Additionally, there are numerous regional and national programs to eliminate local endemic disease, such as onchocerciasis (river blindness) from the Americas and malaria from Hispaniola (island of Dominican Republic and Haiti). Smallpox is the only disease that has been successfully eradicated worldwide and vaccinations were discontinued in 1980.
   For a disease to be eradicated (worldwide) or eliminated (regional), several key principles are present:
   1. Intervention must occur everywhere the disease has the possibility of occurring
   2. The disease and intervention efforts must be monitored closely
   3. Public health officials must be flexible in their approach
   4. There must be focus on the goal of stopping transmission, regardless of cost.

In determining whether or not a disease is eradicable, several considerations should be made. First, eradication should be technologically and biologically achievable. This includes epidemiological susceptibility (i.e., modifiable disease reservoirs) and effective intervention. Furthermore, there must be social and political support for the eradication. For this to happen, the intervention should have desirable economic considerations.

4. **D. Smallpox**
   Smallpox is the only disease that has been successfully eradicated worldwide. The last naturally occurring case was in 1977 and vaccinations were discontinued in 1980.
   Disease eradication was discussed in answer #3 (directly above).
5. **B. Mumps**
   Mumps is a vaccine preventable virus spread by respiratory droplets. Symptomatic patients experience symptoms of malaise, fever, headache, myalgias, fatigue, swelling of the parotid glands, and less commonly swelling of the testicles.
6. **A. Birds**
   West Nile virus (WNV) is an arbovirus transmitted by mosquitos that causes symptoms such as encephalopathy, headache, myalgias, fatigue, weakness, fever, and/or rash in 20%–30% of humans affected. The virus is cycled between mosquitos and birds (> 300 species) which act as amplifier hosts. When an infected mosquito bites a human, horse, or other type of mammal, that mammal becomes a dead-end host. Dead-end hosts do not replicate the virus in such a way that mosquitos become infected by biting them. Humans do not contract *WNV* by handling or consuming birds with *WNV*. Some states and municipalities record information from necropsies of dead birds for epidemiologic purposes.
7. **E. None of the above**
   Although various hemagglutinin and neuraminidase subtypes of influenza A have a large variety of nonhuman hosts (birds, pigs, horses, bats, whales, etc.), both influenza B and influenza C are exclusively human diseases. Influenza D primarily affects cattle and is not known to cause illness in people.
8. **B. Fecal–oral**
   Hepatitis A is usually transmitted through fecal–oral transmission. This may occur either through infected food contact or contact with the anus leading to oral ingestion resulting from sexual activity. Less

directly, it may come from a contaminated food source or preparation surface. The hepatitis A virus is capable of living outside the body for several months. Freezing temperatures do not kill the virus, although boiling does.

The symptoms of the virus may range from subclinical infection all the way to hospitalization. If symptoms do appear, they typically occur two to six weeks after exposure. Those showing symptoms typically recover within several months. Hepatitis A does not cause chronic infection.

9. E. *Vibrio*

Vibrio disease (especially from *Vibrio parahaemolyticus* and *Vibrio vulnificus*) is synonymous with raw shellfish, specifically oysters. Oysters are bottom feeders that may accumulate pathogens. Although *Vibrio* species are naturally occurring in regions that oysters are farmed, they may increase in number during environmental changes like warm waters during summer months, algal blooms and increases in waste products in water. After harvesting oysters, their *Vibrio* concentration may continue to increase due to inappropriate temperature storage and long-storage periods.

To reduce risk of infection from shellfish, oyster beds are monitored for fecal indicator bacteria (FIB). Once harvested, the oysters are typically screened via PCR and growth enrichment media. Oysters intended for raw consumption are given a tag containing information on the date and location of their origin. Restaurants are required to hold onto these tags, and in the case a customer becomes ill from eating the oysters, other oysters farmed from the same location may be identified. The FDA, individual states, and the Shellfish Sanitation Conference (ISSC) are important parties in monitoring oyster safety. The groups recommend the following safeguards to reduce disease transmission: Rapid ultra-low freezing, high hydrostatic pressure, mild heat treatment, and ionizing radiation, followed by frozen storage.

Cooking oysters is sufficient to kill *Vibrio* pathogens. Immunocompromised patients that elect to eat raw shellfish are at the greatest risk of contracting Vibrio infection.

10. C. It occurs while an outbreak is occurring

Ring vaccination is a process where an epidemic is controlled by quarantining and vaccinating contacts of those with a vaccine preventable illness, essentially controlling a ring around them.

Contacts of contacts may also be quarantined and vaccinated. It is an appropriate strategy to use when there is limited vaccine available, the vaccine produces undesirable side-effects, or there is significant immunity in the population (such as in the final stages of disease eradication). It is not suitable when the contacts travel to far away regions, especially out of jurisdiction of the health provider. Because ring vaccination does not immunize a large percentage of the population, it does not build herd immunity.

Preemptive vaccination occurs prior to a disease outbreak. Alternatively, reactive vaccination occurs while an outbreak is occurring. Mass vaccination may be a form of preemptive and/or reactive vaccination, while ring vaccination is strictly reactive. Ring vaccination is more appropriate to use in a rare disease where there is limited immunity in the population. Mass vaccination is a better strategy for a more widespread pathogen.

Ring vaccination was successfully used in smallpox eradication efforts and is endorsed to be the strategy of choice for future potential smallpox outbreaks. It has also been successfully used in Ebola outbreaks in Western Africa.

11. B. Breastfeeding

Vertical transmission of disease occurs when a disease is transmitted to offspring through sperm, ova, placenta, milk, or vaginal fluids during the birthing process. Perinatal HIV transmission may occur during pregnancy, delivery, and through breastfeeding.

Horizontal transmission occurs when a disease is transferred from one host to another person that is not offspring. Types of horizontal transmission include direct transmission (sexual intercourse), common vehicle (used needle from drug use, blood transfusion), respiratory droplets, and vectors such as insects.

HIV is not transmitted through insect bites.

12. B. Cat

Women become immunocompromised during pregnancy. Because of this immunocompromised status, women should avoid specific animal contact. It widely recommended that women reduce their contact with cats during pregnancy due to the risk of *Toxoplasmosis gondii* (*T. gondii*). Specifically, pregnant women should avoid interacting with cat feces. Other zoonotic infections that pose especially high risk to pregnant women are listerosis, Q fever, brucellosis, and leptospirosis. Many of these pathogens are commonly found in animal birthing fluids and membranes. Therefore, it is advised that pregnant women stay away from the animal birthing process and by-products.

Cats are the definitive host of *T. gondii*, a protozoa that crosses the placental barrier. Infection may result in termination of pregnancy or lead to neurological or ocular disease in the infant. It rarely causes symptoms in people with a healthy immune system. If a women with a healthy immune system had previously been infected by *T. gondii*, there is reduced danger to the fetus.

Infected cats shed *T. gondii* a single time in a lifespan, typically for one to two weeks. This is the only time cat is infectious. This shedding period typically occurs after the cat itself is infected. Typically, this occurs after a cat eats undercooked meat. This most often happens in young stray cats. For this reason, pregnant women should not take in new cats. If they have cats from before the pregnancy, chances are that they have already been exposed and built an immunity.

Because *T. gondii* is ubiquitous worldwide, it is more likely that a person will it pick up from soil or food consumption than from a cat.

13. B. *Escherichia coli (E.coli)*

HUS is a disorder defined by the triad of microangiopathic hemolytic anemia, thrombocytopenia, and acute renal failure. Additionally, it is associated with diarrhea, elevated serum creatinine, and oliguria.

HUS is a caused by *enterohemorrhagic E. coli* (*EHEC*), a subset of *STEC*.

14. B. *Escherichia coli (E.coli)*

It is believed that *STEC* acquired the ability to produce shiga toxin from Shigella, of which the only subtype that produces the toxin is *S. dysenteriae*. Once ingested, shiga toxin is cytotoxic to the cells lining the intestines. It may cause cramping, watery diarrhea, emesis, and fever. These symptoms typically resolve after five to seven days. *EHEC* is a subset of *STEC*.

15. D. *Rotavirus*

Worldwide, *rotavirus* is the most common cause of severe gastroenteritis in infants and children. Roughly 500,000 children under five years of age worldwide die of rotavirus annually with over 80% occurring in developing countries. Prior to routine rotavirus vaccination in the United States in 2006, approximately 80% of children had contracted rotavirus gastroenteritis by five years old.

16. C. Erythromycin

The USPSTF gives a grade A recommendation to give topical antibiotics to newborns within 24 hours of birth to prevent gonococcal ophthalmia neonatorum. Erythromycin ointment is the only antibiotic recommended for this indication that is available within the United States.

17. E. Measure the diameter of the induration from one lateral border to the other

The proper way to read a TST is to measure the induration from one lateral border to the other, in the direction of a bracelet or wrist watch. Begin by feeling the lesion and noting where the firm induration begins. Mark this point and then measure where the induration begins on the other side. The width of the induration is then measured in millimeters.

18. B. A 6 mm induration of a healthy person that has a spouse with active TB

Although all of these patients have indurations, the only one that would be positive is the one that has a 6 mm induration and a spouse with active TB.

Criteria for a positive tuberculin skin test (TST)

Criteria for a positive tuberculin skin test (TST)
Induration ≥ 5 mm
HIV +
Contacts of active TB
Fibrotic changes on chest X-Ray
Status post organ transplant
Induration ≥ 10 mm
Immigrant within five years of TB endemic region
Injection drug users
Lab workers exposed to TB
Live/work in congregate settings such as prisons
Past medical history that predisposes to infection or causes immunodeficiency
Patient ≤ five years of age
Induration ≥ 15 mm
Those without TB risk factors

A negative TST does not completely exclude the possibility of TB. Anergy is the phenomenon where the body does not form a reaction to a TST. This typically occurs due to an immunocompromised state which can occur from HIV or weakened immune system from severe TB infection.

19. B. Isoniazid

The drugs listed in this question are typically used to treat TB. When treating patients with isoniazid, pyridoxine (vitamin B6) should be coadministered. This is because isoniazid causes a pyridoxine deficiency leading to peripheral neuropathy, weakness, glossitis, cheilosis, and stomatitis.

Peripheral neuropathy is the most commonly found side-effect from isoniazid which occurs largely due to vitamin B6 deficiency. Peripheral neuropathy incidence is dose related from isoniazid and typically manifests as paresthesias of the hands and feet. It associated with muscle weakness and diminished reflexes that typically appears around the sixth month of isoniazid treatment, but may occur sooner in those with a preexisting pyridoxine deficiency.

20. B. 1–3 weeks

The thought process of a two-step PPD is that the first PPD placed will boost the body's immune response to the second PPD in those that have experienced a waned immune response. It is ideal to have the second PPD read –one to three weeks after the initial Mantoux PPD procedure. However, it is often sufficient to accept a PPD as the first in the series as long as it had been given in the previous 12 months. The three visit, two-step PPD approach is gaining acceptance due to higher compliance. In this approach, the patient is given the PPD on day one. On day seven, the initial PPD is read and a new one is placed if the initial one was negative. The second PPD is read (visit #3) 48–72 hours after being administered. If the first PPD became indurated and resolved prior to identification, it will return in the second PPD to be discovered then.

21. C. 2–8 weeks

After exposure to TB, it takes 2–8 weeks to convert to a positive TST. Therefore, it is recommended to complete the final TST screening eight weeks after last contact with someone that has TB. Likewise, the final IGRA test is recommended 8–10 weeks after last exposure to TB.

22. D. Erythromycin

Diagnosing pertussis infection should begin with careful history taking. If a patient is a close contact of a confirmed case of pertussis, they should receive erythromycin.

Erythromycin is also the drug of choice after confirmation of pertussis from nasopharyngeal secretions or serological testing. Alternative antimicrobials include azithromycin, clarithromycin, and trimethoprim–sulfamethoxazole.

If given too late (typically >three weeks) after contracting *B. pertussis*, antibiotics will not alter the course of illness. Pertussis is infectious from the beginning of the catarrhal stage (rhinorrhea, fever, general URI symptoms) until five days after antibiotic therapy. After several weeks without treatment, an immunocompetent patient will innately clear the infection and stop further disease transmission. If antibiotics are not administered, the disease will continue transmission through the third week after paroxysms (coughing fits).

23. A. Four months

Rifampin should be considered for LTBI patients that cannot tolerate isoniazid or have been exposed to INH-resistant LTBI. Rifampin should be given daily and caution should be used in those also receiving antiretroviral therapy for HIV.

24. B. *Escherichia coli (E.coli)*

UTIs are a significant source of morbidity. Roughly 50%–60% of women will experience a UTI in their lifetime. Men experience significantly less UTIs, mainly due to anatomical differences. *E. coli* is the organism most often responsible for both upper and lower UTIs.

25. E. Tuberculosis

All of the options listed are opportunistic infections that are common in those with poorly controlled HIV. Worldwide, TB is the leading cause of HIV opportunistic infection mortality. TB is also one of the leading infectious causes of death in those with intact immune systems. Comorbid HIV and TB infections are much more common in developing nations than the United States. In many of these underserved nations, TB is a common presenting sign of HIV infection.

Because of the link between HIV and TB, all HIV patients should be screened for TB infection. Screening is especially important because those with HIV and TB do not typically present with classic TB symptoms. People with HIV are at increased risk of activation of latent TB. Therefore, those with active and latent TB should be immediately treated for TB infection. Early initiation of antiretroviral HIV therapy is an important aspect of TB treatment and helps control the spread of TB.

26. D. 80%

Roughly 15%–25% of those that experience an acute hepatitis C virus (HCV) infection will clear it naturally. The remaining 75%–85% people with acute HCV develop chronic HCV infection. HCV is the most common chronic blood-borne pathogen in the United States.

27. **C. Injection Drug Use**
All of the possible answers listed are causes of HCV infection. Injection drug use is the leading contributor to new HCV infections. Research reports that 60% of new HCV infections occur in those who have injected drugs within the prior six months. Hemodialysis and reception of blood products were large contributors of HCV transmission prior to implementation of screening techniques. Fortunately, these methods of HCV transmission are now rare. Sexual contact is not an efficient method of HCV conveyance.

28. **D. Give both a rabies shot and HRIG shot immediately**
Rabies is an RNA virus transmitted by the bite of a mammal. If not treated properly, it leads to progressive encephalomyelitis and death. Rabies postexposure prophylaxis (PEP) is nearly always effective when used appropriately. After initiation of PEP, all otherwise healthy patients quickly develop detectable rabies neutralizing antibodies.

If a previously unvaccinated person is bit by, or comes into close contact with a mammalian species known to carry rabies, it is recommended that they receive HRIG and a rabies vaccination immediately. HRIG is given at the site of injury (bite, scratch) and is highly effective in reducing dissemination of the virus. It is designed to provide rabies virus-neutralizing antibody coverage until the vaccination allows the patient to start producing antibodies. The vaccination is recommended to be given intramuscularly in a four-dose regimen on days zero, three, seven, and 14.

Previously vaccinated individuals should not receive HRIG, as is may inhibit the already formed rabies antibodies. It is recommended that those that have already received the rabies vaccination get a two-dose rabies virus vaccination on days zero and three.

If a person is undergoing rabies vaccination and the source animal is found to be negative for the virus after conducting laboratory testing, the PEP may be discontinued.

29. **C. Pneumonia**
Pneumonia is the most common serious complication of chicken pox (varicella) in healthy adults. Although an exact prevalence is unknown, it is estimated to occur in roughly one of every 400 varicella infections. It is estimated that varicella pneumonia yields a mortality rate between 5% through 33%, depending on factors such as overall health status and access to care.

Risk factors for varicella pneumonia include immunocompromised status, lung disease, smoking, severity of rash and prior exposure to the varicella zoster virus. It has been found that there is increased risk of developing pneumonia if contracting chicken pox as an adult, especially while pregnant.

30. C. Two

The ACIP currently recommends that healthy children receive two varicella vaccinations. The first dose is recommended between the ages of 12–15 months, while the second dose is recommended at 4–6 years. Varicella vaccination was previously only recommended to be a one-time childhood vaccination. Although a single varicella regimen protects most people from chicken pox, it has been found to be associated with breakthrough chicken pox, a disease where wild-type chickenpox causes a mild illness (less than 50 vesicles) in children given the varicella vaccine at least seven weeks prior. Adding a second varicella vaccination reduces most cases of breakthrough chicken pox.

31. B. Doxycycline

The tick-borne illness in this vignette is Rocky Mountain spotted fever (RMSF), the most lethal tick-borne illness in the United States. RMSF is caused by *Rickettsia rickettsii*, an obligate intracellular bacterium found in ticks and rodents. After inoculation, the bacterium disseminates throughout the body. Once in the endothelial cells, it initiates the coagulation cascade, causes vasculitis, increased vascular permeability, and edema. Thrombocytopenia typically occurs. Microvascular thrombosis and hemorrhage may lead to hypovolemia and associated complications.

Somewhere between two to 14 days after inoculation, RMSF symptoms first to occur. The endothelial cell damage first appears as a rash on the ankles and wrists. Soon it spreads centrally and distally to include the palms and soles. Other symptoms include headache, focal neurological symptoms, fever, malaise, anorexia, nausea, abdominal pain, and cough. Bleeding may also occur. Roughly 10% of those affected do not develop a rash.

The preferred drug of choice for RMSF is doxycycline. If a patient is pregnant, chloramphenicol should be given. It is very important not to hesitate when treating RMSF as the disease may progress quickly. RMSF may be diagnosed clinically or with lab work, including indirect fluorescent antibody testing.

32. A. Chicken farmer in Cincinnati, Ohio

Histoplasma capsulatum is the agent responsible for causing histoplasmosis. Although it may occur worldwide, histoplasmosis is most common in the Mississippi and Ohio River Valleys.

H. capsulatum is a fungus that lives in the soil as a mold that may form a thick capsule. It thrives in dirt that is enriched with bird and bat droppings. It becomes aerosolized when the soil is disturbed and may travel in the air for miles. Once causing infection in the body, *H. capsulatum* transforms into a yeast and may become pathologic.

Less than 5% of those infected with *H. capsulatum* develop symptoms of infection. This is because a healthy immune system is usually strong enough to suppress the infection. Because it can stay dormant for many years, symptoms of infection may reveal themselves during periods when one is immunocompromised. Acute pulmonary histoplasmosis is the most common form of infection. It usually presents as a self-limited flu-like illness.

33. E. Oral doxycycline

This patient developed erythema migrans, a skin lesion with a red spot surrounded by a red circle, which is a diagnostic sign of Lyme disease. Lyme disease is caused by the spirochete genus *Borrelia* after a host tick attaches and feeds on human blood for at least 36–48 hours. Lyme disease typically begins to manifest between seven to 14 days of the tick becoming attached. Roughly 80% of those affected with Lyme disease experience an erythema migrans rash. Other symptoms include arthralgias, myalgias, headache, neck stiffness, and neuropathy. Of these symptoms, an erythema migrans rash is the only diagnostic finding that identifies Lyme disease. In the lab, Lyme disease should be initially screened using ELISA. Western immunoblot is then used to confirm the diagnosis.

Doxycycline, amoxicillin, cefuroxime, and ceftriaxone are all considered to be first-line agents in treating Lyme disease. Doxycycline is viewed as the preferred drug of choice due to its ability to neutralize other tick-borne illnesses.

To prevent Lyme disease in areas with higher incidence (US Northeast and Midwest regions), it is preferred that people going into the wilderness wear bug spray and bright clothes that cover the arms and legs.

34. D. Pork

Trichinosis is caused by the helminthic organism, *Trichinella spiralis*. After eating meat of an infected animal, the encysted worm larvae travels from the intestines into muscle tissue and settles between muscle fibers. Once this muscle tissue is eaten by the next host, the cycle repeats. The cycle may be broken by adequately cooking meat prior to consumption. Although the worm lives in many species of carnivorous animals (rats, hyenas, etc.), it is most often acquired in humans from pork. The pigs are infected from eating meat tainted with *Trichinella spiralis*.

Symptoms of trichinosis include fever, myalgias, abdominal pain, and diarrhea. Eosinophilia is commonly associated with trichinosis and diagnosis can be confirmed through muscle biopsy.

Trichinosis may be treated with antiparasitic agents such as albendazole.

35. B. *Listeria monocytogenes*

Listeria monocytogenes is a motile gram-positive bacillus that is the organism responsible for listerosis. This disease is subclinical or mild in healthy adults, causing a sore throat, fever, and diarrhea. Pregnant women have a weakened immune system and are approximately 20 times more likely to contract listerosis than nonpregnant women. This illness can be more severe during pregnancy and even be transmitted to newborns. In newborns, *Listeria* may cause meningitis and death.

Listeria monocytogenes is found in vegetables, meats, cheeses, ice cream, and unpasteurized milk. The bacteria may also find its way into processed foods such as deli meats, hot dogs, and soft cheeses.

36. C. *Salmonella*

Salmonella enterica serotype typhi is responsible for typhoid fever. Meanwhile, *Salmonella enterica* serotypes paratyphi A, paratyphi B, and paratyphi C are causes of paratyphoid fever. Together, these two ailments are called enteric fever. Once diagnosed, patients are to be treated with antibiotics. However, antibiotic resistance is increasing.

These bacteria species have low prevalence in the United States and are common in developing counties, where they may be acquired from exposure to (fecal) contaminated food and water.

The FDA has approved two prophylactic typhoid vaccinations. The first type of vaccination is an intramuscular (IM) capsular polysaccharide that should be given two weeks prior to potential exposure. The second type is a live-attenuated vaccination that is given in pill form on days zero, two, four, and six. It is to be completed one week prior to potential exposure. Antibiotics are not to be given within three days of the live typhoid vaccine. A whole cell typhoid vaccine has been discontinued due to fever and systemic reactions. Meanwhile, protein-conjugated polysaccharide vaccines are currently available in countries outside of the United States.

In the United States, the typhoid vaccine is recommended for travelers going to areas with risk of exposure to *Salmenilla serotype typhi*. It is also recommended for persons with intimate contact to chronic carriers of *Salmonella serotype typhi* and those that work in the laboratory with it.

Travelers should be made aware that the vaccination does not protect them 100% of the time and they should be diligent in what they eat and drink while abroad.

37. A. *Chlamydia trachomatis*

Lymphoganuloma venereum is caused by *Chlamydia trachomatis*. It presents as a painless papule and evolves into painful inguinal lymphadenopathy. It may be transmitted through vaginal or anal intercourse is treated with a three-week course of doxycycline.

Haemophilus ducreyi is the cause of chancroid, a painful ulcer with rugged edges associated with dysuria, inguinal tenderness, and lymphadenopathy. *H. ducreyi* is often spread through sexual contact.

HPV causes cervical cancer, anal cancer, penile cancer, oropharyngeal cancer, and genital warts.

Neisseria gonorrhoeae is the agent that causes gonorrhea.

Treponema pallidum is the agent that causes syphilis.

38. C. HPV

There are roughly 20 million new sexually transmitted infections (STIs) annually in the United States. Of these, roughly half occur in individuals aged 15–24. The annual incidence of new infections is nearly equal among these young men and young women.

HPV is the most common STI. A large majority of sexually active men and women will contract HPV at some point in their lives.

The list below shows the order of annual incidence of STIs in the United States ranked from highest to lowest incidence.

1. HPV
2. Chlamydia
3. Trichomoniasis

4. Gonorrhea
5. Herpes simplex virus–type 2
6. Syphilis
7. HIV
8. Hepatitis B

The list below shows the order of annual prevalence of STIs in the United States ranked from highest to lowest prevalence.

1. HPV
2. Herpes simplex virus – type 2
3. Trichomoniasis
4. Chlamydia
5. HIV
6. Hepatitis B
7. Gonorrhea
8. Syphilis

39. A. Azithromycin

Chlamydia has the highest incidence of any STI in the United States. Because it can cause pathological changes in asymptomatic patients, the USPSTF recommends screening for it in all women $\leq$24 years old and younger (grade A) and women 25 and older with increased risk (grade A). The same goes for pregnant women, but that recommendation has only been given a grade B recommendation. Screening for Chlamydia in asymptomatic men has been given a grade I recommendation.

CDC's recommended treatment for chlamydia is one dose of one gram of azithromycin or doxycycline 100 mg twice daily for one week. Because patients with chlamydia are frequently coinfected with gonorrhea, it is often advised to presumptively treat for both. The recommended treatment for gonorrhea is ceftriaxone 250 mg intramuscularly.

40. B. Two syringes injected (in same visit) intramuscularly

The organism that causes syphilis, *T. pallidum,* is predictably susceptible to penicillin. For this reason, penicillin is the preferred drug of choice for treating all stages of syphilis. *T. pallidum* can reside in sites poorly accessed by all oral penicillins and some intramuscular penicillins. This is why Benzathine penicillin G (Bicillin L-A) is the preferred treatment for syphilis. If a patient is allergic to penicillin, some experts recommend that they undergo penicillin desensitization in order to receive treatment. Some clinicians advocate alternative agents, such as azithromycin, ceftriaxone, doxycycline, and tetracycline. Doxycycline and tetracycline should not be used in pregnant women.

The penicillin G regimen for syphilis varies by the stage of the syphilis infection. The recommended penicillin G regimen for primary, secondary, and early-latent syphilis treatment in adults is 2,400,000 units intramuscularly in a single dose. Treatment in patients with tertiary and late-latent syphilis is 2,400,000 units intramuscularly once weekly for three weeks.

Follow-up serological testing after syphilis treatment should be performed six and 12 months after treatment. The testing should compare the RPR titer from before the treatment and after the treatment. As a rule of thumb, successful treatment is a four-fold titer decline from date of treatment. Roughly 10% of those successfully treated will not experience a fourfold titer decrease.

The patient in this example is asymptomatic and can be defined as having latent syphilis. Because she has had latent syphilis for less than one year (as evidenced by her negative test two months prior), she is further defined as having early latent syphilis. Appropriate treatment for early latent syphilis is 2,400,000 units, which is two syringes of 1,200,000units.

41. A. Desensitization and treatment with penicillin

Penicillin G is the only known treatment for treating syphilis in pregnant women. The CDC recommends that all women with penicillin allergy receive desensitization to penicillin in order to receive penicillin G.

In nonpregnant women, penicillin G is the preferred first agent of choice but there is some debate about using other medication, such as azithromycin, ceftriaxone, doxycycline, and tetracycline.

42. A. OPV, OPV, OPV

Depending on the type used, the OPV contains different combinations of live-attenuated poliovirus strain types one, two, and three. It has been used as the main tool in the worldwide polio eradication effort. Following OPV administration, the polio virus replicates for four to six weeks, allowing an opportunity to develop humoral immunity. Sometimes during this process vaccine associated paralytic polio (VAPP) occurs, as the attenuated virus reverts to a neurovirulent form that causes paralysis. VAPP typically arises in those that have recently taken OPV or are a close contact of someone that has recently taken OPV (contact VAPP). Different OPV poliomyelitis strains are associated with different risk of VAPP. OPV has also been found to lead to vaccine-derived poliovirus (VDPV), which is transmissible like wild poliovirus variants.

Numerous studies have demonstrated that if an IPV is given prior to OPV, the risk of VAPP is significantly reduced. Because many high-income countries have adopted strategies to give IPV prior to OPV, the amount of cases of VAPP have declined. Interestingly, residents of countries with higher levels of circulating poliomyelitis virus are less prone to develop OPV than residents of nonpoliomyelitis endemic countries. This is felt to be due to higher levels of circulating poliomyelitis antibodies. For similar reasons, individuals with immunodeficiencies are at increased risk of developing VAPP after OPV administration.

Vaccination experts recommend that countries introduce at least one dose of IPV into the primary immunization schedule in addition to OPV.

43. B. 10 days

The only FDA-approved cholera immunization is a live-attenuated liquid that provides protection against *Vibrio cholerae* O1 infection. It has been approved for those aged 18–64 to take at least 10 days prior to travel to cholera-affected areas.

As a live vaccine, it may be given along with other live vaccines recommended for travel, but it may not be given within 30 days of another live vaccination if it was not given at the same time.

Even with the cholera vaccination, it is still important for travelers to practice other safety measures such as hand washing, water decontamination, and hygienic food preparation.

44. C. MMR

It is recommended that women considering pregnancy get caught up with the recommended vaccinations prior to conception. With the exception of live vaccines, all forms of immunization are generally considered safe to administer during pregnancy (if a live vaccine is accidentally administered during pregnancy, it is not an indication to terminate the pregnancy). Once a live vaccination is administered to a women of child bearing age, it is recommended that she wait one to three months prior to getting pregnant.

Of the vaccinations listed in the possible answers, MMR is the only live vaccine.

A decision to vaccinate during pregnancy should consider the gestational age, risk of preventable disease exposure, susceptibility of disease, and the risk of the individual vaccination. One benefit of vaccination during pregnancy is the transfer of antibodies from mother to child across the placenta.

Live vaccinations are not contraindicated during breast feeding.

45. C. Intussusception

The rotavirus vaccination is a live-attenuated vaccination that is offered in two forms:

1 Monovalent human rotavirus vaccine, RV1, which is administered orally once at two months of age and once at four months of age.
2 Pentavalent human-bovine reassortant rotavirus vaccine, RV5, which is administered orally at ages two, four, and six months.

The rotavirus vaccine predecessor, a rhesus-based tetravalent vaccine (RRV-TV), was withdrawn from the United States market within one year due to increased risk of intussusception. Because of that, both RV1 and RV5 have been studied excessively for increased risk of intussusception. Neither has shown a statistically significant increase in risk.

The minimum age for RV1 or RV5 administration is six weeks. Due to lack of safety data, the maximum age for the first dose is 14 weeks and six days. Rotavirus vaccinations should not be administered within four weeks of each other and all doses should be administered by eight months of age.

Rotavirus is a gastric illness prominent in children that may cause watery diarrhea, vomiting, and fever. The virus is transmitted through a fecal—oral route and destroys the epithelial surface of the small intestine and blunts the villi. The disease may be subclinical or severe enough to cause dehydration, shock, electrolyte imbalance, and death.

46. B. Hepatitis B

The hepatitis B vaccination is the only vaccine that is routinely recommended at birth. Infants born to mothers with positive hepatitis B surface antigen (HBsAg) should receive hepatitis B vaccination and HBIG within 12 hours of birth. After receiving the first shot at birth, all infants should receive the second hepatitis B vaccination at one to two months of age, followed by the third shot between six to 18 months of age.

The first hepatitis A vaccination is recommended at one year of age.

The first polio vaccination is recommended at two months of age.

The first pneumococcal vaccination is recommended at two months of age.

The first varicella vaccination is recommended at one year of age.

47. D. Seven–year-old girl recently diagnosed with sickle cell disease

The Hib vaccination is recommended to be given at ages two months, four months, and six months, with a booster dose given between 12 to 15 months. At least one proprietary Hib vaccination does not require the six month dose. Unvaccinated healthy children 15 months of age and older only need to receive one dose to catch up.

The Hib vaccine is not routinely recommended for healthy people greater than five years of age. However, the ACIP recommends that all children over of age of five receive it if they have HIV or anatomic/functional asplenia. Sickle cell disease is a main cause of functional asplenia.

If the ten-year-old boy was undergoing an elective splenectomy, it is recommended that he receive a single dose of the Hib vaccine at least 14 days prior to the procedure.

48. D. Virus-like particle (VLP)

Cervical cancer is one of many types of cancer caused HPV. Other cancers caused by HPV include cancers of the vulva, penis, anus, and oropharynx. HPV is a double-stranded DNA virus enclosed in a capsid shell, including capsid proteins L1 and L2.

There are three HPV vaccinations approved by the FDA, two-valent, four-valent, and nine-valent forms. The two-valent form includes protection against HPV subtypes 16 and 18 which are responsible for around two or three of all cervical cancers. The 4-valent form includes protection against HPV subtypes six, 11, 16, and 18. Finally, the nine-valent HPV vaccination affords protection against HPV subtypes six, 11, 16, 18, 31, 33, 45, 52, and 58. It is estimated that there are more than 200 HPV types.

Purified L1 protein are excised from HPV and engineered with harmless microbes to form empty shells that resemble a virus, called VLPs. VLPs are a type of recombinant vaccine (the class that includes vaccines such as hepatitis B).

49. B. Recombinant

The hepatitis B vaccination is a recombinant vaccination completed by inserting the HBsAg gene into yeast and specific types of mammalian cells.

50. E. 24 months

S*treptococcus pneumoniae* (*pneumococcus*) is a bacteria responsible for causing pneumonia, bacteremia, sinusitis, and otitis media. The PCV13 vaccine contains capsular polysaccharide components of 13

Streptococcus pneumoniae serotypes, each attached to a nontoxic diphtheria carrier protein.

The ACIP recommends that the PCV13 vaccine be given to all children aged two to 59 months. PCV13 is recommended as a four-dose series at ages two, four, six, and 12–15 months. It should also be given to children with underlying medical conditions aged 60–71 months, those ≥19 years that have immunocompromised status (including asplenia, CSF leak and cochlear implants) and in all adults ≥65 years old. In adults ≥65 years old that have already received the PPSV23 vaccination, it is recommended to wait at least one year prior to giving the PCV13 vaccination. If the ≥65 year old patient has not received a recent PPSV23 or PCV13 vaccination, it is recommended to give PCV13 and then wait six months to one year until administering PPSV23.

51. **D. PCV13 vaccination followed by PPSV23 in 6–12 months**

PPSV23 and PCV13 vaccinations both provide protection against Pneumococcus. The two vaccinations have many components in common, as PPSV23 contains 12 serotypes in common with PCV13. Both are recommended for adults age ≥ 65.

PPSV23 and PCV13 should not be coadministered. Those age ≥65 that have not previously received a pneumococcal vaccine (or have unknown vaccination history) should receive the PCV13 first, followed by a dose of PPSV23 after 6–12 months. A patient that has been vaccinated with PPSV23 should wait at least one year prior to receiving PCV13.

PPSV23 should not be administered at age 65 if the patient has received it within the prior five years. In this case, it is recommended to give the PPSV23 vaccination five years after the prior vaccination.

52. **C. Nigeria**

The WHO maintains International Health Regulations (IHR) to help countries around the world detect, assess, and report public health events in order to buildup healthcare capabilities. Protection from yellow fever is one of the many capacities in which the IHRs can help increase global health security.

IHR regulations recommend select African and South American countries to require proof of yellow fever vaccination in travelers on a standardized form known as the International Certification of Vaccination or Prophylaxis (ICVP). If a traveler does not have yellow fever vaccination documented on their ICVP, these countries are encouraged to refuse entry, or in some cases quarantine for up to six days. Country entry requirements may vary. Some of the countries require that all travelers show yellow fever vaccination, while others only require travelers to have it if they are coming from yellow fever endemic countries. Exceptions are made for specific groups such as infants and those with a waiver stating they are contraindicated from getting the vaccine.

There is only one yellow fever vaccine that is approved for use in the United States. It is a live vaccine. Other countries around the world use different vaccines. Studies show that there is no substantial difference in immune response between the various vaccines. Because of this, people that received any form of yellow fever vaccine should be considered protected against yellow fever.

53. A. Never

A single dose of yellow fever vaccine provides sustained immunity and lifelong protection against yellow fever. At least, that was the conclusion of a WHO strategic advisory group on immunizations in 2014. Prior to this statement, booster vaccinations were recommended every 10 years. Many countries with yellow fever restrictions followed this recommendation. Thanks to this conclusion, a completed ICVP demonstrating one yellow fever immunization can now be used throughout a lifetime to grant access to countries that require proof of yellow fever vaccination for entry. The CDC also supports giving a single yellow fever vaccine.

Booster doses of the yellow fever vaccine are recommended for specific groups of travelers. The groups include women that were pregnant when they received the vaccine, those that received hematopoietic cell transplants after receiving the vaccine, and people with HIV.

54. C. A major wound with a history of two Tdap vaccinations

The CDC and the ACIP recommend that TIG be administered directly into the wound in those that suffer a major wound and have experienced less than three tetanus vaccinations in their past. Of the listed options, choice C is the only option that fits the criteria.

55. B. Give the Td vaccine now

Tetanus is caused by *Clostridium tetani* spores that are ubiquitous in the environment but are especially common in feces and soil. The spores germinate in injured tissue after entering through broken skin. The deeper the wound, the greater the anaerobic conditions and the higher the likelihood of germination leading to release of the tetanospasmin toxin. This toxin leads to noncommunicable disease defined by body-wide muscle spasms.

Tetanus vaccination begins at a young age, with a five-dose DTaP vaccination series. A routine tetanus booster (Td) is recommended every 10 years for adults. The ACIP recommends that one of these routine tetanus boosters be Tdap, which offers protection against pertussis. If a dirty wound occurs, it is recommended to receive a tetanus booster if the last booster was more than five years prior.

If this patient had endured a deeper injury and not received at least three tetanus injections in the past, it would be recommended that TIG be administered directly into the wound.

TT is only indicated in rare situations, such as when the patient has an allergy to diphtheria toxoid.

56. D. Toxoid

The tetanus component in TT vaccine, tetanus, and Td and all other vaccination combinations of tetanus, diphtheria, and pertussis are composed of toxoid elements. Toxoids are inactivated bacterial/viral toxin elements that are capable of instructing the immune system to develop antibodies to activated toxins.

57. B. Give him the Tdap vaccination now

The Td vaccination does not provide protection against pertussis, the organism that causes whooping cough. The ACIP recommends that close contacts of infants less than one year of age receive at least one Tdap vaccination during their lifetime. This process of protecting everyone around the infant is known as cocooning. It is recommended that unvaccinated potential contacts of the infant receive vaccination at least two weeks prior to contact. Pregnant women should receive the vaccination during every pregnancy (preferably between gestational weeks 27–36).

Td and Tdap may be given on the same day in situations such as this.

58. B. 90%

Although this question may seem obscure, it is intended to bring up two important topics.

Healthy People is a program that sets objective goals to strive for. Rather than being generic in stating something along the lines of, "all healthcare workers should get the flu shot", it provides an objective number that is thought to be attainable.

The second important topic is healthcare worker immunization for all vaccine-preventable diseases. Healthcare workers serve as vectors that transmit infections to the most susceptible population in society. Unvaccinated healthcare workers increase the risk of iatrogenic illness, leading to increases in morbidity, mortality, cost of care, and employee absenteeism. Annual influenza vaccination is typically the most commonly debated immunization for healthcare workers because it is administered every year. As recently as 2008, only 45% of all healthcare workers received annual influenza vaccination. Proposed solutions to lack of influenza immunization include job termination (mandatory immunization), removal of patient contact and/or requirement to wear a surgical mask.

59. A. During every pregnancy

In October 2011, the ACIP first recommended the routine administration of a Tdap during pregnancy as a strategy to protect infants from pertussis.

The vaccination is recommended for all pregnant women to protect the infant from pertussis. The immunity gained from the vaccination is passively passed onto the child. Even with this head start, children are not fully protected against pertussis until they receive several DTaP doses.

It is recommended that all close contacts received the Tdap vaccination to assure the infant is not exposed to pertussis. This process is known as cocooning.

60. E. Two weeks

People may start to develop antibodies as soon as two to six days after vaccination with an inactivated influenza vaccination. Within two weeks of vaccination, roughly 90% of patients have developed protective antibody titers. The response peaks between two to three weeks after influenza vaccination and wanes in time, falling 50% or lower six months after vaccination. The degree of protection from vaccination depends on the match between the vaccine strains and the wild influenza strains circulating in the community, the overall health of the vaccine recipient and previous history of influenza.

An initial peak of IgA and IgM occurs two weeks after vaccination with the live influenza virus vaccination. An IgG reaction peaks between four through twelve weeks and sustains these levels for one year.

61. D. Oseltamivir

Antiviral drugs (including oseltamivir) are the preferred PEP for influenza exposure. Antiviral PEP is recommended for those unimmunized against influenza, those that have received the influenza vaccination within two weeks and those that have been exposed to a variation of the circulating flu that is different from the variant in the vaccine.

Azithromycin is an antibiotic and will not provide protection to influenza which is a virus.

Wearing a surgical mask and washing hands are good ways to prevent the spread of influenza. However, in this vignette, the patient's mother has already been exposed to the flu. The virus may be spread even before causing symptoms.

62. **B. Six months**

The ACIP recommends that children begin to receive their annual flu shot beginning at the age of six months. Furthermore, ACIP recommends that those under eight years of age receiving their first flu shot be given a second shot during that flu season, but no sooner than within four weeks of the first.

63. **A. Six months**

Hepatitis A virus (HAV) is a single-stranded RNA virus that replicates in the liver, is excreted in bile, and is shed in stool. It may cause fever, malaise, anorexia, nausea, abdominal pain, and jaundice. Person-to-person transmission through the fecal–oral route is the most common means of transmission in the United States but it can also be spread through common-source outbreaks from contaminated food or water. Groups at increased risk include travelers, men who have sex with men (MSM), and illicit drug users. After transmission, the average incubation period is 28 days. Peak infectivity occurs two weeks before jaundice or elevation of hepatic enzymes. HAV is diagnosed through HAV IgM antibodies. There is no chronic form of hepatitis A and symptoms are usually self-limited to two months or less.

The hepatitis A vaccine is an inactivated form of the virus that is injected intramuscularly. It may be given alone or in a combination vaccine along with hepatitis B. In the United States, the hepatitis A vaccine is recommended routinely for children aged one year old. After initial vaccination, a second vaccination is recommended in six to 18 months.

64. **B. Hib**

In unimmunized regions, the majority of epiglottitis cases are attributed to *Haemophilus influenzae* species, with sporadic cases from *N.*

meningitides and *Streptococcus* species. The Hib vaccine has dramatically decreased the incidence of epiglottitis in places that immunization programs have been successfully implemented.

Epiglottitis is the result of an infected epiglottis. Symptoms include fever, dysphagia, drooling, altered voice, stridor, and cyanosis. If the swelling gets bad enough, the airway is compromised and respiratory distress ensues.

65. D. VISs are produced by the vaccine manufacturer

VISs are vaccine-specific documents produced by the CDC to educate vaccine recipients about the risks and benefits of the vaccine they are receiving. If the patient is a child, the VIS is to be given to their guardian. Under the National Vaccine Childhood Injury Act, it is mandatory for the vaccine provider to give the appropriate VIS to the patient prior to each and every vaccination, including for each dose in a series. Furthermore, providers are required to record the VIS edition date and the date the VIS was provided.

Information present on the VIS includes information on why to get vaccinated, specific education about each vaccine, information about who should not get the vaccine, risks of vaccine reaction, what to do if there is a reaction, information about the National Vaccine Injury Compensation Program, and resources on how to learn more.

66. B. Chloroquine

Chloroquine is the preferred medication for pregnant women to use as malaria prophylaxis when traveling to regions without chloroquine resistant strains of malaria. Chloroquine can be used for both malaria prophylaxis and malaria treatment in pregnant women and children.

Malarone, doxycycline, and primaquine are not recommended for pregnant women.

If chloroquine is contraindicated, mefloquine can also be used in pregnancy for prophylaxis and treatment. Mefloquine is generally not the primary drug of choice, due to the side-effect profile and availability of other drugs for malaria.

67. C. Microcephaly

Zika virus is transmitted via the *Aedes aegypti* mosquito and sexual contact. Symptomatic people experience self-limited malaise, fever, arthralgia, conjunctivitis, and often times a maculopapular rash. Treatment is generally supportive.

Pregnant women may transmit the virus to their fetus at any stage during pregnancy. Once infected by the virus, microcephaly may occur in the fetus. For this reason, it is recommended that pregnant women with confirmed Zika virus undergo serial ultrasounds to measure cranial development.

68. B. RNA virus

Chikungunya is a mosquito-borne RNA virus. Humans are the primary host. After a three to seven-day incubation period, primary symptoms include fever and arthralgia. Other symptoms include headache, myalgia, conjunctivitis, nausea/vomiting, and maculopapular rash. Antibodies can be detected as soon as four days after illness and PCR can detect it within eight days. Treatment is symptomatic and symptoms usually resolve after seven to 10 days.

69. A. Dysuria

Dengue is an RNA flavivirus that is transmitted by the *Aedes aegypti* mosquito. Illness presents in three main forms: Dengue fever, dengue hemorrhagic fever, and dengue shock syndrome. Dengue is commonly known as "breakbone fever" due to the severe bone pain it causes in conjunction with fever. Pain behind the eyes is a common complaint amongst patients. The disease may progress to include thrombocytopenia and increased vascular permeability, leading to hemorrhagic manifestations and rash. If the disease progresses, it may cause hypotension and shock. The treatment for dengue fever is supportive.

70. D. 60 years old

The herpes zoster (shingles) vaccination is a live-attenuated virus. The FDA approved it for those ≥ 50 years and the ACIP recommends it for those ≥ 60 years. It is approved to reduce the risk of developing shingles and subsequent complications such as postherpetic neuralgia. It is not indicated for the treatment of shingles, postherpetic neuralgia, or the primary prevention of varicella infection.

71. D. Snail

Schistosomiasis is a neglected tropical disease caused by trematode worms of the genus *Schistoma*. This parasite lives a complicated lifecycle through which certain types of freshwater snails act as a host. When the infectious form of a *Schistoma* (known as cercariae) leaves the snail, it contaminates the water and traverses the skin to affect those that come into contact with it.

Once in the body, the parasite matures into adults and lays eggs. Many of the eggs are passed harmlessly through the urine and the stool. It is the eggs that do not leave the body that contribute to the vast majority of morbidity. These eggs travel throughout the body and become lodged in the bladder and intestines, amongst other sites. Once there, they trigger an inflammatory reaction leading to scarring and obstruction. Acute symptoms of schistosomiasis include pruritus, rash, fevers, chills, cough, and myalgia. Children suffering from chronic schistosomiasis may develop anemia, malnutrition, learning disability, or damage to the liver, spleen, intestines, and bladder. Other chronic complications include abdominal discomfort, liver fibrosis, bowel wall perforation, hematuria, and blood in the stool. Praziquantel is the drug of choice to treat schistosomiasis.

Schistosomiasis is endemic in lesser developed countries with poor sanitation. Over 200 million people worldwide are affected by the disease which is spread when contaminated individuals urinate or defecate in water that contains the right type of snail species.

72. A. *Cytomegalovirus*

Roughly two out of every 1000 children born in the United States suffer from profound hearing insufficiency (≥35 dB) and four out of every 1000 suffer from lesser, yet clinically significant hearing loss. Genetic deviations cause between 50%–60% of these cases. Meanwhile, nearly 30% of the remaining cases are due to maternal infection during pregnancy, environmental causes, and complications after birth. These numbers are constantly in flux because maternal infections that cause hearing loss are also in flux. Other common causes of hearing loss include prematurity, postnatal infections, head trauma and pharmacologic toxicity. With significant rubella vaccination rates in the United States, congenital cytomegalovirus infection has become the most prevalent environmental cause of hearing loss in infants.

The Advisory Committee on Heritable Disorders in Newborns and Children (within the Health Resources and Services Administration) recommends a panel of universal screening tests for states to consider when making their own state mandated newborn screening protocols. Screening for hearing loss is one of their core recommendations. When hearing loss is detected early, medical professionals may institute interventions to maintain language, social, and emotional development. Unfortunately, some forms of early-onset hearing loss are not apparent at birth. The Joint Committee on Infant Hearing has identified risk indicators for clinicians to be watchful of. Furthermore, the American Academy of Pediatricians promote periodic hearing screening for every child through adolescence.

73. B. *Chlamydia*

Trachoma is a progressive form of irreversible blindness that is caused by *Chlamydia trachomatis*. Repeated exposure to the organism throughout the lifetime causes scarring of the surface of the eye. It is endemic in developing nations, mainly in areas that are overpopulated with poor hygiene and restricted access to care. The WHO has identified Trachoma as a disease to eradicate.

74. E. Live

BCG vaccination is a type of live-attenuated mycobacterial strain from a descendent of the original-attenuated Bacillus of Calmette–Guerin. It is given in countries with high endemic levels of TB. It is most effective against miliary and meningeal forms of TB. Because it is a live vaccine, it should not be given to immunocompromised persons.

75. E. Nothing

PEP should be given to only those that have prolonged contact with a confirmed case of bacterial meningitis. Prolonged contact usually occurs in daycare/healthcare settings and at home. Casual contacts should not receive PEP but should be monitored closely.

For meningococcal meningitis, PEP includes rifampin for children under one month old, ceftriaxone for patients under 15 years old, and fluoroquinolones for adults. To provide PEP against meningitis from Hib, the appropriate medication is rifampin.

76. C. 24 hours after taking antibiotics

A child with strep throat may return to school after being on appropriate antibiotics and being free of fever for 24 hours. Many clinicians are comfortable shortening this time period. For example, some clinicians may feel comfortable sending children to school the following day, as long as antibiotic therapy was initiated before 3:00 p.m. If a patient is diagnosed with a viral illness and is not prescribed antibiotics, it is still appropriate to refrain from going to school until free of fever for 24 hours.

The American Academy of Pediatrics, American Public Health Association, National Resource Center for Health and Safety in Child Care and Early Education have posted guidelines for parents, teachers, and clinicians to follow regarding the appropriate time to send children back to school. In addition to the common scenarios in the table below, children should refrain from attending school if their ailments make school attendance and learning difficult.

Ailment	When to return to school
Chicken pox	Return once the last lesion has dried and crusted
Diarrhea	No school if ≥ three episodes in past 24 hours or blood in stool. Return once resolved.
Vomiting	No school if ≥ two episodes in past 24 hours. Return once resolved or found to be noninfectious.
Pertussis	Return after five days antibiotics (which should be given 14 days total)
Strep throat	Return after 24 hours of antibiotics and free of fever for 24 hours
Skin infection	Return 24 hours after being placed on antibiotics
Head lice	Return after first treatment
Scabies	Return after treatment has been completed
Conjunctivitis	Return 24 hours after treatment has been initiated

77. A. Immediately

A child that has had ≥ three episodes of diarrhea in past 24 hours, or is experiencing bloody diarrhea regardless of frequency, should not return to school until the diarrhea has ceased. He/she may return to school when he/she has had two or less episodes of diarrhea in a 24 hour period, unless the diarrhea is bloody. If the patient is diagnosed with *Shigella* or *E. coli* 0157:H7, the child should not return to school until diarrhea resolves and two stool cultures are negative.

Guidelines for when to return to school were explained in answer #76 (directly above).

78. E. All of the above

As inappropriate usage of antibiotics continues to increase around the world, the looming threat of entering a postantibiotic era becomes more conceivable. Inappropriate use of antibiotics directly contributes to microbial resistance. Resistant infections take longer to control and are more expensive to treat. If a resistant infection is present, it may contraindicate performing medical procedures such as surgery and chemotherapy. Furthermore, there is a clear association with multidrug resistant organisms and increased mortality.

79. D. Use of diagnostic tests

Microbial antibiotic resistance is a public health global emergency with multifactorial causes, including four out of five possible answers listed. Utilization of diagnostic tests is the only answer listed that helps reduce antibiotic resistance. Diagnostic tests guide health practitioners and educated consumers towards the appropriateness of antibiotics. For example, if a patient with pharyngitis tests positive for mononucleosis, the mononucleosis screening test should direct the clinician away from the use of antibiotics. In addition to determining whether antibiotics are appropriate, diagnostic tests may also shed insight towards the appropriate type of antibiotic (culture and sensitivity).

Antimicrobial agents are commonly added to animal foods to increase the size and health of animals. When animals are administered antibiotics for illness, they run the same risk of resistance as when administered to people. The types of antibiotics used in people and animals are fairly similar to each other. Antibiotic resistance may be transferred from animal pathogens to human pathogens.

Antibiotics are available over the counter in many nations around the world. Lay healthcare consumers are forced to make an uninformed choice about what they feel is the most appropriate antibiotic. These people may also struggle to take the correct dosage in the right frequency for the appropriate duration. In many cases, an antibiotic is not warranted to begin with. These health consumers may also be influenced by external factors such as clearance sales from the pharmacy and advertisement from drug companies.

Even when there is a clinician prescribing antibiotics, there are still considerable factors that lead to antibiotic resistance. Clinicians are often guilty of knowingly or unknowingly overprescribing antibiotics for self-limited bacterial and viral infections. Antibiotics are overprescribed and inadequately used in hospitals around the world, greatly contributing to resistance.

Antibacterial resistance is not only forged through animal and human consumption. Environmental settings, such as wastewater treatment facilities and drainage systems, play a large role in fostering bacterial resistance through horizontal gene transfer. Poor sanitation and globalization are other avenues for antimicrobial resistance.

Worldwide, antibacterial use is on the rise. There have been interventions at all levels of government to reduce progression antibiotic resistance. One of the largest efforts is known as the Global Antibiotic Resistance Partnership. The United States has created an Antibacterial Drug Development Task Force and has passed legislation rewarding discovery of new antibiotics, such as the Generating Antibiotic Initiatives Now (GAIN) Act.

Successful interventions to reduce the rate of bacterial resistance must be as diverse as the causes of bacterial resistance. Access and quality of care (including infection control) must be appropriate. There should be surveillance of antibiotic use and resistance with acceptance of available guidelines for appropriate antibiotic use. Furthermore, hospitals could offer antibiotic stewardship programs. In low and middle-income countries, available antibiotics should be limited and appropriate antibiotic education available.

80. B. GBS

Group B streptococcus (GBS) is a gram-positive bacteria that colonizes the vagina and rectum of nearly a quarter of adult women. Neonatal infection occurs from vertical transmission when the membranes of pregnancy rupture and the newborn comes into contact with the bacteria directly through contact with the vagina, or through the amniotic fluid, which may aspirate into the lungs. Symptoms in a newborn may occur as early as the first week of life, with most infections evident by the third month of life. Infection may lead to sepsis, pneumonia and meningitis. GBS is the leading cause of infectious morbidity and mortality in infants born in the United States. Premature infants are disproportionately affected.

Without intervention, 1%–2% of infants born to mothers colonized with GBS develop symptoms of infection. This number is much lower for children born via cesarean section to mothers with intact membranes. With appropriate prophylaxis, the number of affected children is nearly eliminated. First-line recommended prophylaxis is intrapartum administration of IV penicillin or ampicillin.

Routine GBS screening through vaginal and rectal swabs is recommended between 35–37 weeks of gestation.

The following are indications to initiate GBS prophylaxis:

– Previous newborn with invasive GBS disease
– GBS bacteriuria during current pregnancy

- Positive GBS vaginal–rectal screening in late gestation (ideally between 35–37 weeks)
- Unknown GBS status at onset of labor and any of the following:
  - Delivery at <37 weeks gestation
  - Amniotic membrane rupture ≥ 18 hours
  - Intrapartum temperature ≥ 100.4°F
  - Intrapartum NAAT positive for GBS

The following are not indications to initiate GBS prophylaxis
- Positive GBS status from previous pregnancy
- Negative vaginal–rectal GBS screening
- Cesarean delivery with intact amniotic membranes

81. E. Diarrhea and pneumonia

The global disease mortality for those aged one month through 59 months (five years) is a statistic kept by the WHO. The top four causes of death in this population in descending order are: Pneumonia, diarrhea, malaria, injuries.

82. B. Heterosexual vaginal intercourse

The most common method of HIV transmission is through sexual contact. The major route of HIV transmission worldwide is heterosexual vaginal intercourse. However, the most common route of transmission in the United States is through anal intercourse in MSM. HIV infection is less commonly transferred through oral sex. Factors that increase HIV spread include transmission of the virus during the time period immediately after contracting it, having genital lesions, and concomitant STIs.

The second most common source of HIV infection is through blood contact, including needle sharing for intravenous drug use, accidental needlesticks, and iatrogenic healthcare blood transfusions.

The third most common source of HIV infection is vertical transmission from mother to child.

83. E. *Vibrio cholerae*

This is an outbreak of cholera caused by the bacteria *Vibrio cholera.* Cholera is rare in the United States thanks to elevated hygienic standards, with most cases being imported from international travelers. Although there are oral vaccinations available, the benefits are very short-lived.

Once ingested, *Vibrio cholera* colonizes the intestinal lining and produces the toxin choleragen, which causes influx of intestinal fluids leading to severe diarrhea. Symptoms of cholera occur two to three days after ingestion and include abdominal pain, dehydration, and diarrhea which is often described as "rice water stool". Further complications include hypoglycemia, dehydration, and renal failure. Rehydration is the cornerstone of treatment.

84. C. *Salmonella*

Salmonella is a gram-negative facultative anaerobe that is often found in intestinal flora. It is transmitted through ingestion of contaminated water and foods such as eggs, poultry, beef, vegetables, and dairy products. Reptiles may colonize *Salmonella* and pose as a vector for transmission.

Typhoid fever is caused by S. typhi invading epithelial cells of the small intestine and subsequently leading to bacteremia. Although the infection may be asymptomatic, symptoms can include fever, constipation, diarrhea, emesis, rash and septicemia.

Enteric fever is caused by *S. paratyphi*. The infection may present similarly to typhoid fever, but symptoms are typically less severe. Enteric fever is the least common form of salmonellosis.

Gastroenteritis is caused by >2000 *Salmonella* serotypes. Symptoms are typically self-limited and include nausea, emesis, diarrhea, and dehydration.

85. A. Rat flea (*Xenopsylla cheapis*)

All three types of fleas listed have the capacity to bite humans. Fleas may transmit typhus, salmonellosis, and plague, amongst many other diseases. The rat flea carries the organism that causes the plague, *Yersinia pestis*. Plague is still common in India, occasionally leading to an outbreak. It has also been reported in the American West and Southwest, where the rat flea is carried by prairie dogs, chipmunks, squirrels, and rats.

86. B. Infantile

Clostridium botulinum produces a group of neurotoxins collectively known as botulinum toxin. This toxin blocks release of acetylcholine at the presynaptic membrane in motor neurons, causing general muscle weakness that ultimately leads to paralysis, and impairs respiratory function. Botulinum toxin is one of the most potent substances known, with a lethal effect at 0.0002 μg/kg body weight. Just 100 g could kill nearly all humans on Earth.

C. botulinum is a dormant spore in oxygen rich environments. However, in low-oxygen environments (including canned goods and intestinal tracts), it may germinate and produce botulinum toxin.

There are three naturally occurring forms of botulism: Infantile, foodborne, and wound. Infantile botulism is the most common form of botulism seen in the United States. It is an infectious form of botulism which occurs when infants ingest *C. botulinum* which germinate in intestinal tissue incapable of killing the bacteria. Infants are often exposed to *C. botulinum* through raw honey. Healthy adult intestinal tissue is capable of killing *C. botulinum*. Foodborne botulism is an example of intoxication and occurs after ingestion of tainted food with preformed toxins. These toxins are produced from *C. botulinum* from improper food handling such as canning or not cooking food thoroughly. Wound botulism may occur if bacterial spores are introduced into a wound.

Inhaled botulism is not a naturally occurring phenomenon and is a product of bioterrorism. Due to its extreme potency and lethal ease of dissemination, it is listed as a category A bioterrorism agent.

87. E. Five years old

A TST is nearly always preferred over IGRA for children less than five years old. Little data exists for the use of IGRA in children under five.

88. D. IGRA should be taken during time of live vaccination or >four weeks after live vaccination

Similar to the case with the TST, live virus administration may affect IGRA results. For this reason, it is recommended that IGRA be taken on the day of live vaccination administration, or four weeks status post live vaccination administration.

89. D. Two-step PPD

According to the "same-day rule", two or more live vaccinations may be given at the same time. After receiving a single live vaccination, the recipient must wait four weeks to receive another live vaccination. This is due to theoretical immunocompromised status, where the immune system will not mount the typical reaction. The same rule holds true for PPDs. A PPD may be given on the same day as a live vaccination or four weeks after the administration of one, but not within that time period.

In this case, a single PPD could be given on the same day as the live vaccination, but a two-step PPD requires a second PPD placement after the first. It is preferred to give the second PPD between one through three weeks after getting the initial PPD, but it may be given as late as one year after the first PPD placement.

Unfortunately, this student needs to have the second step completed within two weeks.

Drawing blood prior to receiving the varicella vaccination or on the same day as vaccination for IGRA testing would provide a TB test that would be free of adulteration from the body's response to a live vaccine.

90. C. *Meningococcal*

The annual Hajj in Saudi Arabia is the largest mass gathering in the world. Muslims are required by their faith to make this trip at least once in his or her lifetime.

The Saudi government recommends that all visitors and locals in the area be up to date with routine immunizations, including hepatitis A, hepatitis B, polio, and influenza. The only vaccination required to obtain a Hajj visa is the meningococcal vaccine. Everyone >two years old must have received a single dose of quadrivalent A/C/Y/W-135 vaccine and be equipped with proof of immunization status on a certificate. Hajj pilgrims are required to have had the Meningococcal Vaccine no less than 10 days before the visit and no more than three years prior to the trip.

91. B. Hepatitis B

Hepatitis B, hepatitis C, and HIV are the infections that carry the most concern in needlestick injuries. Hepatitis B leads to a positive seroconversion in $\geq 30\%$ of needlestick injuries if the source patient is positive for HBsAg. Meanwhile, hepatitis C and HIV transmission occur at a rates of $\sim 1.8\%$ and 0.3%, respectively. A larger inoculum typically increases the risk of disease transmission. Therefore, risk factors for transmission include further progression in the natural course of the disease in the source patient at time of needle contamination, visible blood from the source patient present on the needle, using a larger needle, the depth of the needle stick and visible bleeding as a result of the needle stick.

After a needlestick, instruct patients to immediately wash the wound with soap and water. There is no evidence to show that "squeezing out" the inoculum helps reduce the risk of infection. If possible, collect blood from the source patient to evaluate for hepatitis B, hepatitis C, and HIV.

All healthcare workers and those with potential exposure to a needlestick injury should receive the hepatitis B vaccination series (initial dose, one month after initial dose, and six months after initial dose). If a needlestick-exposed patient has not received the hepatitis B vaccination, it is recommended to receive HBIG and begin the hepatitis B vaccination series as quickly as possible. If a needlestick-exposed patient is currently in the midst of the vaccination series, he/she should receive HBIG immediately and continue the vaccination as scheduled. If the needlestick patient has already received the hepatitis B vaccination series, the practitioner may wish to check titers. No medical intervention is warranted if the titers are positive.

There is no vaccination available to protect against hepatitis C. Immunoglobulins and antivirals to prevent hepatitis C are not indicated after needlestick exposure. After a patient has been exposed to hepatitis C, clinicians should initiate serial hepatitis C and liver enzyme testing. More frequent tests lead to treatment in the easier to treat acute phase.

Although there is no HIV vaccination available, needlestick patients may receive PEP medications to prevent the transmission of HIV. Serial testing is indicated to monitor HIV status.

92. **E. Both answers C and D**

All healthcare workers and those with potential exposure to a needlestick injury should receive the hepatitis B vaccination series (initial dose, one month after initial dose, and six months after initial dose).

Please see question #91 (directly above) for further explanation.

93. **A. Administering a reduced amount of the recommended dose of a vaccine**

Fractional dosing occurs when smaller than recommended vaccine doses are administered. It is widely discouraged for clinicians to give fractioned doses. However, under extenuating circumstances, fractional dosing has been recommended by governing bodies. For example, fractional dosing was encouraged during a yellow fever outbreak in Africa when there was a shortage of available vaccine. It was thought that dividing the amount of available vaccine to a larger number of people would yield greater benefit than administering limited doses to a limited number of people.

94. **A. Electronic Disease Notification (EDN)**

The Department of Health and Human Services (DHHS), specifically the CDC (within DHHS) is the agency responsible for preventing the spread of communicable diseases from foreign countries to the United States. It works together with the US Department of State and US Department of Homeland Security to screen every immigrant coming to the United States for infectious diseases. Once the immigrants are screened, the CDC utilizes the EDN system to notify state public health agencies of the health status of all newly immigrated people.

The EDN provides health departments with timely and accurate notifications on immigrants that are to follow up with physicians. It also monitors diseases of public health significance and provides information that is used by international healthcare organizations.

95. C. Hepatitis A

Vaccines may be broken down into two groups, live or inactivated.

Live vaccines are derived from the viruses and bacteria that cause disease. These agents are attenuated (weakened) to become less harmful to people. After injection into the body, live viruses replicate and cause an immune response similar to that produced by a natural infection. The body learns how to respond and forms a strong, lasting immune response. It is theoretically possible for all live vaccines to revert to their wild and harmful form but only the OPV has been shown to do so (can be limited by giving IPV first). In roughly 1% of varicella vaccine recipients, the vaccine has been shown to produce a very mild form of the disease, consisting of ~five vesicles. Examples of live vaccines include oral typhoid, BCG, varicella, and yellow fever.

Inactivated vaccines may either be whole or broken down into subunits. They are incapable of replicating in the host and therefore stimulate a weaker immune response that wanes over time and requires multiple doses. Whole cell inactivated vaccines include hepatitis A and rabies. Meanwhile, subunit vaccines may be either protein or polysaccharide based. Protein-based subunit vaccines include toxoids and protein subunits, including the tetanus and diphtheria vaccinations. Subunit polysaccharide vaccines include acellular pertussis.

Conjugate vaccines may be considered a type of inactivated subunit vaccine. A conjugate vaccination is made by attaching polysaccharides of a harmful pathogenic bacterium to a harmless antigen that is easily recognized by the immune system. The immune system learns to recognize the harmful antigen and generates immunity. Hib is a conjugate vaccine.

Recombinant vaccines are products of genetic engineering, where a harmless agent such as yeast, is programed to produce antigens of harmful pathogens. Similar to conjugate vaccines, the immune system learns to recognize the harmful antigen and generates immunity. Hepatitis B is a recombinant vaccine. When the recombinant vaccine places the antigen of interest into a virus shell, this type of recombinant vaccine is known as a virus-like particle, VLP. The HPV vaccination is a VLP.

96. **A. Haemophilus influenza**

A conjugate vaccination is made by attaching polysaccharides of a harmful pathogenic bacterium to a harmless antigen that is easily recognized by the immune system. The immune system learns to recognize the harmful antigen and generates immunity. It is especially useful for harmful capsular bacteria that are not easily recognized by the immune system. Hib is a conjugate vaccine.

- Hepatitis B is a recombinant vaccine
- Rabies is an inactivated whole vaccine
- Shingles is a live-attenuated vaccine
- Tetanus is a toxoid vaccine—a type of protein subunit

Please see question #95 (directly above) for further explanation.

97. **B. *Mycobacterium***

Mycobacterium leprae is the organism responsible for causing Hansen's disease (aka leprosy), an illness that targets the skin, respiratory mucosa, eyes, and the peripheral nervous system. The prevalence of this disease has dropped >99% in the past three decades, with 95% of new cases being found in 14 select countries (India #1).

Although the mechanism of transmission is not known, it is thought to be spread through direct contact, and possibly respiratory droplets. It is known to have a low infectious rate and is typically only spread to close contacts. Once infected, the incubation period can last anywhere from nine months to 20 years, with an average of four to eight years.

The disease asymmetrically targets peripheral nerve trunks, causing nerve thickening, and subsequent effects of nerve damage. These effects include loss of sensation in the skin, muscle weakness, and curling of the fingers. One of the more visible signs of Hansen's disease is lepromas, disfiguring tumor-like lesions that typically form on the trunk. Hansen's disease may also cause blindness, collapsing of the nose, and thickening of the earlobes.

Diagnosis may be made from either positive skin smears or clinical findings of skin sensory loss and skin lesions consistent with Hansen's disease (with or without thickened nerves). Characteristic skin lesions may appear as macules, papules, or nodules (lepromas). Patients with negative smears are said to have paucibacillary leprosy

(PB) while those with positive smears are diagnosed with multibacillary leprosy (MB). Close contacts of over five years should also be investigated to restrict spread of the disease.

Early treatment of Hanson's disease averts most disabilities. Because monotherapy leads to drug resistance, multidrug therapy is the mainstay of treatment. The WHO recommends dapsone, rifampicin, and clofazimine for those with MB and dapsone and rifampicin for those with PB. Hanson's is also associated with a negative stigma, leading to isolation and psychological damage. To address the stigma, the name of the disease was changed from leprosy and efforts have been made to provide public education.

98. E. Depends on multiple factors

Not only are vaccinations beneficial to individuals, but potentially to the population as a whole. Once enough of the population is immunized, herd immunity occurs to reduce transmission to the point that an epidemic is nearly impossible.

Herd immunity is different for all diseases and depends on multiple factors. These factors include the type of illness (immunogenicity, infectivity, and secondary attack rate), social patterns of the population (type and duration of contact), population density, and weather.

Some examples of herd immunity thresholds for different diseases are as follows: Pertussis 92–94%, rubella 83–85%, diphtheria 85%, and influenza 33–44%.

BIBLIOGRAPHY

[1] Plotkin SA, Orenstein WA, Offit PA. Vaccines: expert consult. 5th ed Philadelphia, PA: Elsevier Health Sciences; 2008. p. 19.

[2] Conway MJ, Colpitts TM, Fikrig E. Role of the vector in Arbovirus transmission. Annu Rev Virol 2014;1(1):71–88.

[3-4] Hopkins DR. Disease eradication. N Engl J Med 2013;368(1):54–63.

[5] CDC. Mumps. ⟨http://www.cdc.gov/mumps/⟩ [accessed 20.10.2016].

[6] CDC. Prevention & control. ⟨http://www.cdc.gov/westnile/prevention/index.html⟩ [accessed 20.10.2016].

[7] WHO. Animal influenza. World Health Organization. ⟨http://www.who.int/zoonoses/diseases/animal_influenza/en/⟩ [accessed 20.10.2016].

[8] CDC. Hepatitis A Information. ⟨http://www.cdc.gov/hepatitis/hav/index.htm⟩ [accessed 20.10.2016].

[9] Morris GJ, Potter ME. Foodborne infections and intoxications. 4th ed Oxford: Academic Press; 2013. p. 120–2.

[10a] Lagrange SR. Vaccine issues. Hauppauge, NY: Nova Publishers; 2004. p. 27–8.

[10b] Zubay G. Agents of bioterrorism: Pathogens and their weaponization. Columbia University Press; 2005. p. 248–50.

[11] Macera CA, Shaffer R, Shaffer PM. In: Stoskopf C, editor. Introduction to epidemiology: Distribution and determinants of disease. Clifton Park, NY: Cengage Learning; 2013. p. 186–7.

[12] Davis RG. Animals, diseases, and human health: Shaping our lives now and in the future. Santa Barbara. CA: ABC-CLIO 2011;127–66.

[13] Liu D. Manual of security sensitive microbes and toxins. Boca Raton, FL: CRC Press; 2014. p. 520–5.

[14] Jenkins WD. Public health laboratories: Analysis, operations, and management. Sudbury, MA: Jones & Bartlett Learning; 2011. p. 117.

[15] Cortese MM, Parashar UD. Centers for Disease Control and Prevention (CDC). Prevention of Rotavirus Gastroenteritis Among Infants and Children: Recommendations of the Advisory Committee on Immunization Practices [ACIP]. MMWR 2009;58[No. RR-02].

[16] Clinical summary: Ocular Prophylaxis for Gonococcal Ophthalmia Neonatorum: Preventive medication. ⟨http://www.uspreventiveservicestaskforce.org/Page/Document/ClinicalSummaryFinal/ocular-prophylaxis-for-gonococcal-ophthalmia-neonatorum-preventive-medication⟩ [accessed 25.10.2016]

[17-18] Testing for Tuberculosis Infection. CDC. ⟨http://www.cdc.gov/tb/education/corecurr/pdf/chapter3.pdf⟩ [accessed 25.10.2016].

[19] Boullata JI, Armenti VT. Handbook of drug-nutrient interactions. 2nd ed New York, NY: Springer Science & Business Media; 2010. p. 326–7.

[20-21] Testing for Tuberculosis Infection. CDC. ⟨http://www.cdc.gov/tb/education/corecurr/pdf/chapter3.pdf⟩ [accessed 25.10.2016].

[22] Pertussis (Whooping Cough) Treatment. CDC. ⟨http://www.cdc.gov/pertussis/clinical/treatment.html⟩ [accessed 25.10.2016]..

[23] Latent tuberculosis infection: A guide for primary health care providers. CDC. ⟨http://www.cdc.gov/tb/publications/ltbi/treatment.htm#treatmentRegimens⟩ [accessed 25.10.2016]..

[24] Barratt J, Topham P, Harris K, editors. Oxford desk reference: Nephrology. New York, NY: Oxford University Press; 2009.

[25] Dierberg K, Chaisson R. Human immunodeficiency virus-associated tuberculosis: Update on prevention and treatment. Clin Chest Med 2013;34(2):217–28.

[26] Hepatitis C FAQs for health professionals. CDC. ⟨http://www.cdc.gov/hepatitis/hcv/hcvfaq.htm⟩ [accessed 25.10.2016].

[27] Chou R, Cottrell EB, Wasson N, Rahman B, Guise J-M. Screening for Hepatitis C Virus Infection in Adults. Comparative Effectiveness Review No. 69. (Prepared by the Oregon Evidence-based Practice Center under Contract No. 290-2007-10057-I.) AHRQ Publication No. 12(13)-EHC090-EF. Rockville, MD: Agency for Healthcare Research and Quality. November 2012.

[28] Rupprecht CE, Briggs D. Brown C, et al. Centers for Disease Control and Prevention (CDC). Use of a Reduced (4-Dose) Vaccine Schedule for Postexposure Prophylaxis to Prevent Human Rabies: Recommendations of the Advisory Committee on Immunization Practices. MMWR, March 19, 2010, Vol 59, #RR-02. 59(RR02);1-9.

[29] Alanezi M. Varicella pneumonia in adults: 13 years' experience with review of literature. Ann Thorac Med. 2007;2(4):163.

[30-31] Bope ET, Kellerman RD. Conn's current therapy: 2014, 170. Philadelphia, PA: Saunders; 2014.

[32] Bope ET, Kellerman RD. Conn's current therapy: 2014. Philadelphia, PA: Saunders; 2014. p. 365–8.

[33] Bope ET, Kellerman RD. Conn's current therapy: 2014. Philadelphia, PA: Saunders; 2014. p. 131–5.
[34] Lucas AO, Gilles HM. Short textbook of public health medicine for the tropics. 4th ed Boca Raton, FL: CRC Press; 2002. p. 84.
[35] Krasner RI, Shors T. The microbial challenge. 3rd ed Burlington, MA: Jones & Bartlett Learning; 2013. p. 215.
[36] Jackson BR, Iqbal S, Mahon B. Centers for Disease Control and Prevention (CDC). Updated recommendations for the use of typhoid vaccine. MMWR Morb Mortal Wkly Rep 2015;64(11):305–8.
[37] CDC. Lymphogranuloma Venereum (LGV). 2015 Sexually Transmitted Diseases Treatment Guidelines. ⟨http://www.cdc.gov/std/tg2015/lgv.htm⟩ [accessed 25.10.2016].
[38] Incidence, Prevalence, and Cost of Sexually Transmitted Infections in the United States. CDC. ⟨http://www.cdc.gov/std/stats/sti-estimates-fact-sheet-feb-2013.pdf⟩ [accessed 25.10.2016].
[39] CDC. Chlamydial infections. 2015 Sexually Transmitted Diseases Treatment Guidelines. ⟨http://www.cdc.gov/std/tg2015/chlamydia.htm⟩ [accessed 25.10.2016].
[40-41] CDC. Syphilis. 2015 Sexually Transmitted Diseases Treatment Guidelines. ⟨http://www.cdc.gov/std/tg2015/syphilis.htm⟩ [accessed 25.10.2016].
[42] Platt LR, Estivariz CF, Sutter RW. Vaccine-associated Paralytic Poliomyelitis: A review of the epidemiology and estimation of the global burden. J Infect Dis 2014;210(Suppl 1):S380–9.
[43] FDA approves vaccine to prevent cholera for travelers. FDA. ⟨http://www.fda.gov/NewsEvents/Newsroom/PressAnnouncements/ucm506305.htm⟩ [accessed 25.10.2016]. Published June 10, 2016.
[44] Arora M, Sharma A. A practical guide to First trimester of pregnancy. New Delhi, India: JP Medical; 2014. p. 60–6.
[45] Cortese MM, Parashar UD. Centers for Disease Control and Prevention (CDC). Prevention of Rotavirus Gastroenteritis Among Infants and Children: Recommendations of the Advisory Committee on Immunization Practices [ACIP]. MMWR 2009;58[No. RR-02].
[46] Guidance for Developing Admission Orders in Labor & Delivery and Newborn Units to Prevent Hepatitis B Virus Transmission. Immunization Action Coalition. ⟨http://www.immunize.org/catg.d/p2130.pdf⟩ [accessed 25.10.2016].
[47] CDC. Child and adolescent schedule. Immunization Schedules. ⟨http://www.cdc.gov/vaccines/schedules/hcp/imz/child-adolescent.html⟩ [accessed 25.10.2016].
[48] Petrosky E, Bocchini JA, Hariri S, et al. Centers for Disease Control and Prevention (CDC). Use of 9-valent human papillomavirus (HPV) vaccine: Updated HPV vaccination recommendations of the advisory committee on immunization practices. MMWR Morb Mortal Wkly Rep 2015;64(11):300–4.
[49] WHO. Hepatitis B. World Health Organization. ⟨http://www.who.int/biologicals/vaccines/Hepatitis_B/en/⟩ [accessed 25.10.2016].
[50a] Centers for Disease Control and Prevention. CDC. Licensure of a 13-Valent Pneumococcal Conjugate Vaccine (PCV13) and Recommendations for Use Among Children --- Advisory Committee on Immunization Practices (ACIP). MMWR 2010;59(No. RR-9).
[50b-51] Tomcyzk S, Bennett NM, Stoecker C, et al. Centers for Disease Control and Prevention (CDC). Use of 13-Valent Pneumococcal Conjugate Vaccine and 23-Valent Pneumococcal Polysaccharide Vaccine Among Adults Aged ≥65

Years: Recommendations of the Advisory Committee on Immunization Practices (ACIP). MMWR 2014;63(No, RR-37).

[52a] Yellow fever vaccination requirements and recommendations; Malaria situation; and other vaccination requirements. World Health Organization. ⟨http://www.who.int/ith/ITH_country_list.pdf?ua = 1&ua = 1⟩ [accessed 25.10.2016].

[52b] Centers for Disease Control and Prevention. CDC Health Information for International Travel 2016. New York: Oxford University Press; 2016.

[53] Centers for Disease Control and Prevention. CDC Health Information for International Travel 2016. New York: Oxford University Press; 2016.

[54-55] Tiwari TSP. VPD Surveillance Manual. 5th ed. Publisher unknown; 2011:chap 16. ⟨http://www.cdc.gov/vaccines/pubs/surv-manual/chpt16-tetanus.pdf⟩ [accessed 25.10.2016].

[56] Plotkin SA, Orenstein W, Offit PA. Vaccines. 6th ed Elsevier Health Sciences; 2013. p. 746.

[57] Ask the experts about vaccines. Immunization Action Coalition. ⟨http://www.immunize.org/askexperts/experts_per.asp⟩ [accessed 28.10.2016].

[58] Increase the Percentage of Health Care Personnel Who are Vaccinated Annually Against Seasonal Influenza. Healthy People. ⟨https://www.healthypeople.gov/node/4668/data_details⟩ [accessed 28.10.2016].

[59] Housey M, Zhang F, Miller C, Lyon-Callo S, et al. Centers for Disease Control and Prevention (CDC). Vaccination with tetanus, diphtheria, and acellular pertussis vaccine of pregnant women enrolled in medicaid—Michigan, 2011–2013. MMWR Morb Mortal Wkly Rep 2014;63(38):839–42.

[60] 122. Cox RJ, Brokstad KA, Ogra P. Influenza virus: Immunity and vaccination strategies. Comparison of the immune response to inactivated and live, Attenuated influenza vaccines. Scandinavian Journal of Immunology. 2004;59 (1):1–15. doi:10.1111/j.0300-9475.2004.01382.x.

[61] Plotkin SA, Orenstein W, Offit PA. Vaccines. 6th ed Elsevier Health Sciences; 2013. p. 1307–8.

[62] Grohskopf LA, Olsen SJ, Sokolow LZ, et al. Centers for Disease Control and Prevention (CDC). Prevention and control of seasonal influenza with vaccines: recommendations of the Advisory Committee on Immunization Practices (ACIP)—United States, 2014–15 influenza season. MMWR Morb Mortal Wkly Rep 2014;63:691–7.

[63] Fiore AE, Wasley A, Bell BP. Centers for Disease Control and Prevention (CDC). Prevention of hepatitis A through active or passive immunization: recommendations of the Advisory Committee on Immunization Practices (ACIP). MMWR 2006;55(No. RR-7).

[64] Hay WW, Levin MJ, Deterding RR. Current diagnosis and treatment pediatrics. 22nd ed New York, NY: McGraw-Hill Education; 2014. p. 546–7.

[65] CDC. Facts about VISs. Vaccines & Immunizations. ⟨http://www.cdc.gov/vaccines/hcp/vis/about/facts-vis.html⟩ [accessed 28.10.2016].

[66] Bope ET, Kellerman RD. Conn's current therapy: 2014. Philadelphia, PA: Saunders; 2014. p. 139–45.

[67] Petersen EE, Staples E, Meaney-Delman D. Centers for Disease Control and Prevention (CDC). CDC. Interim Guidelines for Pregnant Women During a Zika Virus Outbreak — United States, 2016. MMWR 2016:early release volume 65.

[68] Chikungunya. Information for healthcare providers, CDC. ⟨http://www.cdc.gov/chikungunya/pdfs/CHIKV_Clinicians.pdf⟩ [accessed 28.10.2016].

[69] WHO. Dengue. International Travel and Health. ⟨http://www.who.int/ith/diseases/dengue/en/⟩ [accessed 28.10.2016].

[70] Hales CM, Harpaz R, Ortega-Sanchez I, Bialek SR. Centers for Disease Control and Prevention (CDC). Update on recommendations for use of herpes zoster vaccine. MMWR Morb Mortal Wkly Rep 2014;63(33):729–31.

[71] 146 Colley DG, Bustinduy AL, Secor WE, King CH. Human schistosomiasis. Lancet. 2014;383(9936):2253–64. Available from: http://dx.doi.org/10.1016/s0140-6736(13)61949-2.

[72] Morton CC, Nance WE. Newborn hearing screening — A silent revolution. N Engl J Med 2006;354(20):2151–64.

[73] WHO. Visual impairment and blindness. World Health Organization. ⟨http://www.who.int/mediacentre/factsheets/fs282/en/⟩ [accessed 28.10.2016].

[74] WHO. Tuberculosis. World Health Organization. ⟨http://www.who.int/ith/vaccines/tb/en/⟩ [accessed 29.10.2016].

[75] Greenlee JE. Acute bacterial Meningitis. Merck Manual. ⟨http://www.merckmanuals.com/professional/neurologic-disorders/meningitis/acute-bacterial-meningitis⟩ [accessed 29.10.2016].

[76-77] American Academy of Pediatrics, American Public Health Association, National Resource Center for Health and Safety in Child Care and Early Education. Caring for Our Children: National Health and Safety Performance Standards: Guidelines for Out-of-Home Child Care Programs. 2nd ed. Elk Grove Village, IL: American Academy of Pediatrics; 2002. Also available at http://nrckids.org

[78-79] Laxminarayan R, Duse A, Wattal C, et al. Antibiotic resistance—the need for global solutions. Lancet Infect Dis 2013;13(12):1057–98.

[80] Centers for Disease Control and Prevention (CDC). Prevention of Perinatal Group B Streptococcal Disease Revised Guidelines from CDC, 2010. MMWR 2010;59[No. RR-10].

[81a] Causes of Death Among Children Under 5 Years, 2015. 2014. ⟨http://www.who.int/gho/child_health/mortality/child_health_004.jpg?ua = 1⟩ [accessed 28.10.2016].

[81b] Department of Health Statistics and Information Systems (WHO, Geneva) and WHO-UNICEF Child Health Epidemiology Reference Group (CHERG). CHERG-WHO Methods and Data Sources for Child Causes of Death 2000-2013. 2014: Global Estimates Technical Paper WHO/HIS/HIS/GHE/2014.6.2

[82] Rom WN, Markowitz SB. Environmental and occupational medicine. 4th ed Philadelphia, PA: Lippincott Williams & Wilkins; 2007. p. 745–7.

[83-84] Moore GS. In: Boca Raton FL, Press CRC, editors. Living with the earth: Concepts in environmental health science. 2nd ed. 2002. p. 341–2.

[85] Moore GS. In: Boca Raton FL, Press CRC, editors. Living with the earth: Concepts in environmental health science. 2nd ed. 2002. p. 229–30.

[86] Richards IS, Bourgeois M. Principles and practice of toxicology in public health. 2nd ed Burlington, MA: Jones & Bartlett; 2014. p. 63–4.

[87-88] Mazurek GH, Jereb J, Vernon A, et al. Centers for Disease Control and Prevention (CDC). Updated Guidelines for Using Interferon Gamma Release Assays to Detect Mycobacterium tuberculosis Infection, United States. MMWR 2010; 59 (No.RR-5).

[89] Zuckerman JN, Jong EC. Travelers' vaccines. 2nd ed Shelton, CT: PMPH-USA; 2010. p. 93.

[90] Bowron CS, Maalim SM. Saudi Arabia: Hajj pilgrimage. CDC Travelers' Health. ⟨http://wwwnc.cdc.gov/travel/yellowbook/2016/select-destinations/saudi-arabia-hajj-pilgrimage⟩ [accessed 29.10.2016]. Published July 10, 2015. Chapter 4 - 2016 yellow book.

[91-92] Schlossberg D. Clinical infectious disease. 2nd ed Cambridge: Cambridge University Press; 2015. p. 703–7.

[93] Marshall GS. The vaccine handbook: A practical guide for Clinicians: The purple book. 3rd ed West Islip, NY: Professional Communications; 2010. p. 141.

[94] Lee D, Philen R, Wang Z. Centers for Disease Control and Prevention (CDC). Disease Surveillance Among Newly Arrived Refugees and Immigrants-Electronic Disease Notification System, United States, 2009. MMWR Morb Mortal Wkly Rep 2013;62(SS07):1–20.

[95] Centers for Disease Control and Prevention. Epidemiology and Prevention of Vaccine-Preventable Diseases. Hamborsky J, Kroger A, Wolfe S, eds. 13th ed. Washington D.C.Public Health Foundation, 2015.

[96] Centers for Disease Control and Prevention. Epidemiology and Prevention of Vaccine-Preventable Diseases. Hamborsky J, Kroger A, Wolfe S, eds. 13th ed. Washington D.C.Public Health Foundation, 2015.

[97a] Tulchinsky TH, Varavikova EA. The New Public Health. 3rd ed San Diego, CA: Elsevier Academic Press; 2014. p. 216–7.

[97b] WHO. Leprosy. World Health Organization. ⟨http://www.who.int/lep/en/⟩ [accessed 31.01.2016].

CHAPTER EIGHT

Emergency Preparedness

8.1 EMERGENCY PREPAREDNESS QUESTIONS

1. What is the most appropriate initial step after a natural disaster has occurred?
   A. Address the public
   B. Conduct a needs assessment
   C. Dispatch health care providers
   D. Request financial services
   E. Coordinate a plan of action
2. What is the purpose of the National Incident Management System (NIMS)?
   A. Grant the national guard with jurisdiction to control all emergencies nationwide
   B. Provide a template and structure to manage all emergencies nationwide
   C. Serve as central command to oversee each emergency nationwide
   D. Allow recruitment of people into leadership positions during an emergency
   E. Create a national system for payment of emergency operations
3. Which of the following answers lists three components of the Incident Command System (ICS)?
   A. Clinical, academic, logistics
   B. Command, operations, finance/administration
   C. Operations, logistics, clinical
   D. Finance/administration, planning, communication
   E. Communication, command, logistics
4. As the new director of a county health department, you are looking to evaluate some of the county's emergency preparedness plans. What is the fourth and final phase of emergency management?

Board Review in Preventive Medicine and Public Health.
DOI: http://dx.doi.org/10.1016/B978-0-12-813778-9.00008-6

A. Activity
B. Mitigation
C. Preparedness
D. Recovery
E. Response

5. What is the name of the complex system of laboratories that allow for rapid testing and notification of public health emergencies?
A. American Laboratory Notification Network (ALNN)
B. Clinical Laboratory Improvement Amendments (CLIA)
C. Laboratory Response Network (LRN)
D. Public Health Emergency System (PHES)
E. Public Health Laboratory Notification System (PHLNS)

6. Which of the following potential biological agents is considered a category A biological agent?
A. Botulism
B. Brucellosis
C. *Escherichia coli* 0157:H7
D. Ricin toxin
E. Psittacosis

7. Which of the following potential biological agents is considered a category B biological agent?
A. Anthrax
B. Ebola
C. Glanders
D. Plague
E. Smallpox

8. Which of the following is classified as a category C bioterrorism agent?
A. Anthrax
B. *Cytomegalovirus*
C. Nuclear radiation
D. Multidrug resistant tuberculosis
E. Tularemia

9. The Federal Emergency Management Agency (FEMA) is located within which organization?
A. Department of Defense
B. Department of Health and Human Services
C. Department of Homeland Security
D. Department of Housing and Urban Development
E. Department of Justice

10. Which of the following is endorsed by the Department of Homeland Security to outline how government and community partners organize themselves in response to catastrophic events such as natural disasters and terrorist attacks?

A. American Disaster Response (ADR)
B. Federal Emergency Management Structure (FEMS)
C. National Protection Framework (NPF)
D. National Response Framework (NRF)
E. There is no national structure. Emergency response is coordinated by individual states.

11. Which antimicrobial agent should be disseminated during a bioterrorism attack to counteract the effects from terrorists using an aerosolized strain of the organism responsible for pneumonic plague?

A. Amoxicillin
B. Azithromycin
C. Doxycycline
D. Nitrofurantoin
E. Trimethoprim–sulfamethoxazole

12. Which supplement is given to people exposed to nuclear radiation as a result of accident at a nuclear power plant?

A. Aluminum chloride
B. Glucose
C. Magnesium citrate
D. Metoprolol
E. Potassium iodide (KI)

13. Which of the following options is not a component of the National Planning Frameworks as described within the National Preparedness System?

A. National Prevention Framework
B. National Protection Framework
C. National Mitigation Framework
D. National Response Framework
E. All of these options are National Planning Frameworks within the National Preparedness System

14. A semitruck carrying hundreds of gallons of pesticides is involved in a crash, resulting in the truck emptying its contents on a busy urban road. Local health officials are concerned about rainwater runoff washing the pesticide into a water reservoir that is used for drinking water. When ingested in sufficient quantities, the pesticide may cause seizures. Which stage of risk assessment includes recognition of the lowest observable effect level (LOEL)?

A. Dose-response assessment
B. Exposure assessment
C. Hazard identification
D. Risk characterization
E. Resolution analysis

15. A biological weapon was dispersed over an event with 100,000 participants. Health care providers in this municipality are trained to use the START triage method for those affected. What do the letters in START stand for?

A. Same triage action responsibilities to all
B. Simple triage and rapid treatment
C. Sort the ailment by ranking therapy
D. Standard assignment of rendering treatment
E. Start treatment and react timely

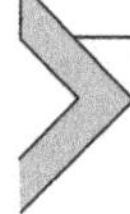

8.2 EMERGENCY PREPAREDNESS ANSWERS

1. **B. Conduct a needs assessment**
   The first priority in an emergency is conduct a needs assessment. This includes gathering information pertaining to damages, needs, and vulnerabilities of the affected population.
2. **B. Provide a template and structure to manage all emergencies nationwide**
   The NIMS is a model that provides a template for seamless response to emergency situations by all agencies across the United States. NIMS accomplishes this by standardizing the chain of command, individual roles, personal responsibilities, and use of technology. With this coordination, NIMS can better prevent, protect against, respond to, recover from, and mitigate the effects of emergencies.
3. **B. Command, operations, finance/administration**
   The ICS is the standard incident management model for the NIMS. It leverages facilities, equipment, and personnel to operate within a common structure to aid in incident management activities in both the private and public sectors. The five components of the ICS are command, operations, planning, logistics, and finance/administration.
4. **D. Recovery**
   In order from first to last, the four stages of comprehensive emergency management are mitigation, preparedness, response, and recovery.

   During the mitigation phase, efforts are made to prevent future emergencies and reduce their potential effects. Creating building codes that require structures be able to resist hurricanes and earthquakes is an example of mitigation.

   During the preparation stage, plans and preparations are made to handle an emergency before the emergency occurs. This includes actions such as stocking food and water during a hurricane.

   During the response stage, actions are taken for an imminent disaster or during a disaster. Closing a dangerous bridge before an ice storm and opening a homeless shelter before a blizzard are examples of the response stage.

During the recovery stage, efforts are taken to prevent an emergency situation from worsening and bring back normalcy. Providing mental health counseling after a mass shooting is an example of recovery, as is obtaining finances to pay for repairs.

5. **C. Laboratory Response Network (LRN)**

The LRN is regarded as a national security asset that is comprised of integrated domestic and international laboratories. These laboratories are prepared to quickly respond to biological, chemical, and radiological threats amongst many other types of public health emergencies. It provides rapid testing and timely notification of laboratory results. It may also be used in nonemergent situations.

The CLIA is a set of regulatory standards that assures quality laboratory testing. It is regulated by the Centers for Medicare and Medicaid Services (CMS). To certify CLIA compliance, laboratories review documentation and reporting of their results, scrutinize personnel credentialing, and partake in quality control efforts. To confirm accuracy, laboratories undergo proficiency testing, where the lab receives an unknown sample and are tasked with reaching the appropriate lab measurement/diagnosis. CLIA covers both private and public health laboratories.

Outside of LRN and CLIA, the other answer options do not exist.

6. **A. Botulism**

Botulism is the only category A biological agent listed. The rest of the answers are category B agents. Category A biological agents pose the highest threat of all biological agents to the United States due to their high mortality and potential to wreak havoc in the health care/public health systems. They require special action for preparedness. These agents have the potential to be easily disseminated and transmitted.

The list of category A biological agents include:

- Anthrax (*Bacillus anthracis*)
- Botulism (*Clostridium botulinum* toxin)
- Plague (*Yersinia pestis*)
- Smallpox (*Variola major*)
- Tularemia (*Francisella tularensis*)
- Viral hemorrhagic fevers
  - Filoviruses
    - Ebola
    - Marburg

- Arenaviruses
  - Lassa
  - Machupo

7. C. Glanders

Glanders is the only category B biological agent listed. The rest of the answers are category A agents.

Behind category A biological agents, category B biological agents are the second highest public health preparedness priority. These agents are moderately easy to disseminate and have moderate morbidity and mortality rates. They require special attention so boost diagnosis and surveillance capabilities.

The list of category B biological agents include:

– Brucellosis
– Food safety threats
  - *Salmonella* species
  - *Escherichia coli* O157:H7
  - *Shigella*
– Glanders (*Burkholderia mallei*)
– Melioidosis (*Burkholderia pseudomallei*)
– Psittacosis (*Chlamydia psittaci*)
– Q Fever (*Coxiella burnetii*)
– Ricin Toxin from *Ricinus communis* (castor beans)
– *Staphylococcal enterotoxin* B
– Typhus fever (*Rickettsia prowazekii*)
– *Viral encephalitis*
  - (Alphaviruses)
– Water safety threats
  - *Vibrio cholerae*
  - *Cryptosporidium parvum*

8. D. Multidrug resistant tuberculosis

Category C bioterrorism agents reclassify more frequently than agents in categories A and B. This classification includes newly emerging diseases that can potentially be exploited to cause widespread illness. The list of category C agents changes with disease outbreak circumstances around the world. Some diseases that have been classified as category C include drug resistant tuberculosis, yellow fever, chikungunya, rabies, and novel influenza strains.

9. C. Department of Homeland Security

The FEMA is one of 16 components within the Department of Homeland Security.

10. D. National Response Framework (NRF)

The NRF is part of the National Preparedness System. It outlines how government and community organizations unify to achieve the national preparedness goal and respond to disasters of all proportions. NRF describes specific authorities and recommends best practices for disaster management. It is built upon the flexible concepts outlined in the NIMS and is part of one of the national planning frameworks. The other national planning frameworks include the National Prevention Framework, National Protection Framework, National Mitigation Framework, and the National Disaster Recovery Framework.

11. C. Doxycycline

Plague is a caused by the gram-negative bacteria *Yersinia pestis*. Depending on the route of exposure, it may cause one of four types of plague:

- Bubonic plague—caused from bites of infected animals, including fleas
- Pharyngeal/ocular plague—occurs when eye or mucosal surface exposure leads to local infection
- Septicemic plague—occurs when the bacteria becomes disseminated through the blood
- Pneumonic plague—occurs when *Y. pestis* invades the lungs through inhalation

Due to ease of dissemination and high risk of mortality (~100% if treatment is not initiated in the first 24 hours), pneumonic plague is most likely to be weaponized. After inhalational exposure, symptoms typically occur in two to four days. Initial symptoms include fever, chills, myalgias, and lethargy. Later symptoms include tachypnea, chest pain, and productive cough. Finally, pulmonary necrosis and hemorrhage may lead to pulmonary abscess and cavitary lesions.

Pneumonic plague is very infectious as each cough aerosolizes large numbers of *Y. pestis* which can infect nearby people. Those diagnosed with pneumonic plague should be immediately isolated and suspected contacts should be quarantined for at least 72 hours after beginning antibiotics. Gentamicin is the drug of choice for *Y. pestis*. However, during mass treatment and postexposure prophylaxis (PEP), it is recommended to either give doxycycline 100 mg twice daily or ciprofloxacin 500 mg twice daily. The duration of treatment is 10 days for confirmed cases and seven days after the last potential exposure for PEP.

12. E. Potassium iodide (KI)

It is recommended that potassium iodide (KI) be given to those within a 10 mile emergency planning zone. KI is a stable salt containing iodide that is absorbed by the thyroid gland to block future absorption of radioactive iodide by saturating iodide receptors. By blocking uptake of radioactive iodide, KI protects the thyroid from potential radiation injury. If the thyroid absorbs radioactive iodine first, the damage may become irreversible. KI does not protect the rest of the body from other radioactive elements that result from a nuclear accident.

13. E. All of these options are National Planning Frameworks within the National Preparedness System.

The National Preparedness System outlines processes of emergency preparedness to achieve the national preparedness goal: "A secure and resilient nation with the capabilities required across the whole community to prevent, protect against, mitigate, respond to, and recover from the threats and hazards that pose the greatest risk." The national preparedness goal categorizes components of national preparedness into five mission areas: Prevention, protection, mitigation, response, and recovery.

The six parts of the National Preparedness System are known as the National Planning Frameworks. They include the following:

National Prevention Framework—Describes how the community works together to prevent the occurrence and potential damage resulting from emergencies.

National Protection Framework—describes how the community works together to obtain protection from emergencies.

National Mitigation Framework—describes how the community works together to mitigate effects of emergencies.

National Response Framework—describes how the community works together to form response for the smallest to the largest of emergencies. It defines principles, roles, and structures in the emergency response process.

National Disaster Recovery Framework—describes how the community works together to restore and revitalize themselves after an emergency. This includes considerations to health, social, economic, and environmental components of the community.

14. A. Dose-response assessment

Risk assessment is customarily broken down into four stages: Hazard identification, dose-response assessment, exposure assessment, and risk characterization.

Hazard identification involves collection of data to determine the toxic properties of the agent in question.

The dose-response assessment (also known as hazard evaluation) evaluates the level of exposure that produces adverse health or environmental effects. Recognizing the LOEL would be included in the dose-response assessment.

Exposure assessment recognizes the exposure setting then calculates the amount, frequency, and route of exposure.

Risk characterization compiles the data from the previous three steps to reach an objective conclusion about the overall risk in question.

15. B. Simple triage and rapid treatment

Emergency medical responders and preparedness officials should be familiar with standardized triage and treatment protocols that provide coordination during catastrophic events. Limited resources and decreased access to care are common obstacles faced in times of mass health emergencies. During these times, protocols differ from the usual standard of care. These protocols strive to deliver optimal survival benefit to all patients and provide the greatest good for the greatest number of people.

The START triage process stands for simple triage and rapid treatment. This protocol records four observations to categorize the sick and wounded into one of four categories of medical urgency. The four observations are the ability to walk, respiratory rate (>30/min or <12/min), quality of blood perfusion (radial pulse or capillary refill), and mental status. Based upon these variables, patients are categorized as either minor, delayed, immediate, or expectant (non-survivable). JumpSTART is the pediatric version of START.

Other triage processes are starting to gain ground. For many groups, the Sort-Assess-Lifesaving Interventions-Treatment/Transport (SALT) is being advanced as a new standard for many groups.

BIBLIOGRAPHY

[1] Kapur GB, Smith JP. Emergency public health: Preparedness and response. Sudbury, MA: Jones & Bartlett Learning; 2011. p. 185.

[2-3] Landesman LY. Public health management of disasters: The practice guide. 3rd ed Washington, DC: American Public Health Association Publications; 2011. p. 43.

[4] Phillips BD. Disaster Recovery. 2nd ed Boca Raton: CRC Press; 2016. p. 5–6.

[5] Laboratory response network (LRN). CDC- Emergency Preparedness and Response. ⟨https://emergency.cdc.gov/lrn/⟩ [accessed 27.10.2016].

[6-7] Clements BW, Casani JA. Disasters and Public Health: Planning and Response. 2nd ed Cambridge, MA: Elsevier Science; 2016, chapter 6.

[8] Ryan J. Biosecurity and Bioterrorism: Containing and Preventing Biological Threats. 2nd ed Cambridge, MA: Elsevier Science; 2016. p. 113–32.

[9] Organizational chart. U.S. Department of Homeland Security. ⟨https://www.dhs.gov/organizational-chart⟩ [accessed 27.10.2016].

[10] National response framework. Federal Emergency Management Agency. ⟨https://www.fema.gov/national-response-framework⟩ [accessed 28.10.2016] Updated June 16, 2016.

[11] Public health emergency preparedness & response: Principles & practice. 1st ed. Landover, MD: PHS COF; 2010:132–136.

[12] Emergency Preparedness at Nuclear Power Plants. Nuclear Regulatory Commission- Backgrounder. ⟨http://www.nrc.gov/reading-rm/doc-collections/fact-sheets/emerg-plan-prep-nuc-power.pdf⟩ [accessed 15.10.2016]. Published April 2014.

[13] National preparedness system. FEMA. ⟨https://www.fema.gov/national-preparedness-system⟩ [accessed 15.12.2016]. Updated 2016.

[14] Torres J, Bobst S. Toxicological risk assessment for beginners. New York, NY: Springer; 2015. p. 2–4.

[15] Carmona RH, Darling RG, Knoben JE, Michael JM. The U S Public Health Service Mission. Public health emergency preparedness & response: Principles & practice. 1st ed Landover, MD: Public Health Service; 2010. p. 270–6.

INDEX

A

Absolute risk reduction (ARR), 136
Absorbed radiation dose, 217, 237–238
Accountable care organizations (ACOs), 24, 45–46
Active surveillance, 100–101, 149
Acute radiation syndrome, 217–218, 238
Administration for Children and Families (ACF), 58
Administration for Community Living (ACL), 58
Adverse selection, 24, 47
Advisory Committee on Childhood Lead Poisoning Prevention (ACCLPP), 204
Advisory Committee on Heritable Disorders in Newborns and Children, 293, 332–333
Advisory Committee on Immunization Practices (ACIP), 355, 379
Affordable Care Act (ACA), 28, 36, 52, 66–67
Age Discrimination in Employment Act (ADEA), 36, 66
Agency for Healthcare Research and Quality (AHRQ), 21, 23–24, 41–42, 45–46, 58
Agency for Toxic Substances and Disease Registry (ATSDR), 31, 58, 194, 211
Agent, 187, 197
Agriculture Research Service (ARS), 57
Air pollutants, 187, 197
Alcohol
  abuse, risk factor for, 291, 330
  as cause of liver cancer, 278, 310
  dehydrogenase (ADH), 251
Alcoholism, 228, 251
Alternative hypothesis, 88, 130
Alzheimer's dementia, 289, 326
American College of Occupational and Environmental Medicine (ACOEM), 235, 264
American Conference of Governmental Industrial Hygienists (ACGIH), 31, 56–57
American Psychiatric Association, 288, 325
American Public Health Association (APHA), 31
Americans with Disability Act (ADA), 228–229, 231, 251–252, 255
Analysis of covariance (ANCOVA), 95, 141
Analysis of variance (ANOVA), 95, 98, 140–141, 146
Anencephaly, 293, 335–336
Animal and Plant Health Inspection Service (APHIS), 57
Annual incidence of an illness, 102, 152
Antibiotic resistance, 365, 397–399
Antibiotics, 364–365, 396–397
Antigenic shift, 115, 175
Anxiety, 288, 326
Approval of new drugs, 32, 59
Arsenic intake, 191, 205
Arthritis, 284, 319–320
Ashkenazi Jewish heritage, 293, 334–335
Aspirin
  risk of colon cancer and, 271, 300
  USPSTF recommendation for, 300
Assessment Protocol for Excellence in Public Health (APEX PH), 5, 16
Asthma, 276, 306
Attack rate ratio, 103, 153
Attorney General of the United States, 39, 72
Attributable risk percentage (AR%), 111, 168–169
Autism, 290, 327–328
Autonomy, 34, 63–64
Average inpatient length of stay, 23, 44–45
Azithromycin, 357, 362, 383, 391–392

B

Bacille Calmette–Guérin (BCG) vaccine, 364, 396
Back pain, 284, 319
Bagassosis, 221, 243
Basal cell carcinoma (BCC), 314
Basal energy expenditure, 295, 338–339
Basal metabolic rate (BMR), 338–339
Bayes' theorem, 83, 122–123
Beef consumption and colorectal cancer, 278, 310
Behavioral Risk Factor Surveillance System (BRFSS), 108, 161–162
Beneficence, 34–35, 63–64
Benzathine penicillin G (Bicillin L-A), 357, 383–384
Benzene exposure, 222, 244
Beryllium toxicity, 221, 243
Binge drinking, 291, 330
Binomial distribution, 84, 96–97, 123–124, 143
Biological agents
  category A, 414, 418–419
  category B, 414, 419
  category C, 414, 419
Biological exposure indices (BEIs), 56–57, 264
Biological oxygen demand (BOD), 195, 212–213
Bioterrorism agent, 414, 419
Bladder cancer, 277, 308
  risk of, 227, 250
Blood alcohol content (BAC), 189–190, 202
Blood lead level (BLL), 190–191, 204
Blood pressure management, 274, 304
Body mass index (BMI), 85, 96, 125, 142–143, 282, 316
Botulism, 367, 401–402, 414, 418–419
Boyle's law, 232, 257–258
Breast cancer, 277, 308
Breastfeeding, 286–287, 322–323
  HIV transmission, 373
Bronchoprovocation testing, 224, 247
Byssinosis, 226, 249

C

CAGE Questionnaire, 289, 326
Calcium channel blocker, 274, 304
Cancers, 79, 118, 276, 307
  kidney, 281, 315
  lifetime risk
    of being diagnosed with, 276–277, 308
    of dying from, 277, 308
  lung, 279, 312
  melanoma, 280, 314
  risk factor for
    bladder cancer, 277, 308
    breast cancer, 277, 308
    cervical cancer, 278, 310
    colorectal cancer (CRC), 310
    liver cancer, 278, 310
    oral cancer, 279, 312
    ovarian cancer, 279–280, 312–313
    pancreatic cancer, 278, 311
    stomach cancer, 278, 311
    thyroid cancer, 281, 315
  screening, 273, 303
  skin cancer, 280, 314
  testicular cancer, 281, 314–315
Capitated payment, 28, 52–53
Carbohydrates, 282, 316
Carbon monoxide, 260
Carcinogenesis, 307
Carcinogenic radiation, 217, 237
Cardiovascular disease, 274, 304
  leading cause of mortality, 275, 304–305
Carpal tunnel syndrome, 227, 250
Case-control study, 79, 112, 118, 170
Cataracts, 285, 320–321
Causal association between two variables, 92
Causative organism of the sexually transmitted infection, 357, 382–383
Center for Nutrition Policy and Promotion (CNPP), 57
Centers for Disease Control and Prevention (CDC), 29, 55, 58, 101, 149
  Division of Community Health, 6

Centers for Medicare and Medicaid Services, 58
Central hearing loss, 219, 240
Centralized health department classification system, 30, 56
Central limit theorem, 84, 123–124
Certificates of Need (CON), 71
Cervical cancer, 278, 310
  screening, 273, 303
  vaccination to prevent, 279, 291
Charles' law, 257–258
Chief financial officer (CFO), 25–26
Chikungunya, 363, 394
Child abuse
  cases, 293, 335
  common form of, 294, 337
  victimization rate of child maltreatment, 294, 337
Childhood obesity prevention, 283, 317
Child nutrition programs, 55
Chi-squared test, 93–94, 97, 137–139, 143–144
Chlamydia, 357, 383
Chlamydia trachomatis, 356, 382
Chlorine level in swimming pool, 195, 213
Chloroquine, 363, 393
Cholera immunization, 358, 385
Cholesterol drug, 114, 174
Chromium, 221, 242
Chronic kidney disease (CKD), 282, 315
Chronic Obstructive Pulmonary Disease (COPD), 276, 306–307
Chronic silica inhalation, 226, 249
Cigarette, 290, 329
  exposure, 306–307
  smokers, 291, 330
Cirrhosis, 113, 173
Clean Air Act (CAA), 187, 197–198
Clean Water Act (CWA), 188, 198–199
Clear-cell adenocarcinoma, 277, 309
Clinical Laboratory Improvement Amendments (CLIA), 37, 67–68, 414, 418
Clinical trial testing, 116, 176
Clostridium botulinum, 401–402
Cohort study, 79, 118
Coinsurance, 28, 53–54
Colorectal cancer (CRC), 278, 310
  screening, 270, 292, 297–299, 331
Commercial airplanes
  expired air, likelihood of sharing, 235, 263–264
  sickle cell anemia, 235, 264
Commercial insurance. *See* Private insurance
Community Health Assessment and Group Evaluation (CHANGE)
  sectors, 4, 15
  tool, 4, 14–15
Community Preventive Services Task Force, 5, 16
Comprehensive Environmental Response, Compensation, and Liability Act (CERCLA), 194, 210–212
Conduct internal utilization review, 23, 45
Conductive hearing loss, 219, 240
Confidence intervals, 86, 89, 126–127, 132
Confounding, 82, 121
  alcohol as confounder, 82, 122
Congenital hypothyroidism, 293, 334
Conjugate vaccine, 369, 406
Consolidated Omnibus Budget Reconciliation Act (COBRA), 35–36, 65–66
Contact dermatitis, 222, 244
Controlled Substances Act (CSA), 33, 61–62
Copayment, 28, 53–54
Core Functions. *See* Public health, core functions of
Coronary artery disease (CAD) screening, 273, 303
Cost, access, and quality, 27, 50
Cost-benefit Analysis (CBA), 26–27, 50
Cost-effectiveness analysis (CEA), 26, 48–49
Cost-minimization analysis (CMA), 25–26
  calculations, 26, 50
Cost-sharing, 27–28, 53–54
Cost-utility Analysis (CUA), 29, 54–55
Council of State and Territorial Epidemiologists, 101, 149

Critical access hospital, 23, 44–45
Cross-sectional studies, 79, 118
Crude rate, 109, 163
Cryptosporidium, 196, 213–214
removal of, 187, 198
Current value of money, 48
Cystic fibrosis, 293, 334–335
Cytomegalovirus, 364, 395–396

D

Dalton's law, 257–258
DASH diet, 283, 317
Deaths
due to gun usage, 195, 213
leading cause of, 109, 164, 274, 304
aged one to 59 months, 366, 400
heart disease, 274, 304
unintentional injuries, 189, 202
from medical errors, 38, 69–70
from opportunistic infections, 354, 377
Decibels (dB), 223, 245–246
Deductible, 28, 53–54
Degrees of freedom (df), 93, 137
Delaney Clause, 188, 200, 211–212
Demographic gap, 100, 148
Demographic rating, 24
Dengue, 363, 394
Dental hygienist, 32, 59–60
Department of Agriculture, 29–30, 55–56, 193, 209
functions of, 31, 57
Department of Health and Human Services (DHHS), 23, 31, 45, 58, 264
Department of Justice (DOJ), 31, 58
Department of Labor (DOL), 29
Department of Veterans Affairs, 31, 58
Department of Vital Statistics, 79, 118
Dependent variable, 90, 133–134
Depression, 271–272, 301
Desensitization and treatment with penicillin, 357, 384
Diabetes mellitus (DM)
as leading cause of CKD, 282, 315
prevalence of, 282, 315
screening, 270, 299
Diagnostic and Statistical Manual of Mental Disorders (DSM) guidelines, 288, 325
Diagnostic-related groups (DRGs), 23, 45
Diarrheal illness, leading cause of, 196, 213–214
Dietary cholesterol, maximum daily limit of, 283, 317
Dietary iodine, 296, 340
Diethylstilbestrol (DES), 277, 309
Differential misclassification, 173–174
Diffusion of Innovation Model, 4, 13
Diphtheria toxoids (Td) vaccine, 360, 389–390
Disability adjusted life years (DALYs), 290, 328
Disability among adults, common cause of, 284–285, 319–320
Discounted price, 48
Disease management program, 296, 340–341
Dose-response assessment, 416, 422
Doxycycline, 355, 379, 415, 420–421
oral, 356, 380
Dracunculiasis, 349, 370–371
Drug Abuse Warning Network (DAWN), 117, 178–179
Dysuria, 363, 394

E

Early and Periodic Screening, Diagnostic, and Treatment (EPSDT) program, 22, 43
Eight Joint National Committee (JNC 8), 274, 304
blood pressure management for Black patients, 274, 304
Elder abuse, common perpetrator of, 294, 337
Electronic Disease Notification (EDN), 368–369, 404–405
Emergency management, 413–414, 417–418
Emergency Medical Treatment and Active Labor Act (EMTALA), 17, 36, 67
Employee assistance programs (EAP), 229, 252–253
Employee Retirement Income Security Act (ERISA), 35, 65
Endemic, 115, 175–176
Enzootic, 115, 175–176
Epidemic, 115–116, 176
Epidemic curve, 102–103, 152
Epidemic triangle, 187, 197

Epizootic, 115, 175–176
Ergonomics, 222–223, 244–245
Erythromycin, 352–353, 375–377
Escherichia coli, 351, 353, 374, 377
Essential hypertension, 275, 305
Experimental study, 108, 162
Expired air in airplanes, likelihood of sharing, 235, 263–264
Exposure (risk factor), 112–113, 171–172
External validity, 137

F

False-negative error rate, 105, 157
False-positive error rate, 105–106, 157–158
Family Medical Leave Act (FMLA), 230, 254
Fat consumption, recommended, 295, 339
Federal Aviation Administration (FAA), 234, 261
Federal Emergency Management Agency (FEMA), 189, 201, 414, 420
Federal Insecticide, Fungicide, and Rodenticide Act (FIFRA), 194, 211–212
Federally Qualified Health Centers (FQHCs), 22–23, 44
Federal Medical Assistance Percentage (FMAP), 25, 48
Federal Motor Carrier Safety Administration (FMCSA), 232, 256
Fee-for-service arrangement, 28, 53
Fetal death, 100, 148
  rate, 160
Filtration, 187, 198
Firearm violence, 192, 206
Fisher's exact test, 97, 143–144
"Five A's" of smoking cessation, 291, 330
Fixed-shift work schedules, 223–224, 246
Flexible spending account (FSA), 27, 51–52
Fluoride, drinking water, 188, 199
Folic acid, recommendation, 287, 323–324
Food, Drug, and Cosmetic Act, 211–212
Food additives, 188, 200
Food and Drug Administration (FDA), 37, 58, 68–69, 193, 208–209
  role in performing food inspections, 193, 209
Food and Nutrition Service (FNS), 57
Food fortification, 340
Food Safety and Inspection Service (FSIS), 57
Food Safety Modernization Act (FSMA), 208–209
"Four Ps" of health-care social marketing, 6, 18
Freshwater, atmospheric pressure forces in, 233, 259
F-test, 94, 140
Fumes, 217, 237

G

Gantt charts, 62–63, 104, 155
Gastroenteritis, 351–352, 374
GBS, 365, 399–400
Genetically modified organisms (GMOs), 37, 68–69
Geographic information systems (GIS), 104, 155
Geometric mean, 84, 125
Glanders, 414, 419
Gleason score, 280, 313
Graduate medical education (GME), 21, 42
Gravity forces (G-forces), 232, 257
Gun ranges, 222, 243–244

H

Haddon matrix, 190, 202–203
Haemophilus influenzae type B (Hib) vaccine, 359, 362, 369, 387, 392–393, 406
Hand-arm vibration syndrome (HAVS), 245
Hand hygiene, 38, 69
Hansen's disease, 369, 406–407
Hawthorne effect, 82, 84, 120–121, 123–124
Hazardous Materials Transportation Act (HMTA), 200–201
Health behavior model, 3
  reciprocal determinism, 3, 12
Health Belief Model, 3, 12

Healthcare, 113, 173
plan, performance indicators of, 34
quality measures, 27, 51
Healthcare associated infection (HAI), 55
Healthcare Effectiveness Data and Information Set (HEDIS), 23, 34, 45, 62
Healthcare expenditure
Government financed programs, 22, 43
payer contribution to, 22
percentage of Medicare dollars spent on last year of life, 23, 45
recovery of financial losses, 24
Health economics, 38, 70–71
Health insurance, 22, 42–43
premium, 24, 46
Health Insurance Portability and Accountability Act (HIPAA), 35, 64, 253–254
Health maintenance organizations (HMOs), 23–24
Health outcomes, 27, 29, 51, 54
Health Planning and Resources Development Act, 71
Health professional shortage areas (HPSAs), 22, 43–44
Health resources, evaluation of appropriate use of, 24–25, 47
Health Resources and Services Administration (HRSA), 2, 10, 23, 29, 45, 55, 58
Health status rating, 24, 46
Healthy People 2020, 361, 390–391
Healthy People objectives, 2, 10–11
Hearing loss, screening for, 395–396
Heart disease, 274, 304
Heat cramps, 239
Heat edema, 239
Heat exhaustion, 239
Heat stroke, 218, 239
Heat syncope, 239
Hemolytic uremic syndrome (HUS), 374, 377
Henry's law, 233, 257–258
Hepatitis A, 362, 369, 392, 405–406
transmission, 350, 371–372
Hepatitis B, 368, 403–404
vaccination, 358–359, 368, 386–387, 404
Hepatitis C, 354, 377–378
Herd immunity, 369, 407
Heroin addiction, 252
Herpes zoster (shingles) vaccination, 363, 394
High blood pressure. *See* Hypertension
Hill–Burton Act, 35, 65
Histoplasma capsulatum, 355, 380
Histoplasmosis, 355, 380
Homicide from firearm use, 192, 206
Horizontal transmission of disease, 373
Hormone replacement therapy, 277, 308
Hospital Consumer Assessment of Healthcare Providers and Systems (HCAHPS), 21, 35, 41–42, 65
Human Immunodeficiency Virus (HIV), 351, 366, 373, 400
Human Papillomavirus (HPV), 277, 309, 357, 382–383
Humanitarian outcomes, 29, 54
Human rabies immunoglobulin (HRIG), 354, 378
Hydrocodone Cough Syrup, 33, 61–62
Hypertension, 274–275, 304–305
common form of, 275, 305
highest risk of, 275, 305
stage two, 275, 305
Hypothermia, 193, 209–210

I

Idiopathic scoliosis screening, 272–273, 303
Inactivated influenza vaccination, 361, 391
Inactivated polio vaccine (IPV), 358
Inactivated whole cell vaccine, 369
Incidence density, 83, 123
of influenza, 83, 123
Incident Command System (ICS), 413, 417
Independent Payment Advisory Board (IPAB), 28, 52
Independent Practice Association (IPA), 24, 45–46

Indian Health Services (IHS), 31, 58
Individual mandate, 24, 47
Infant mortality rate (IMR), 107, 160–161
congenital malformations, 117, 179
Inferential statistics, 84, 123–124
Infertility, 287–288, 325
Influenza
incidence density of, 83, 123
vaccination, 362, 392
Influenza B, 350, 371
Information bias, 82, 121
Inhalation hypoxia, 227, 250
Institutional Review Board (IRB), 116, 177
Intelligence quotient (IQ), 84, 124
Interferon-gamma release assay (IGRA), 367, 402
Internal validity, 137
International Agency for Research on Cancer (IARC), 235–236, 264–265
International Certification of Vaccination or Prophylaxis (ICVP), 388–389
International Classification of Diseases (ICD), 107, 159–160
Interquartile range, 97, 144–145
Interval scale, 96, 142
Intimate partner violence (IPV), 294, 336–337
Intramuscular injection of promethazine, 234, 259–260
Intussusception, 358
"Iron triangle", 27
Ischemic heart disease, 295, 337–338
Ishikawa diagram, 17, 34, 62–63, 104, 155
Isocyanates, 234, 260
Isolation, 153–154, 406–407
Isoniazid, 352, 376

J

Jet lag, 224, 247

K

Kaplan–Meier function, 84, 123–124
Kaplan–Meier method of survival analysis, 91, 135
Kappa ratio, 94, 139–140
Kelvin scale, 96, 143
Kidney transplant, 276, 307
Koebner phenomenon, 228, 251
Kruskal–Wallis one-way test, 99, 146

L

Laboratory Response Network (LRN), 37, 67–68, 414, 418
Laser exposure, 226, 249
Latent TB infection (LTBI), treatment for, 353, 377
Leachate, 194, 210
Lead-time bias, 81, 120
Legionella, 188, 200
Legionellosis, 200
Length bias, 80, 119
Leprosy. *See* Hansen's disease
Leukemia, 279, 311–312
Licensing health-care facilities, 1, 9
Life expectancy, 100, 109, 148–149, 163
Lipid screening, 270, 299–300
Listeria monocytogenes, 356, 381
Liver cancer, 278, 310
Local health departments, 30, 56
Logistic regression, 99, 147
Long-term care expenses, 22, 42
Low-protein diet, 278, 311
Lung cancer, 279, 312
incidence of, 223, 245
radon exposure, 279, 312
Lyme disease, 380
Lymphocyte proliferation test, 226, 249

M

Malaria, 349, 370
Malpractice, 36–37, 67
Mandated Purchase of health insurance. *See* Individual mandate
Mandatory reportable diseases, 101, 150
Manganese, 222, 244, 260
Mann–Whitney U test, 97, 99, 144, 146
Maternal mortality rate, 100, 148, 160
Maximum contaminant levels (MCLs), 264
Maximum medical improvement (MMI), 232, 256

McNemar's test, 97, 144
Mean, 84, 125
Measles, mumps, rubella (MMR), 358, 385–386
Median, 85–86, 126
Medicaid, 22, 36, 39, 42, 67, 71–72
Title of the Social Security Act (SSA), 32, 59
Medical licensure, 32, 59–60
Medical nutrition therapy (MNT), 269, 297
Medical paternalism, 34, 63–64
Medicare, 21, 25, 36, 42, 48, 67
Affordable Care Act (ACA) spending, 28, 52
payment of physicians in, 25, 47–48
Title of the Social Security Act (SSA), 32, 58
Medicare Access and Children's Health Insurance Program (CHIP) Reauthorization Act (MACRA) of 2015, 73
Melanoma, 280, 314
ABCs of detecting, 280–281, 314
Meningococcal, 368, 403
Mental disorder, lifetime prevalence of a, 288, 326
Mental illness and tobacco smoking, 291, 330
Mercury
inorganic, 204–205
organic, 203–204
toxicity in seafood, 190, 203–204
Merit-Based Incentive Payment System (MIPS), 40, 73
Meta-analysis, 92, 135–136
Metabolic syndrome, 275, 305
Metal fume fever, 227, 249–250
Methacholine, 224, 247, 260
Methyl mercury, 191, 203–205
Microcephaly, 363, 393–394
Mixed hearing loss, 219, 240
Mobilizing Action for Planning and Partnerships (MAPP), 5, 16–17
Mode, 125–126
Mortality rate, 108–109, 162–163
Multidrug resistant tuberculosis, 414, 419
Multiple regression, 91, 135
Multistage Process of Carcinogenesis Theory, 276, 307
Multivariate analysis of variance (MANOVA), 99, 147–148
Mumps, 350, 371
Myalgias, 106, 158

N

National Ambient Air Quality Standards (NAAQS), 197
National Association of County and City Health Officials (NACCHO), 30, 56
National Center for Health Statistics (NCHS), 80, 118
National Committee for Quality Assurance (NCQA), 23, 45
National Environmental Policy Act (NEPA), 194, 210–211
National Health and Nutrition Examination Survey (NHANES), 106, 159
National Healthcare Safety Network (NHSN), 29
National health expenditure in United States, 37, 67
National Health Information Center (NHIC), 32, 60
National Implementation and Dissemination for Chronic Disease Prevention, 18
National Incident Management System (NIMS), 413, 417
National Institute for Occupational Safety and Health (NIOSH), 30–31, 56–57, 219, 235, 241, 264
National Institutes of Health, 58
National Notifiable Diseases Surveillance System (NNDSS), 101, 149
National Planning Frameworks, 415, 421
National Practitioner Data Bank (NPDB), 29, 55
National Preparedness System, 415, 421
National Radon Action Plan, 212
National Response Framework (NRF), 415, 420

National Rural Health Association (NRHA), 31
National Survey on Drug Use and Health (NSDUH), 117, 178
National Transportation Safety Board (NTSB), 234, 261
National Vaccine Injury Compensation Program, 24, 46
National Vital Statistics System, 106, 158
Natural disaster response, 413, 417
Negative Predictive Value (NPV), 105, 156–157
Neonatal mortality, 287, 324
  rate, 107, 160–161
Newborn screening tests, 293, 332–333
Neyman bias, 82, 121
Nickel exposure, 222, 244, 260
Nitrate exposure, 220, 242
Nitrogen, 192, 207
Noise, 189, 201
  limits, 224–225, 247–248
Noise-induced hearing loss, 218, 240
Nominal scale, 142
Nondifferential error/bias, 80–81, 113–114, 119–120, 173–174
Nonmaleficence, 34, 63–64
Norovirus outbreak, epidemic curve of, 102, 151
Nosocomial infections, reducing, 38, 69
Nuclear Regulatory Commission (NRC), 189, 201
Nuclear waste, 189, 201
Null hypothesis, 88–89, 130, 132
Number needed to harm (NNH), 136
Number needed to treat (NNT), 136
Nutrition, 2, 10

O

Obesity, 284, 318
  environmental factors contributing to, 196, 214
  screening, 272, 302
Occupational injuries and fatalities, 219, 240–241
Occupational Safety and Health Administration (OSHA), 29, 55, 219, 229–230, 235, 241, 253–254, 264
  duration for maintaining employee medical records, 231, 255
  General Duty Clause, 219–220, 241
  hearing conservation program, 231, 254–255
  reporting an incident, 231, 255
Odds ratio, 112, 170–171
Office of Disease Prevention and Health Promotion (ODPHP), 32, 60
Oral polio vaccine (OPV), 358, 384–385
Ordinal scale, 96, 142–143
Organophosphate toxicity, 227–228, 250
Organ Procurement and Transplantation Network (OPTN), 282, 315
Oseltamivir, 362, 391–392
Osteoporosis, 284, 318
Ovarian cancer, 279–280, 312–313
Overweight, 282, 316
Oxygen saturation, 235, 262
Ozone, 192, 208

P

Paired t-test, 94, 98, 140, 145
Pancreatic cancer, 278, 311
  survival rate, 278–279, 311
Pandemic, 115, 175–176
Parens patriae, 35, 65
Partnerships to Improve Community Health (PICH), 18
Pasteurization, 192, 206–207
PATCH program, 4, 15
Patient-Centered Medical Homes (PCMHs), 23
Patient perception of healthcare, 21
Patient Self-Determination Act (PSDA), 35, 65
Pay-for-performance model, 38, 40, 70, 73
Pearson correlation coefficient, 90–91, 95, 134–135, 141
Pellagra, 284, 318–319
Penicillin G, 384
PEPFAR, 2, 10
Percentile calculation, 85–86, 126
Perinatal mortality rate, 161
Peripheral neuropathy, 376

Permanent partial disability, 226, 248
Permissible exposure limits (PELs), 56–57, 219, 235, 241, 264
Personal health information (PHI), 64
Pertussis infection, treatment for, 353, 376–377
PEST analysis, 33, 60
Phenylketonuria (PKU), 288, 325
Phosphorus, 220, 241
Physical activity recommendation, 292, 331
Physician compensation systems, 28, 53
Physician hospital organizations (PHOs), 24, 45–46
Physician payment reimbursements, 28, 52–53
Physician Quality Reporting System (PQRS), 40, 73–74
Physicians, legislation preventing referring his/her patients, 25, 47
Physician Self-referral Act (Stark Law), 25
Plague (*Yersinia pestis*), 366, 401
Plan, do, study, act (PSDA), 33, 60–61
Planned Approach to Community Health (PATCH) program, 5–6, 17–18
Pneumococcal conjugate vaccine, 359–360, 387–388
Pneumoconiosis, 243
Pneumonia, 354, 378–379
Pneumonic plague, 420–421
Pneumothorax, 234–235, 261–262
Point source pollution, 191, 205
Police powers, 32, 59
Polychlorinated biphenyls (PCBs), 192, 208
Population attributable risk percent (PAR %), 111, 168–169
Population growth of the population pyramid, 110, 166
Positive linear relationship, 90–91, 134
Positive Predictive Value (PPV), 104, 155–156
Post-exposure prophylaxis, 364, 396
Potassium, adequate intake (AI) of, 295, 338
Potassium iodide (KI), 415, 421
PRECEDE–PROCEED Model, 17
Preferred provider organizations (PPOs), 24, 45–46
Pregnancy Medication Exposure Registry, 106–107, 159
Pregnancy Risk Assessment Monitoring System (PRAMS), 112, 170–171
Prescription drug monitoring programs (PDMPs), 39–40, 73
Preterm birth and related complications, 287, 324
Preterm pregnancy, 287, 324
Prevalence, 89, 132
Primary care physician
  number per geographic population, 22, 43–44
Prison population in the United States
  common ailment among, 3, 11
  leading cause of death among, 3, 11–12
Private insurance, 21–22, 41, 43
Private managed care plans, 25
Probabilities, 87, 129
Prospective Payment System (PPS), 21, 41–42
Prostate-specific antigen (PSA) screening, 272, 302
Psoriatic lesion, 228, 251
Publication bias, 100, 148
Public health, core functions of, 1, 7
  relations with essential public health services, 1, 7–9
Public Health Accreditation Board (PHAB), 21, 41–42
Pulmonary function test (PFT), 226, 249
Purified protein derivative (PPD) tuberculosis test, 88–89, 131
P-values, 86–88, 127–128, 130–131
Pyridoxine (vitamin B6), 352, 376

Q

Quality Adjusted Life Years (QALY), 29, 54–55
Quality control stages of healthcare organizations, 33–34, 62
Quarantine, 104, 153–154
Quasi-experimental study, 108, 162

R

Racial and Ethnic Approaches to Community Health (REACH) program, 18
Radon exposure, 279, 312
  limit, 195, 212
Randomized controlled trial (RCT), 80, 119
Rate, defined, 162–163
Recall bias, 82, 121
Receiver operating characteristic (ROC) curve, 106, 158
Recommended exposure limits (RELs), 56–57
  for hazardous chemicals, 219, 241
Regression analysis, 82, 121
Relative risk, 111–112, 169–170
Renal cell carcinomas, 281, 315
Reserved Powers Doctrine, 59
Resource Conservation and Recovery Act (RCRA), 194, 210–211
Respiratory infection, 286, 322
Resting energy expenditure (REE), 295, 338–339
Returning to school, 365, 397
Ring vaccination, 351, 372–373
Risk, defined, 162–163
Risk calculation, 136
Rocky Mountain spotted fever (RMSF), 379
Rotavirus, 352, 374
  vaccination, 358, 386
Rothman community organization models, 4, 15–16
Ryan White HIV/AIDS Program, 2, 10

S

Safety data sheet (SDS), 224, 247
Salmonella, 366, 401
Salmonella enterica, 356, 381–382
Sample size, calculation of, 89–90, 132–133
Sarcoidosis, 221, 243
Schedule I drugs, 39, 72–73
Schedule II drugs, 39, 72–73
Schedule III drugs, 39, 72–73
Schedule IV drugs, 39, 72–73
Schedule V drugs, 39, 72–73
Schistosomiasis, 364, 394–395
Screening mammogram, 269–270, 297
Screening mammography, 116–117, 177–178
Screening recommendations, 150
Secondary prevention, 292, 331–332
Secure and Responsible Drug Disposal Act, 33, 61–62
Sensitivity, 101, 104–105, 150, 155–157
Sensitivity analysis, 114, 174
Sensorineural hearing loss, 219, 240
Sensory function, 233, 258
Sick building syndrome (SBS), 111, 166–167
Sickle cell disease, 387
Sign test, 98, 145
Simple triage and rapid treatment (START), 416, 422
Six sigma, 33
Skewed distribution, 85, 125–126
Skin cancer, 280, 314
Smallpox, 349, 371
  vaccine, 103–104, 153–154
Social Cognitive Theory, 3, 12
Social Security Act, 36, 67
Sodium recommendations, 283, 316–317
Space motion sickness (SMS), 259–260
Spearman rank correlation coefficient, 98, 146
Specific, measurable, attainable, relevant, and time-oriented (S.M.A.R.T.), 17, 33, 61
Specificity, 104, 155–156
Squamous cell carcinoma (SCC), 314
Stage two hypertension, 275, 305
Standard deviation, 87, 128
Standard error (SE), 92, 136–137
Standardized mortality ratio (SMR), 87, 129
  indirect adjustment, 88, 129–130
Stannosis, 221, 243
Stark Laws. *See* Physician Self-referral Act
State governments, 39, 71

State health departments, 30, 56
Statistical inference and validity, 92–93, 137
Statisticians, 80, 119
Stomach cancer, 278, 311
Strategic Advisory Group of Experts (SAGE) on Immunization, 38–39, 71
Stratification, 82, 121
Substance Abuse and Mental Health Services Administration (SAMHSA), 58, 117, 178–179
Sudden infant death syndrome (SIDS), 286, 322
Suicide
  highest "success" rate of, 289, 326
  leading method among men, 292, 330–331
  men *vs* women, 289, 327
Sulfur-oxides, 188, 199
Superfund Amendments and Reauthorization Act (SARA), 194, 210–211
Supplemental Nutrition Assistance Program (SNAP), 30, 55–56
Supplemental Nutrition Program (SNAP), 55
Surgeon General, functions of, 1, 9
Surveillance, Epidemiology, and End Results (SEER) Program, 118
Survival analysis, 91, 135
Survival rate, 99, 147
SWOT analysis, 32, 60

T

Temporary partial disability, 225, 248
Tertiary prevention, 26, 48–49
Testicular cancer, 281, 314–315
  screening, 272, 302
Tetanus, diphtheria, acellular pertussis (Tdap) vaccination, 360–361, 389–391
Tetanus immunoglobulin (TIG), 360
Tetanus vaccination, 360–361, 389–390
The Joint Commission (TJC), 21, 41–42
Theory of Reasoned Action, 3, 12
Thiazide diuretic, 274, 304
"Three-legged stool", 27
Threshold limit values (TLVs), 56–57
Thyroid cancer, 281, 315
Tick-borne illness, 355, 379
Time period of observation, 123
Time-series analysis, 96, 141–142
Tin, 221, 243
Title IV of Public Health Law of the Healthcare Quality Improvement Act, 55
Tobacco-free zones, 26–27
Tobacco product, use among students, 290, 329
Tobacco smokers, lowest prevalence of, 290–291, 329
Total energy expenditure (TEE), 338–339
Toxic Substances Control Act (ToSCA), 194, 210–212
Toxoplasmosis gondii, 373–374
Toxoplasmosis infection, 351, 373–374
Trachoma, 364, 396
Traffic fatalities, 189–190, 202
Traffic-related air and noise pollution, 194–195, 212
Transfatty acids, 295–296, 340
Transplantation, 292–293, 332
  kidney, 276, 307
Transportation of hazardous wastes, 189, 200–201
Transtheoretical Model, 3–4, 12–13
Traumatic brain injury, 285, 320
Trichinosis, 356, 381
Trichloroethylene (TCE), 228, 251
Troposphere, 191, 205–206
TST. *See* Tuberculin skin test
T-test, 140–141
Tuberculin skin test (TST), 352, 367, 375, 402
Tuberculosis (TB), 354, 377
  outbreak investigation, 353, 376
  purified protein derivative (PPD), 353, 376
Two-step PPD, 367, 402–403
Type I error, 88, 131
Type II error, 89, 131

U

Unintentional injuries, 189, 202
United Network for Organ Sharing (UNOS), 282, 315
United States, legal authority to isolate and quarantine, 104, 154
Unplanned pregnancies, 286, 321–322
Urinary tract infections (UTIs), 38, 69, 353, 377
US Census Bureau, 107–108, 161
US Department of Agriculture (USDA). *See* Department of Agriculture
US Department of Health and Human Services. *See* Department of Health and Human Services (DHHS)
US Outpatient Influenza-like Illness Surveillance Network (ILINet), 106, 158
US Preventive Services Task Force (USPSTF), 24, 46, 269–271, 297, 301
  cancer screening, 273, 303
  cervical cancer screening, 273, 303
  colorectal cancer (CRC) screening, 270, 297–299
  coronary artery disease (CAD) screening, 273, 303
  depression screening, 271–272, 301
  diabetes mellitus (DM) screening, 270, 299
  financial costs, 271, 300
  genetic screening for risk of ovarian cancer, 279–280, 312–313
  idiopathic scoliosis screening, 272–273, 303
  interventions to reduce tobacco use in children and adolescents, 272, 302
  lipid screening, 270, 299–300
  obesity screening, 272, 302
  prostate-specific antigen (PSA) screening, 272, 302
  recommendation for aspirin, 270–271, 300
  recommended treatment for conjunctival gonorrhea, 352, 375
  sodium recommendations, 283, 316–317
  testicular cancer screening, 272, 302
Utilization management, 24, 47

V

Vaccination fractional dosing strategy, 368, 404
Vaccine Adverse Event Reporting System (VAERS), 116, 176–177
Vaccine-associated paralytic polio (VAPP), 358, 384–385
Vaccine efficacy percentage, 111, 114, 167–168, 174
Vaccine information statement (VIS), 362–363, 393
Vaccine Injury Compensation Trust Fund, 46
Vanadium, 221, 242
Varicella (chicken pox) vaccination, 355, 379
Vertebral compression fractures (VCFs), 284
Vertical transmission of disease, 373
Vibrio cholerae, 366, 400–401
Vibrio disease, 350–351, 372
Victimization rate of child maltreatment, 294, 337
Virulence, 115, 175
Virus-like particle (VLP), 279, 291
Vitamin A, 340
  deficiency, 285–286, 321
Vitamin D deficiency, 295, 338
Vitamin deficiencies, 318–319
Vitamin E supplementation, 288, 325
Volume of the breathing apparatus, 232, 256–257

W

Water fluoridation, 2, 9–10
Well water, 191, 205
West Nile virus (WNV), 350, 371
Wet bulb globe temperature (WBGT), 218, 238–239
Women, Infants, and Children (WIC) program, 2, 10, 29–30, 55
Workers' compensation, 230–232, 253–256

Working during pregnancy, 228, 250–251
World Health Organization (WHO), 107, 159–160

Y

Years lived with disability (YLD), 290, 329
Years of potential life lost (YPLL), 109–110, 164–165
Yellow fever vaccination, 349, 360, 370, 388–389
Youth Risk Behavior Surveillance System (YRBSS), 108, 161

Z

Zika virus, 363, 393–394
Z-score, 84, 124

Printed in the United States
By Bookmasters